Superconductive Tunnelling and Applications

Superconductive Tunnelling and Applications

L. Solymar

Fellow of Brasenose College and
Lecturer in Engineering Science
Oxford University

WILLEY – INTERSCIENCE

A Division of John Wiley & Sons, Inc.
New York – London – Sydney – Toronto

© 1972 L. Solymar
First published in Great Britain 1972
by Chapman and Hall Ltd
11 New Fetter Lane, London EC4P 4EE
Published in the U.S.A. by Wiley-Interscience Division,
John Wiley & Sons, Inc. 605 Third Avenue, New York, N.Y. 10016
Library of Congress Catalog Card Number: 72 752
Wiley ISBN: 0 471 81270 6
Chapman and Hall SBN: 412 10210 2
Printed in Great Britain

Contents

III. JOSEPHSON TUNNELLING

IV. ADDITIONAL MATERIAL

V. APPENDICES

To Lucy, Gillian and Marianne

Preface

Superconductive tunnelling is an offshoot of the general studies of tunnelling in solids which has recently acquired an independent status. It is of great interest to pure scientists because it provides a wealth of information about various properties of the superconducting state, and it is important for the applied scientist because it comprises the basis for a varied set of remarkable devices.

This book aims to review the present state of the art without the use of microscopic theory. The experimental results and applications are discussed on the basis of macroscopic theories. Part I gives a brief introduction to super-conductivity and tunnelling, Parts II and III are concerned with normal and Josephson tunnelling respectively. The purpose of Part IV is (i) to include those results which came out between the submission of the manuscript and the arrival of the proofs and (ii) to make an attempt at the complete coverage of the published literature. Chapters 21 and 22 need to be mentioned separately. The necessity for the former one arose because a number of phenomena defied all my attempts at classification, so I grouped them under the title Further Topics. Chapter 22 is a brief summary of the human aspects of Josephson tunnelling. It shows, with the aid of a few graphs, how a new branch of science breaks new ground in the second half of the twentieth century.

The units used are that of the rationalised MKS system which is used by engineers and is increasingly being adopted by physicists. The only allowance is made to the unit of magnetic flux density which is always given in gausses. I felt that Teslas (and mainly their abbreviation as T) are not yet widely accepted and some readers might have had difficulties in recognising the unit. The angstrom, however, was given no reprieve; there seemed no point in giving two dimensions of a thin film in millimeters and the third one in angstroms. The nanometer was introduced instead which, for our purpose, is a more convenient unit anyway (a junction is just a few nanometers thick).

Having discussed the organisation of the book it seems worthwhile to mention the audience I had in mind.

1. *Final year undergraduates* in physics and electronic engineering who are interested in low temperature physics and would like to have some idea what the various branches can offer. The introductory chapters of Parts II and III and most of the device chapters are within the grasp of such an audience.

2. *Postgraduate students* who have started to do research in superconducting tunnelling or in a neighbouring branch of low temperature physics.
3. *Applied scientists* who are drawn to the field of superconductive tunnelling because of the device potentialities.
4. *Specialists* in tunnelling who would like to read up an easy account of the properties of superconductive tunnelling.
5. *Experimenters* who are specialists in some branch of superconductive tunnelling and would like to see what has been done in the field as a whole.

I greatly benefited from preprints of the publications of Messrs. K. Aihara, R. F. Averill, W. T. Band, D. Blaugher, J. Bostock, D. J. Brassington, S. A. Buckner, J. T. Chen, J. Clarke, G. B. Donaldson, T. Ezaki, G. Faraci, J. R. Gavaler, G. Giaquinta, W. D. Gregory, N. Hara, M. A. Janocko, C. K. Jones, R. A. Kamper, V. E. Kose, D. N. Langenberg, L. Leopold, A. Longacre, Jr., M. L. A. MacVicar, N. A. Mancini, T. Matsushita, M. Mitani, S. I. Ochiai, R. L. Peterson, A. B. Pippard, I. F. Quercia, D. Repici, R. M. Rose, S. Shapiro, D. B. Sullivan, W. E. Tennant, J. Vrba, J. R. Waldram, S. B. Woods, D. Woody, K. Yamafuji, J. E. Zimmerman, to whom I wish to extend my thanks.

I am greatly indebted to Dr. G. B. Donaldson for his comments and criticism of the first version of the manuscript and for reading Chapter 6 of the final version. My thanks are also due to Dr. G. R. S. Seraphim and Mr. K. Overson for reading various parts of the manuscript. Dr. H. Motz helped me considerably by constantly demanding further clarification when I gave a series of lectures (bearing the same title as this book) in the Department of Engineering Science, Oxford, in Michaelmas term, 1970. For the illustrations I am indebted to Mrs. J. Takacs.

Finally I wish to thank my wife Marianne for her patience and forbearance during the time this book was written.

Note on the spelling of 'Tunnelling'.
The original spelling contained within references to books and articles is retained. Thus 'Tunneling' appears frequently as the spelling most favoured by authors from the U.S.A.

I. A review of superconductivity and tunnelling

1.   Superconductivity

1.1 Introduction

In this chapter we shall review briefly the basic theories of superconductivity needed in later chapters. The aim is rather to *refresh* the memory of the reader than to acquaint him with a new discipline. Accordingly, the following sections represent more of a summary than an introduction. Since the major part of the book is based on macroscopic theories there is more emphasis put on the Ginzburg–Landau theory and on the macroscopic Schrödinger equation but an attempt is made to give the main assumptions and the more important results of the microscopic theories as well.

The BCS ground state will be discussed in Section 1.10, the elementary excitations (E_k diagram) in Section 1.11 and 'dirty' superconductors in Section 1.13. The semiconductor model will be introduced in Section 1.12 where the difference between quasiparticles and normal electrons will also be briefly discussed.

1.2 Historical review

Superconductivity was discovered in 1911 by Kamerlingh Onnes. He found that below a certain critical temperature (about $4 \cdot 2°K$) the resistance of mercury dropped to an unmeasurably small value. A number of other superconducting materials were soon discovered and various properties of the superconducting state revealed. It was found for example that a sufficiently large magnetic field or transport current could destroy superconductivity, and interestingly (at least as far as the subject of this book is concerned) it was proved as early as 1924 (by Kamerlingh Onnes [1] and Tuyn [2]) that a persistent current could be induced in a ring consisting of two different superconductors in contact.

More rapid development of superconductivity started in 1933 when Meissner and Ochsenfeld [3] discovered that a superconductor cooled below its critical temperature would expel the magnetic field. This discovery was quickly followed by the two-fluid model of Gorter and Casimir [4] and the elctromagnetic theories of the London brothers [5, 6]. Subsequent major landmarks in describing the phenomenology of the superconducting state were made by Ginzburg and Landau [7] and by Pippard [8].

In contrast to the steadily improving macroscopic theories progress in setting up a microscopic theory was slow. The first indications that the electron–phonon interaction was responsible for superconductivity came as late as 1950 with Frohlich's [9] theoretical model (describing an attractive electron–electron interaction mediated by phonons) and the discovery of the isotope effect [10, 11] (showing that the critical temperature depends on the mass of the nucleus). Having found the basic mechanism it took a few more years to find a satisfactory mathematical solution. Using Cooper's [12] result that in the presence of an attractive interaction electron pairs are formed, Bardeen, Cooper and Schrieffer [13] succeeded in 1957 in formulating a theory (known as the BCS theory) which was capable of accounting for a surprising number of the experimental results. The next important development was due to Bogoliubov *et al.* [14] and Gorkov [15] who rederived the BCS results using different formulations and a very elegant mathematical technique. The following rapid development is a story in itself which we shall not review in this section. Some parts of it will be discussed in more detail in Chapter 6 in connection with the role of superconductive tunnelling in revealing the properties of the superconducting state.

1.3 Effect of a magnetic field

A magnetic field applied to a superconductor cannot penetrate into it. This phenomenon is perfectly well described by the laws of electromagnetism. The magnetic field will set up eddy currents which will cancel the magnetic field inside the superconductor. Since superconductors have no resistivity the eddy currents do not decay and so the magnetic field cannot penetrate. On the same basis we may expect that if a normal conductor were in an external magnetic field before it became superconducting, the internal flux would not change. It was, however, found by Meissner and Ochsenfeld that superconductors do *not* behave this way; when the critical temperature is reached a superconductor will set up its own circulating current and expel the magnetic field. This unexpected effect (known as the Meissner effect) provided the first evidence that classical electromagnetism was insufficient for explaining the properties of superconductors.

The effect of a magnetic field can be described in relatively simple terms for a class of superconductors called Type I.* As the magnetic field is decreased it is found that at a certain magnetic field, $H_c(T)$, superconductivity is destroyed. The function $H_c(T)$ depends on the properties of the material but may in general be approximated by the expression

$$H_c(T) = H_c(0)(1 - t^2) \tag{1.1}$$

where $H_c(T)$ is known as the *critical magnetic field*, and $t = T/T_c$ is the *reduced*

---

*The division into Type I and Type II superconductors will be discussed in Section 1.5.

temperature. $H_c(T)$, plotted as shown in Fig. 1.1, may be looked upon as representing a boundary between a superconducting and a normal phase of the material. So one may expect to describe the transition between the two phases

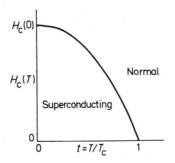

Fig. 1.1 The critical magnetic field as a function of reduced temperature.

with the aid of thermodynamics. This was indeed one of the earliest theoretical approaches. At constant temperature the change in the Gibbs free energy per unit volume may be described as [16]

$$dG = -\mu_0 M \, dH \qquad (1.2)$$

where M is the magnetisation related to the flux density by $B = \mu_0(H + M)$ and μ_0 is the free space permeability. For a diamagnetic material M is negative so that the Gibbs function increases with increasing magnetic field. Assuming that the superconductor is a perfect diamagnet, $M = -H$ and we get

$$G_S(H) = G_S(0) + \tfrac{1}{2}\mu_0 H^2. \qquad (1.3)$$

When $H = H_c$ there is a transition to the normal phase; the Gibbs free energies must be equal, $G_S(H_c) = G_N(H_c)$. Assuming further that the normal phase is nonmagnetic, $G_N(H_c) = G_N(0) = G_N$ we may write for the difference of the Gibbs free energies at zero magnetic field

$$G_S(0) - G_N = -\tfrac{1}{2}\mu_0 H_c^2. \qquad (1.4)$$

A number of other properties may be derived in a similar manner. Our main purpose was to derive Equations (1.3) and (1.4) which will be needed for the discussion of the Ginzburg–Landau theory.

1.4 The London equations

The first phenomenological theory describing the properties of the superfluid in the presence of electromagnetic fields was formulated by the London brothers [5, 6]. We shall follow here (in a much-abridged form) the account given by F. London [17].

. The starting point is the equation of motion for a nonviscous charged fluid

$$\frac{d\mathbf{v}}{dt} = \frac{e}{m}(\mathscr{E} + \mathbf{v} \times \mathbf{B}) \tag{1.5}$$

where e and m are the charge and mass of the particles comprising the fluid, $\mathscr{E}$ the electric field and $\mathbf{v}$ the particle velocity. Changing from Lagrangian to Eulerian representation and using a few vector identities we get

$$\frac{\partial \mathbf{w}}{\partial t} = \nabla \times (\mathbf{v} \times \mathbf{w}) \tag{1.6}$$

where

$$\mathbf{w} = \nabla \times \mathbf{v} + \frac{e}{m}\mathbf{B}. \tag{1.7}$$

Now Equation (1.6) has the property that if $\mathbf{w} = 0$ at $t = 0$ then $\mathbf{w}$ is identically zero for all subsequent time. It turns out that by making this rather arbitrary assumption the expulsion of the magnetic field from a superconductor may be explained. Hence our equation is

$$\nabla \times \mathbf{v} = -\frac{e}{m}\mathbf{B} \tag{1.8}$$

or introducing the vector potential, $\mathbf{A}$, it reduces to

$$\mathbf{v} = -\frac{e}{m}\mathbf{A}. \tag{1.9}$$

Further, using the relationship

$$\mathbf{j} = \rho e \mathbf{v} \tag{1.10}$$

where $\mathbf{j}$ is current density, ρ the number of charge carriers per unit volume, and Maxwell's equation (without the displacement current)

$$\nabla \times \mathbf{H} = \mathbf{j} \tag{1.11}$$

we get (in the gauge $\nabla\cdot\mathbf{A} = 0$)

$$\nabla^2 \mathbf{A} = \lambda_L^{-2}\mathbf{A} \tag{1.12}$$

where

$$\lambda_L^2 = \frac{m}{\rho e^2 \mu_0}. \tag{1.13}$$

Assuming a half-infinite superconductor and specifying the vector potential to be $A(0)$ in the plane $z = 0$, the solution of Equation (1.12) is

$$\mathbf{A} = \mathbf{A}(0)\,e^{-z/\lambda_L} \tag{1.14}$$

that is, λ_L may be regarded as the penetration depth into the superconductors (it is usually referred to as the London penetration depth). Taking $\rho = 10^{28}/m^3$ and identifying e and m with the charge and mass of an electron we get the numerical value $\lambda_L = 0.53 \ 10^{-7}$ m which is about the right magnitude. Note that ρ, the number of superconducting charge carriers, is a function of temperature, hence λ_L is also dependent on temperature.

1.5 The Ginzburg–Landau theory

The most successful phenomenological theory of superconductivity was proposed by Ginzburg and Landau [7]. The starting point is to express the difference of the Gibbs free energies between the normal and superconducting phases in terms of the order parameter ψ where $|\psi|^2$ is the density of superconducting electrons. It follows from the Landau–Lifshitz theory [18] of phase transitions that in the vicinity of T_c the required relationship may be obtained by expanding in terms of $|\psi|^2$, leading to

$$G_S(0) = G_N + \alpha|\psi|^2 + \frac{\beta}{2}|\psi|^4 \qquad (1.15)$$

where $G_S(0)$ is the Gibbs free energy in the superconducting state in the absence of a magnetic field; G_N is the Gibbs free energy in the normal state (assumed nonmagnetic), and α and β are constants. At a given temperature the density of superconducting electrons, $|\psi_0|^2$, which minimises $G_S(0)$ may be obtained from the condition $\partial G_S(0)/\partial|\psi|^2 = 0$, yielding

$$|\psi_0|^2 = -\frac{\alpha}{\beta} \qquad \text{and} \qquad H_c^2 = \frac{\alpha^2}{\mu_0 \beta} \qquad (1.16)$$

where Equation (1.4) has also been used.

For a finite magnetic field there is a further contribution to the Gibbs free energy due to the expulsion of the magnetic field. For an applied field H_a the amount expelled is $H_a - H$ and the increase in the Gibbs free energy is $\frac{1}{2}\mu_0(H_a - H)^2$.

Finally (and this is the essence of the Ginzburg–Landau argument), there is a contribution to the Gibbs free energy due to the gradient of ψ. This means that sudden variations in the order parameter cost energy. Another way of looking at it is to regard ψ as a kind of wave function and then $\nabla\psi$ is related to the kinetic energy density. The form chosen to satisfy gauge invariance is

$$\frac{1}{2m}|-i\hbar\nabla\psi - 2eA\psi|^2. \qquad (1.17)$$

We put here $2e$ for the charge of the superfluid particle anticipating the microscopic theory. Taking account of all contributions, the Gibbs free energy is

$$G_S = G_N + \alpha|\psi|^2 + \frac{\beta}{2}|\psi|^4 + \frac{1}{2}\left(\frac{1}{\mu_0}\nabla \times \mathbf{A} - H_a\right)^2 + \frac{1}{2m}|-i\hbar\nabla\psi - 2e\mathbf{A}\psi|^2. \quad (1.18)$$

Now we are looking for the functions ψ and $\mathbf{A}$ which minimise the total Gibbs free energy for a given volume. The solution belongs to the realm of variational calculus which provides the following differential equation to be satisfied

$$\frac{1}{2m}[-i\hbar\nabla - 2e\mathbf{A}]^2\psi + \alpha\psi + \beta|\psi|^2\psi = 0 \quad (1.19)$$

and

$$\nabla^2\mathbf{A} = \frac{(2e)^2\mu_0}{m}|\psi|^2\mathbf{A} + \frac{i\hbar 2^2\mu_0}{2m}(\psi^*\nabla\psi - \psi\nabla\psi^*). \quad (1.20)$$

In a weak field ψ may be expected to stay constant, $\psi \approx \psi_0$, $\nabla\psi \approx 0$ and Equation (1.20) reduces to Equation (1.12) with

$$\lambda^2 = \frac{m}{(2e)^2|\psi_0|^2\mu_0} \quad (1.21)$$

which, apart from the factor 2 is identical with the London penetration depth.

Let us now investigate Equation (1.19) in one dimension for the case when $\psi \ll \psi_0$. Then the cubic term may be neglected and we get

$$\frac{d^2\psi}{dz^2} = -\frac{\kappa^2}{\lambda^2}\left(1 - \frac{A^2}{2H_c^2\lambda^2\mu_0^2}\right)\psi \quad (1.22)$$

where the new parameter κ is defined by

$$\kappa = \lambda^2\frac{2^{3/2}eH_c\mu_0}{\hbar}. \quad (1.23)$$

Now make the assumption that the magnetic field *can* penetrate the superconductor, $H = H_a$ and (choosing the vector potential zero at $z = 0$)

$$A(z) = \mu_0 H_a z. \quad (1.24)$$

Substituting the above equation into Equation (1.22) we get the well-known differential equation of the quantum harmonic oscillator which has non-diverging solutions in ψ only when

$$H_a = \frac{\kappa\sqrt{2}}{2n+1}H_c \quad (1.25)$$

where n is an integer. Clearly, the largest value of H_a occurs when $n = 0$. If $\kappa > 1/\sqrt{2}$ then $H_a > H_c$. This means that under the circumstances investigated, a magnetic field larger than H_c (defined previously for the perfect diamagnet) may exist in the superconductor. The new limit is usually denoted by H_{c2} and the old one came to be called the *thermodynamic critical field*. The relationship between them is

$$H_{c2} = \kappa\sqrt{2}\,H_c. \tag{1.26}$$

We may now distinguish two different regions as shown in Fig. 1.2. Up to H_{c1} the superconductor is perfectly diamagnetic, between H_{c1} and H_{c2} there is an increasing penetration of the magnetic field, the superconductor is said to be in the *mixed* state; at H_{c2} superconductivity is destroyed and the material

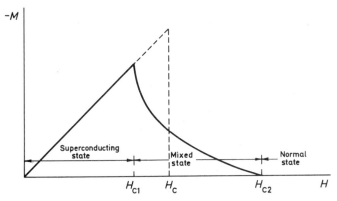

Fig. 1.2 The magnetisation curve of a Type II superconductor. The magnetic flux starts to penetrate the superconductor at H_{c1}; superconductivity disappears at H_{c2}. The thermodynamic critical field corresponding to perfect diamagnetism is H_c.

becomes nonmagnetic. We may now refer to superconductors having a mixed state as Type II while those having a diamagnetic state only are classified as Type I. From Equation (1.26) the borderline is given by $\kappa = 1/\sqrt{2}$. Interestingly this classification agrees very well with that suggested by Pippard [8] on the basis of somewhat different considerations. As it was shown by Gorkov [19, 20] (who derived the Ginzburg–Landau equations from microscopic theory) the parameter κ may be expressed as

$$\kappa(T) = \frac{\lambda(T)}{\xi(T)} \cong \frac{\lambda_L(0)}{\xi_0} \tag{1.27}$$

where

$$\lambda(T) = \frac{1}{\sqrt{2}}\lambda_L(0)(1-t)^{-1/2} \tag{1.28}$$

and

$$\xi(T) = 0.74(1-t)^{-1/2}\xi_0. \tag{1.29}$$

In the above equation $\xi(T)$ is called the temperature-dependent coherence length and ξ_0 is a parameter (also called coherence length) introduced originally by Pippard [8]. It may be expressed as

$$\xi_0 = 0 \cdot 18 \frac{\hbar v_F}{k T_c} \tag{1.30}$$

where v_F is the Fermi velocity.

The physical significance of the temperature-dependent coherence length $\xi(T)$ is that it determines the range in which the order parameter may change due to a perturbation (e.g. a boundary). With the aid of the coherence length the classification of superconductors is as follows: Type I when $\lambda_L(0) \ll \xi_0$ and Type II when $\lambda_L(0) \gg \xi_0$.

1.6 The macroscopic Schrödinger equation

One of the reasons why the mechanism of superconductivity was so difficult to unravel was that the problem could not be tackled by one-electron wave mechanics. However, once the condensation mechanism was found and the nature of the ground state established it became possible (and desirable) to find a simple model which would contain some grains of truth – enough to understand a certain set of phenomena. Such a model was introduced by Feynman [21] designed for undergraduate consumption but as shown by Mercereau [771] (and as we shall see later when discussing Josephson junctions) it has quite wide applicability. According to this model the superconducting electrons are all in the same state and may therefore be expected to obey a *macroscopic* Schrödinger equation where $|\Psi|^2$ (note that it is now time dependent) gives the density of superconducting electrons. Hence the differential equation to solve is

$$i\hbar \frac{\partial \Psi}{\partial t} = \frac{1}{2m}(-i\hbar\nabla - 2e\mathbf{A})^2 \Psi + 2eU\Psi \tag{1.31}$$

where U is a scalar potential. An advantage of this formulation is that we can immediately apply the solutions of the one-electron Schrödinger equation known from elementary quantum mechanics. Thus for example Equation (1.20) now follows straight from the definition of the probability current

$$\mathbf{j} = \frac{i\hbar 2e}{2m}(\Psi^* \nabla \Psi - \Psi \nabla \Psi^*) + \frac{(2e)^2}{2m}|\Psi|^2 \mathbf{A} \tag{1.32}$$

and from Maxwell's

$$\nabla \times H = \mathbf{j} \tag{1.33}$$

There is also certain similarity between Equations (1.19) and (1.31) though of course the former is independent of time. It was shown by Frohlich [22] that a proper choice of U (equivalent to introducing a pressure term) leads to a quantum fluid picture that contains the Ginzburg–Landau equations as a special case.

The wavefunction Ψ is complex. In many cases it is preferable to work in terms of real functions. We shall therefore introduce the superfluid density ρ and the phase of the wavefunction v as new variables by the relation

$$\Psi = \rho^{1/2} \exp iv. \tag{1.34}$$

The current may then be expressed in the form

$$\mathbf{j} = \frac{2eh}{m}\left(\nabla v - \frac{2e}{h}\mathbf{A}\right)\rho. \tag{1.35}$$

1.7 Fluxoid quantisation

Let us now investigate the relationship between current density, superfluid density, phase and vector potential for a superconducting ring. From Equation (1.35)

$$\nabla v = \frac{\mathbf{j}m}{2eh\rho} + \frac{2e}{h}\mathbf{A}. \tag{1.36}$$

Let us integrate both sides of Equation (1.36) along a closed curve in the middle of the ring

$$\oint \nabla v\,d\mathbf{s} = \frac{m}{2eh}\oint \frac{\mathbf{j}}{\rho}\,d\mathbf{s} + \frac{2e}{h}\oint \mathbf{A}\,d\mathbf{s}. \tag{1.37}$$

But

$$\oint \mathbf{A}\,d\mathbf{s} = \Phi \tag{1.38}$$

is the magnetic flux enclosed by the ring and

$$\oint \nabla v\,d\mathbf{s} = 2\pi n, \tag{1.39}$$

due to the fact that v can be different by an integral multiple of 2π as it gets back to the same point round the ring (only $|\Psi|^2$ must be the same). Hence

$$\Phi + \frac{m}{(2e)^2}\oint \frac{\mathbf{j}}{\rho}\,ds = \frac{2\pi h}{2e}n. \tag{1.40}$$

The left-hand side was called by London a *fluxoid* and it may be seen from Equation (1.40) that such a fluxoid is quantised in integral multiples of

$$\Phi_0 = \frac{2\pi h}{2e}. \tag{1.41}$$

If the thickness of the ring is large in comparison with the penetration depth λ then the current density in the middle of the ring is zero and the second term on the left-hand side of Equation (1.40) disappears. In that case the fluxoid is equal to the flux, that is we find that the magnetic flux threading a thick superconducting ring must be an integral multiple of a basic unit, Φ_0. Experimental verification of flux quantisation was done in 1961 by Doll and Näbauer [23] and by Deaver and Fairbank [24].

1.8 Surface energy and Abrikosov's vortex structure

It was first noted by London [17] that in the presence of a magnetic field the lowest energy would be achieved by a laminar structure of superconducting and normal layers. The thickness of the superconducting layer is roughly the penetration depth (to lower the energy by admitting the magnetic field) whereas the thickness of the normal layer is even smaller so as not to contribute much to the total energy. Since this state was not observed, London concluded that a positive surface energy must exist. Forming a superconducting-normal surface costs energy; hence a laminar structure cannot represent the lowest energy state. This argument proved to be correct for Type I but not for Type II superconductors. Type II superconductors possess a negative surface energy and the break-up into normal and superconducting regions does indeed occur above the lower critical field H_{c1} as shown by Abrikosov [25, 26]. The favoured structure is, however, filamentary and not laminar. The superconductor is pierced by a number of filaments which are regularly spaced parallel to the external field. The maximum value of the field is in the middle of the filament and decays according to the penetration depth, λ. The centre of the filament is normal, that is $|\psi|^2 = 0$ increasing to its maximum value in about a coherence length, $\xi(T)$. The magnetic field is surrounded by circulating superconducting currents creating vortex lines. The flux and the currents associated with a single isolated vortex line extend over a distance of about λ. Thus no appreciable interaction occurs until the separation of vortices becomes less than λ. Hence at $H = H_{c1}$ a density of vortices corresponding to a separation λ comes immediately into existence accounting for the infinite slope of the magnetisation curve (Fig. 1.2) at H_{c1}. As the magnetic field increases the density of vortex lines increases approaching a separation $\xi(T)$ as H tends to H_{c2}.

The magnetic flux per unit cell of the vortex lattice must be an integral multiple of the flux quantum. The existing theories and experimental observations seem to suggest that one flux quantum per vortex is the optimum configuration.

According to Abrikosov the vortex lines form a square lattice. Later calculations [27, 28] indicate that a triangular lattice has smaller energy and that is supported by experimental results obtained by neutron diffraction [29] and by deposition of ferromagnetic particles on the surface of the specimen [30].

For thin films it was shown by Tinkham [31, 32] that in a perpendicular magnetic field there is always a mixed state independently of whether the superconductor is Type I or Type II.

1.9 The electron–phonon interaction and Cooper-pairs

The main difficulty in finding a microscopic theory of superconductivity resided in the smallness of the interaction energy. Taking a few hundred gauss for the critical magnetic field of a typical Type I superconductor, the difference between the energies of the normal and superconducting states is $\frac{1}{2}\mu_0 H_c^2$ which gives about 10^{-8} eV per atom. This is very small indeed in comparison with the relevant Fermi energy (say about 10 eV).

The basic mechanism as suggested by Frohlich [9] is the electron–phonon interaction though later it became clear that only the exchange of virtual phonons* by a pair of electrons need to be considered. It may be looked upon physically as an electron being affected by the lattice deformation caused by another electron. In terms of wave vectors this means that an electron of wave vector $\mathbf{k}$ emits a phonon $\mathbf{q}$ which is absorbed by an electron having a wave vector $\mathbf{k}'$. The nature of the resulting electron–electron interaction turns out to be attractive for sufficiently small phonon energy, $\hbar\omega_q$.

It was shown by Cooper [12] that in a thin shell around the Fermi energy two electrons (known since as a Cooper pair) can form a bound state in the presence of an attractive interaction. The lowest energy can be achieved by pairing electrons of opposite momenta and spins.

1.10 The BCS ground state

The first successful microscopic theory was formulated by Bardeen, Schrieffer and Cooper in 1957 and is known as the BCS theory. It is remarkable that such a relatively simple model could account for so many experimental results.

The main assumptions in the derivation of the ground state are as follows:

(i) The superconducting ground state can be expressed solely in terms of Cooper pairs so that the states $(\mathbf{k}, -\mathbf{k})$ are occupied or empty simultaneously.**

(ii) The various interactions may be taken identical in the normal and superconducting states and only the phonon and screened Coulomb interactions need to be separated for attention.

---

* It means that energy need not be conserved because of their very short lifetime.
** It is assumed that the spins are antiparallel and remain antiparallel after scattering so there is no need to use spin indices.

(*iii*) The difference between the phonon and screened Coulomb interactions is $-V_{kk'}$ which may be expressed in the simple form*

$$V_{kk'} = V \quad \text{for} \quad |\varepsilon_k|, |\varepsilon_{k'}| \leqslant k\theta_D$$

and

$$V_{kk'} = 0 \quad \text{otherwise.} \tag{1.42}$$

where the energy ε_k is measured from the Fermi surface and θ_D is the Debye temperature.

In accordance with the above assumptions the ground state energy of the superconducting state at $T = 0°K$ (relative to the energy of the normal state) may be written as

$$W_S = 2 \sum_k \varepsilon_k v_k^2 - V \sum u_k v_k u_{k'} v_{k'} \tag{1.43}$$

where v_k^2 is the probability of the state $(\mathbf{k}, -\mathbf{k})$ being occupied and u_k^2 the probability that it is empty, consequently

$$v_k^2 + u_k^2 = 1. \tag{1.44}$$

The first term in Equation (1.43) gives the difference of kinetic energy between the superconducting and normal phases at zero degree.

The mathematical problem is now to minimise the ground state energy (Equation (1.43)) with respect to the probability v_k. The result is

$$v_k = \left[\frac{1}{2}\left(1 - \frac{\varepsilon_k}{E_k}\right)\right]^{1/2} \tag{1.45}$$

where

$$E_k = [\varepsilon_k^2 + \Delta_k^2]^{1/2} \tag{1.46}$$

and Δ_k may be determined from an integral equation.

Assuming further that

$$\Delta_k = \Delta = \text{constant} \qquad |\varepsilon_k| < k\theta_D$$
$$\text{for}$$
$$\Delta_k = 0 \qquad\qquad |\varepsilon_k| > k\theta_D \tag{1.47}$$

and

$$\Delta \ll k\theta_D, \qquad N_N(\varepsilon_k) \cong N_N(0) \tag{1.48}$$

(where $N_N(0)$ is the density of states at the Fermi surface in the normal state) a particularly simple expression may be obtained for the ground state energy

$$W_S = -\tfrac{1}{2}N_N(0)\Delta^2. \tag{1.49}$$

---

* This is an obvious oversimplification which completely disregards the details of the interaction. It gives a good approximation because the superconducting properties (in reduced coordinates) are hardly dependent on crystal structure and normal state properties.

The functions v_k^2 and E_k for this case are plotted in Fig. 1.3. It may be seen that the distribution of electrons is smeared even at absolute zero temperature. This distribution has a higher energy than the Fermi state but of course the total energy will be reduced below that of the normal state if the negative interaction energy is included as well.

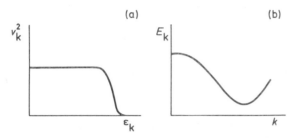

Fig. 1.3 (a) The occupation of electrons in the ground state, v_k^2 as a function of energy. (b) The excitation energy as a function of k.

1.11 Elementary excitations and the energy gap

Having determined the ground state we shall now enquire into the nature of the elementary excitations, which came to be called quasiparticles in modern theories. In a superconductor a given state k is partially occupied with probability v_k^2 and partially empty with probability u_k^2. Thus if we wish to create a quasiparticle with definite k-vector and spin we must partially *create* an electron and partially *destroy* one. This argument was put in mathematical form by Bogoliubov. In that theory the ground state (equivalent to the BCS ground state) can be obtained by requiring no quasiparticles.

Using the same simplifying assumptions as discussed in the previous section for the BCS ground state the excitation energy of one quasiparticle turns out to be

$$\text{excitation energy} = W_S + E_k. \tag{1.50}$$

Hence the minimum energy of excitations is Δ which may now be identified with an energy gap.

The density of states may also be obtained in the form

$$N_S(E) = N_N(E)\left[\frac{dE_k}{d\varepsilon_k}\right]^{-1} = N_N(E)\frac{E_k}{\sqrt{(E_k^2 - \Delta^2)}}. \tag{1.51}$$

To see the physical significance of the energy gap and compare the excitations in a superconductor with those in a normal metal and in a semiconductor, we shall follow here the description of Douglass and Falicov [33] in a somewhat modified form. Fig. 1.4 (a) and (b) show the excitation diagrams of a normal metal and insulator (semiconductor) respectively. Holes (denoted by dotted lines) are also plotted upwards because a certain excitation energy is required to create a hole. In a normal metal an infinitesimal amount of energy is sufficient to excite an electron-hole pair whereas in an insulator a minimum energy $E_g = 2\Delta$ is required. Hence it follows that a normal metal has finite conductivity and an insulator zero conductivity at absolute zero temperature.

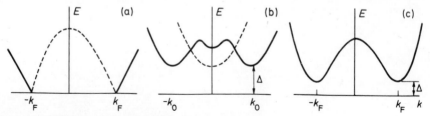

Fig. 1.4 Excitation diagrams of (a) a metal, (b) an insulator (semiconductor) and (c) a superconductor. The dotted lines signify holes (not shown for the superconductor because there the distinction between holes and electrons is somewhat blurred).

In a superconductor the distinction between electrons and holes is somewhat blurred so the excitation diagram (Fig. 1.4 (c)) should be understood in terms of quasiparticles.

In the context of these diagrams the main difference between an insulator and a superconductor is that in an insulator the minima of $E(k)$ are determined essentially by the lattice, and are fixed with respect to the lattice frame. In a superconductor the $E(k)$ curve is essentially determined by the electron inter-actions so it is possible to displace the electron distribution uniformly, displacing the minima of $E(k)$ with it. This corresponds to a superconductor carrying a current. Hence the effect of the energy gap is entirely different in the two cases. For an insulator it leads to zero conductivity while for a superconductor it may be considered responsible for the infinite conductivity.

We wish to mention here a further difference between insulators and super-conductors and that is the strong temperature dependence of the energy gap in the latter. As may be expected the energy gap is zero at the critical temperature and increases monotonically to its maximum value at zero temperature. This is shown in Fig. 1.5 in reduced coordinates. The full line is the BCS theoretical result, the circles represent experimental points for aluminium.

Finally, we shall give here the zero temperature value of the energy gap derived by BCS in the form

$$\Delta(0) = 1{\cdot}76 \, kT_c = 2k\theta_D \exp(-1/N_N(0)V). \tag{1.52}$$

This is simple enough. One may conclude immediately that elements with an even number of valence electrons per atom will have lower critical tempera-tures than elements with an odd number of valence electrons because an even number of electrons is more likely to fill a Brillouin zone so that $N_N(0)$ is small. However, as far as V is concerned Equation (1.52) should be regarded as being only qualitatively valid. Since V was introduced in such a simplified form it is difficult to assign any definite numerical value to it (unless one works backwards from the measured value of T_c). For an accurate determination of the critical temperature much more of the normal state properties are needed.

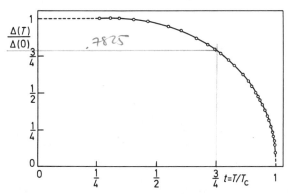

Fig. 1.5 The reduced energy gap as a function of reduced temperature according to the BCS theory. The circles show the experimental points (obtained by tunnelling measurements on an Al–I–Al junction) of Blackford and March.

1.12 Quasiparticles versus normal electrons; the semiconductor model

There is no doubt that Bogoliubov's quasiparticles are different from ordinary electrons. Thus instead of the normal and superconducting electrons of the two-fluid model we should now be talking about quasiparticles and Cooper pairs. But there is a lot of similarity between the quasiparticle description of a superconductor and the familiar semiconductor picture of electron-hole excitations. The minimum energy needed to create two quasiparticles (they must be created in pairs by breaking up a Cooper pair) is 2Δ in analogy to the energy needed to create an electron-hole pair in a semiconductor (denoting the half gap by Δ). Quasiparticles may also recombine and emit the energy in the form of phonons or photons just as electron-hole pairs do. The only difference is in the relative probabilities of phonon and photon emissions. Quasiparticles (like electrons) are fermions and follow the Fermi–Dirac distribution. So why not forget quasiparticles and just adopt the familiar electron-hole picture? It can be done and it has often been done. The main advantage of it is that the tunnelling picture used in the past for semiconductors and metals can be immediately applied to superconductors.

In Chapter 4 we shall treat superconductive tunnelling with the aid of the semiconductor model. It gives the correct quantitative answer too as will be shown in Section 4.6. The only modification needed is to account for the super-conducting density of states as given by Equation (1.51). The density of states has a singularity just below and above the gap as shown in Fig. 1.6. At zero temperature (Fig. 1.6(a)) all states are filled up to $E_F - \Delta$ and all states above the gap are empty. At finite temperature some of the electrons are excited above the gap leaving behind the same number of holes (shown schematically in Fig. 1.6(b)).

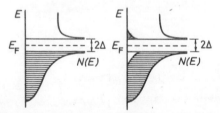

Fig. 1.6 The density of states as a function of energy in the semiconductor representation. Note that the energy scale is very much distorted. The Fermi energy is by several orders of magnitude greater than the gap energy. (a) No excited states at zero temperature, (b) electrons above the gap and holes below the gap at finite temperature.

1.13 'Dirty' superconductors

The superconducting properties depend on impurity concentration as it was first recognised by Pippard [8]. A Type I superconductor can for example, be turned into a Type II superconductor by adding impurities. However, a large amount of impurity scattering may be expected to lead to much more drastic consequences. A collision time of $\tau = 10^{-14}$ sec implies an uncertainty in energy of the order of $\hbar \tau^{-1}$, that is about 0·1 eV which is well above the usual energy gap of a few meV. Thus one could expect the disappearance of the gap whereas experimentally little change is observed provided the impurities are nonmagnetic.*

Anderson [34] explained this effect (or rather the absence of the effect) by studying impurity scattering first and then considering the interaction leading to superconductivity. The essential result is that the energy gap becomes isotropic (without *assuming* the isotropy as done in the BCS model). The Anderson theory becomes applicable when the energy uncertainty $\hbar \tau^{-1}$ is comparable to Δ or equivalently when $l \approx \xi_0$, where l is the mean free path.

Superconductors with a high scattering rate came to be called 'dirty' superconductors. For these superconductors the penetration depth and coherence length are given by the following expressions

$$\lambda(T) = 0.615 \, \lambda_L(0) \left(\frac{\xi_0}{l}\right)^{1/2} (1-t)^{-1/2} \tag{1.53}$$

and

$$\xi(T) = 0.85 \, (\xi_0 l)^{1/2} (1-t)^{-1/2}. \tag{1.54}$$

*Magnetic impurities destroy pairs (and hence depress the energy gap) by affecting the spins. Experimental results obtained by tunnelling will be discussed in Section 6.4.

2.   Tunnelling

2.1 Historical review and basic concepts

There are many quantum mechanical phenomena in conflict with common sense but tunnelling is perhaps the most striking among them. Classically, a barrier is a barrier; if there is not sufficient energy available it cannot be scaled. A ball will not get to the other side of a mountain unless given enough kinetic energy, otherwise it rolls back. Quantum mechanics maintains that if the mountain is thin enough the ball can get to the other side as if there was a tunnel bored into the mountain.

The concept of tunnelling was born nearly simultaneously with modern quantum mechanics. It followed from Schrödinger's equation, and indeed was used soon after the publication of Schrödinger's original paper to explain a number of phenomena, viz. the field ionisation of atomic hydrogen by Oppenheimer [35] (1928), field emission from a free-electron metal by Fowler and Nordheim [36] (1928), alpha decay by Gamow [37] (1928), metal–vacuum–metal junctions by Frenkel [38] (1930), metal–insulator–metal junctions by Sommerfeld and Bethe [39] (1933). There were a few near misses, e.g. Wilson's explanation of rectification in a metal–semiconductor junction [40] and Zener's explanation of dielectric breakdown [41]. In both cases the models set up were not comprehensive enough and the results were not in agreement with experiments. Wilson predicted rectification in the wrong direction and Zener's mechanism did not have a chance to act (it is gratifying to note that the type of breakdown suggested by him (called 'Zener breakdown' nowadays) was indeed found much later in certain p–n junctions).

Interestingly, after nearly thirty years existence as the sole playing ground of physicists, the phenomenon was brought to the attention of engineers in 1957 by Esaki's invention of the tunnel diode [42]. This led to considerable amount of further work, improved technology and above all helped to spread an interest in the difficult concept of tunnelling among those concerned with applications.

Superconductive tunnelling came in the 1960s with the measurements of Giaever [43, 44, 45] and Nicol et al. [46] on normal metal–insulator–superconductor (NIS) and superconductor–insulator–superconductor (SIS) sandwiches, and by Josephson's prediction [47] of a number of exciting effects in the SIS configuration.

After this brief historical summary* we shall now review the simplest formulation of the tunnelling problem. An electron moving with a kinetic energy E in zero potential is incident upon a potential barrier $U_2 > E$ as shown in Fig. 2.1.

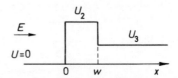

Fig. 2.1 One-dimensional potential barrier.

The probability of the electron appearing at the other side of the barrier may be worked out from the time-independent Schrödinger equation

$$\left(-\frac{\hbar^2}{2m}\nabla^2 + U\right)\psi = E\psi. \tag{2.1}$$

We shall be concerned with the one-dimensional solution only. For constant potential we get in region 1

$$\psi_1 = A \exp ik_1 x + B \exp(-ik_1 x) \tag{2.2}$$

where

$$k_1 = \frac{1}{\hbar}(2mE)^{1/2}. \tag{2.3}$$

The solutions in regions 2 and 3 are as follows

$$\psi_2 = C \exp(-\kappa x) + D \exp \kappa x \tag{2.4}$$

and

$$\psi_3 = F \exp ik_3 x \tag{2.5}$$

where

$$\kappa = \frac{1}{\hbar}[2m(U_2 - E)]^{1/2} \quad \text{and} \quad k_3 = \frac{1}{\hbar}[2m(E - U_3)]^{1/2} \tag{2.6}$$

and A, B, C, D, F are constants. We can determine them from the boundary conditions that at $x = x_1$ and $x = x_2$ both ψ and $d\psi/dx$ should be continuous. The quantity of interest is $|F/A|$, that is the amplitude relationship between the output and input. After a lot of algebra (and assuming that $\exp(-\kappa w)$ is negligible in comparison with $\exp \kappa w$) we get

$$\left|\frac{F}{A}\right| = \frac{4k_1\kappa}{(k_1^2 + \kappa^2)^{1/2}(k_3^2 + \kappa^2)^{1/2}} \exp(-\kappa w) \tag{2.7}$$

*The history of superconductive tunnelling will be discussed in considerably more detail in Chapters 3 and 8.

whence the ratio of current densities is

$$\frac{j_3}{j_1} = \frac{k_3}{k_1}\left|\frac{F}{A}\right|^2 = \frac{16\,k_1 k_3 \kappa^2}{(k_1^2+\kappa^2)(k_3^2+\kappa^2)}\exp\left(-2\kappa w\right). \qquad (2.8)$$

Quite obviously, the above expression is dominated by the exponential factor. The wider is the barrier the smaller is the current which can flow across. Practical barriers in solids may be of the order of one electron volt; substituting this value for $U_2^{\varnothing}-E$ into Equation (2.6), taking the free electron mass and a current transmission of 10^{-6}, we get for the width of the barrier

$$w = \frac{1}{2\kappa}\ln 10^6 \approx 10^{-9}\ \text{m}. \qquad (2.9)$$

This happens to be the right order of magnitude. In order to observe tunnelling the width of the barrier should be of the order of 1nm.

2.2 Metal–insulator–metal junctions

The best introduction to superconductive tunnelling is to investigate first the case when both metals are still in their normal states. We shall give a heuristic derivation starting with a simple model.

In thermal equilibrium the Fermi levels of the two metals match as shown in Fig. 2.2 (a). If a negative voltage V is applied* to the metal on the left-hand side

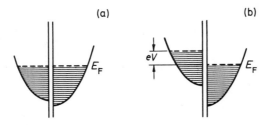

Fig. 2.2 Energy diagram for a metal–insulator–metal junction (a) at thermal equilibrium, (b) at a potential difference V.

then all the electrons will have an energy, eV higher and consequently the energy diagram will appear as shown in Fig. 2.2 (b). Let us now calculate the current flowing across the junction. We shall not consider here the proportion of electrons incident upon the junction but shall simply argue that the number of electrons which will move from left to right in an energy interval dE must be proportional to the number of occupied states on the left, that is to

$$N_1(E-eV)f(E-eV)\,dE \qquad (2.10)$$

* All the junctions treated in this book behave independently of the sign of the applied voltage. We chose the sign in such a way that the net electron flow is from left to right.

where N_1 is the density of states, f is the Fermi function, and note that energy is measured from the Fermi level of the metal on the right. But the electrons can move to the right only if there are unoccupied states there, that is the current must be proportional to

$$N_2(E)[1-f(E)] \tag{2.11}$$

as well. Including finally the probability of transition across the barrier, $P_{12}(E)$, the current flowing from left to right is

$$I_{l \to r} \sim P_{12}(E)N_1(E-eV)N_2(E)f(E-eV)[1-f(E)]\,dE. \tag{2.12}$$

Similar arguments lead to a current from right to left

$$I_{r \to l} \sim P_{21}(E)N_1(E-eV)N_2(E)f(E)[1-f(E-eV)]\,dE. \tag{2.13}$$

Assuming further that $P_{12}(E) = P_{21}(E)$ (an electron has just as much chance to tunnel from left to right as the opposite way) and integrating for all energies we get the net current in the form

$$I \sim \int P_{12}(E)N_1(E-eV)N_2(E)[f(E-eV)-f(E)]\,dE. \tag{2.14}$$

Now one of the usual approximations is to take $P_{12}(E)$ independent of energy and take it outside the integral sign. This seems perfectly justified for small applied voltage. Similarly, we may argue that the density of states is a slowly varying function and take its value at the Fermi level, i.e.

$$N_1(E-eV) \cong N_1(E) = N_1(0) \qquad \text{and} \qquad N_2(E) \cong N_2(0). \tag{2.15}$$

With these approximations Equation (2.14) reduces to

$$I = AN_1(0)N_2(0)\int [f(E-eV)-f(E)]\,dE \tag{2.16}$$

where A is a constant incorporating both P_{12} and the geometry of the junction.

For small voltages the Fermi function may be expanded to give

$$f(E-eV)-f(E) = -eV\frac{df}{dE}. \tag{2.17}$$

If the temperature is not too high $-df/dE$ may be approximated by a delta function, giving for the current

$$I = AN_1(0)N_2(0)eV \tag{2.18}$$

that is, a metal–insulator–metal junction obeys Ohm's Law; the relationship between current and voltage is linear.

2.3 Many-body formulation of tunnelling

The theoretical formulation given in the previous section was regarded as being highly respectable until the early 1960s when the wandering attention of theoretical physicists fell upon the problem of tunnelling. The simple model came under scrutiny and tunnelling became part of the many-body problem.

The first query one might raise in connection with our simple treatment is the nature of P_{12}, which we called the probability of transition (or tunnelling). Intuitively one feels that it must be related to the transmitted current given by Equation (2.8). However, in order to apply the well-developed machinery of modern quantum mechanics we should have some basic sets of states, a perturbing Hamiltonian and instead of a vaguely defined probability of transition a proper matrix element connecting two states in the presence of the perturbation. This problem was first attacked by Bardeen [48]; we shall follow here the more detailed exposition of Kane [49].

Let us consider the barrier region again. As we stated before, Equation (2.4) represents an exact solution of Schrödinger's equation in the region $x_1 < x < x_2$; there are two exponential functions declining in the positive and negative directions respectively. We shall introduce a new notation and denote these two functions by

$$\psi_1 = C_1 \exp(-\kappa x) \quad \text{and} \quad \psi_r = C_r \exp \kappa x. \tag{2.19}$$

There is no change so far. But now we are going to resort to an approximation. We say that ψ_1 is a solution not only for $x_1 < x < x_2$ but for the whole region $x > x_1$ as well (for $x < x_1$ it is supposed to be matched to the correct solution). By doing this we commit only a small error because ψ_1 is already very small at $x = x_2$ and declines further exponentially. Similarly, we shall regard ψ_r as a solution for $x < x_2$.

Next we assume that an electron is initially in the state ψ_1 and work out its transition rate into ψ_r. We may write then for the complete time-dependent wavefunction

$$\Psi = q(t)\psi_1 \exp\left(-i\frac{E_1}{\hbar}t\right) + r(t)\psi_r \exp\left(-i\frac{E_r}{\hbar}t\right) \tag{2.20}$$

where $q(t)$ and $r(t)$ are independent of the spatial coordinate. Substituting Equation (2.20) into the time-dependent Schrödinger equation

$$H\Psi = i\hbar\frac{\partial\Psi}{\partial t} \tag{2.21}$$

we get after a number of operations and simplifying assumptions (see Appendix 1)

$$i\hbar\frac{dr}{dt} = \int \psi_r^*(H - E_1)\psi_1\, dx \exp\left(i\frac{E_r - E_1}{\hbar}t\right). \tag{2.22}$$

If we now write the Hamiltonian H in the form $H_0 + H_1$ and note that $H_0\psi_1 = E_1\psi_1$, we have

$$\int \psi_r^*(H - E_1)\psi_r dx = \int \psi_r^* H_1 \psi_1 dx; \tag{3.23}$$

so we may define the effective matrix element for tunnelling by

$$T_{rl} = \int \psi_r^*(H - E_l)\psi_l \, dx. \qquad (2.24)$$

The essential assumptions in deriving this expression were that ψ_l and ψ_r are good approximate solutions of the exact Hamiltonian H, and exact solutions of the hypothetical 'unperturbed' Hamiltonian, H_0.

After a few more mathematical operations (Appendix 2) we get

$$T_{rl} = \frac{\hbar^2 \kappa}{m} C_r^* C_1. \qquad (2.25)$$

Having got the matrix element we may now determine the current (number of particles transferred per unit time due to the perturbation) by Fermi's golden rule. Performing the calculations we obtain the same expression for the transmitted current as before. Thus we have shown (at least for the case when the barrier has sharp boundaries) that Bardeen's approach gives the same result as the simple one-electron calculation.

The way was now open to describe tunnelling in the language of quantum field theory. It is only one step from Equation (2.25) to arrive at the formula

$$H_T = \sum_{r,l} (T_{rl} C_r^\dagger C_l + T_{lr} C_l^\dagger C_r) \qquad (2.26)$$

first introduced by Cohen *et al.* [50]. H_T is here the so-called effective tunnelling Hamiltonian (or simply tunnelling Hamiltonian) in second quantised formulation, $C_r^\dagger$ and C_l are creation and annihilation operators, and the sum is over all the left-hand and right-hand states. The transition is described as destroying an electron on the left and creating it on the right, and vice versa.

A number of subsequent theoretical papers (including Josephson's original paper [47]) made use of the above approach which has become known in the literature as the tunnelling Hamiltonian method.

2.4 A note on the density of states

In Section 2.2 we took the probability of transition and the density of initial and final states as being independent of energy though, clearly, they all depend on energy. Our aim is here to find the energy dependence of the product that is of $|T_{rl}|^2 N_1(E) N_2(E)$. The solution of this problem was attempted by Harrison [51] using the WKB approximation (for a discussion see Ref. [53]). In that approach, instead of assuming sharp boundaries k is regarded as a continuous function of the spatial coordinates, and Schrödinger's equation is solved approximately. Unfortunately, the approximate solution diverges at the classical turning points where $k = 0$. It is possible to overcome this difficulty but the necessary mathematical technique is rather laborious so we shall not go into details. Harrison's conclusions may be summarised as follows:

(*i*) in the case of specular reflection at the boundaries there is a selection rule which conserves the spin and the transverse components of the *k* vector,

(*ii*) the matrix element may be written as*

$$|T_{rl}|^2 = k_r k_l |D_{rl}|^2 \tag{2.27}$$

where D_{rl} is independent of energy.

Since the one-dimensional densities of states are inversely proportional to *k* it follows from Equation (2.27) that

$$|T_{rl}|^2 N_1(E) N_2(E) \tag{2.28}$$

is independent of energy. Thus the approximations of Section 2.2 are perfectly justified; the expression shown in (2.28) can be taken out of the integrand of Equation (2.14).

It follows that no tunnelling measurement can give information* about the density of states. This conclusion is often quoted in the literature and experimental results seem to confirm it. It should, however, be noted that not all approaches lead to this conclusion. If instead of the WKB approximation the problem is solved exactly for sharp boundaries, the density of states does *not* cancel [52]. Since practical boundaries are never ideal we may assume that this latter calculation is not applicable.

It should be further noted that Harrison's calculations give, in general, matching conditions (selection rules) for the component of group velocity normal to the barrier. Dowman *et al.* [52] claim that the relevant quantity is not the group velocity but the *k*-vector in the repeated-zone scheme of the metal normal to the barrier. They stress the importance of the insulator structure and of the insulator–metal interface in interpreting the measured tunnelling characteristics.

---

* This applies only to the density of states in a normal metal or semiconductor; for superconductors the density of states does not cancel.

II. Normal electron tunnelling

3.   Introduction

3.1 The discovery of superconductive tunnelling

The first measurements on superconductive tunnelling were performed by Giaever in 1960 when he measured the current–voltage characteristics of a normal metal–insulator–superconductor sandwich. He observed that as one of the metals became superconductive the resistance of the junction drastically increased. He explained the result by reference to the superconducting energy gap which reduced the electron flow by not accepting electrons with small excitation energies. It was a simple experiment giving a large amount of information.

We shall briefly describe in this Introduction how superconducting tunnelling developed from a promising start into the most sensitive probe of the superconducting state. But let us pause for a moment and consider first a philosophical question. Why was superconductive tunnelling discovered in 1960 and not earlier? Could it have been discovered in the 1930s at the same time (or a little later) as tunnelling between normal metals? It was certainly not beyond the means of the experimenters to perform such an experiment. All that had to be done was to drop a junction exhibiting tunnelling into liquid helium and measure its current–voltage characteristic. Did anyone think of it? Two experts, one on superconductivity (Meissner) and one on tunnelling (Holm) did in fact combine forces to measure the resistance of pressure contacts at low temperatures [54]. The work was done in 1932, perhaps a little too early. At a time when the theory of metals was just being born and the concept of band structure was only two years old it would have needed more than human insight to recognise the superconducting energy gap.

Meissner and Holm did certainly observe Josephson tunnelling. They did report explicitly that at the onset of superconductivity the contact resistance of the insulating layer disappeared. It seems very likely that they observed normal electron tunnelling between superconductors as well, as the following quotation shows: 'We found in our measurements Sn–Sn and Pb–Sn contacts which did not become superconducting and had large resistances'. Their aim was to measure the small contact resistance so they were not particularly interested in samples with large resistances.

It is not very susprising that the experiments were not repeated in the 1930s; why repeat a fairly ordinary experiment when a large number of pressing problems were crying for attention? But the situation changed as the clues about the energy gap slowly accumulated. By the early fifties there were a number of convincing proofs for the existence of an energy gap. So by then there was a real need for a simple experiment which could measure the gap. The time was ripe but still nobody thought of doing the experiment. Then in 1957 Esaki invented the tunnel diode, Glover and Tinkham [55] measured the drop in far infrared absorption, and Bardeen, Cooper and Schrieffer [13] established the BCS theory. After this any physicist could have hit upon the idea but it was left to Giaever, a mechanical engineer by training, to perform the first experiment and make a major impact upon the subject. His conversion to Physics started in 1958 and by 1959 he was measuring tunnel currents on evaporated-film Al–Al_2O_3–Al junctions at room temperature. In 1960 his studies in physics reached superconductivity. He learned of the superconducting energy gap, recognised its effect upon the tunnelling current and immediately set forth to measure the tunnelling characteristics at low temperatures. After a few attempts he found the effect.*

This is not a usual story of discovery. Luckily, it happens a few times; otherwise the history of science would be very dreary indeed.

3.2 An historical review

Giaever's experiments were quickly followed by others aimed first at the determination of the energy gap but rapidly extending to more and more aspects of superconductivity and also finding a variety of applications.

Photon-assisted tunnelling was first investigated by Dayem and Martin [57] and explained theoretically by Tien and Gordon [58]. The first experiments on phonon-assisted tunnelling were done by Goldstein and Abeles [59] and Lax and Vernon [60]. The controversial two-particle tunnelling was first reported by Taylor and Burstein [61] and Adkins [62]. Subharmonic tunnelling belongs to the same class (at least as far as the ambiguity of the explanation is concerned); it was also reported by Taylor and Burstein, observed by Rowell [63] and reported in more detail by Yanson et al. [64]. An interesting geometrical resonance effect was discovered by Tomasch [65].

The most remarkable single story is that of the role of tunnelling in the development of a strong-coupling theory of superconductivity. It started with an innocent little kink in the tunnelling measurements of Giaever et al. [66] on lead–insulator–magnesium (lead superconductor, magnesium normal) junctions at a voltage roughly corresponding to the Debye energy in lead. This gave the essential clue that the reason for some of the different superconducting properties of lead (e.g. $2\Delta(0)/kT_c \cong 4\cdot3$ instead of the BCS value $3\cdot52$) should be sought in the strong electron–phonon coupling. There followed a fascinating parallel

*A more detailed version of the circumstances leading to the discovery of superconductive tunnelling is given by Schmitt [56].

development of theories and experimental techniques culminating in the so-called gap inversion technique of McMillan and Rowell [67]. They used part of the tunnelling data to determine the normal state parameters in Eliashberg's theory [68] and then compared the predictions of the theory with the rest of the experimental data. They were able to conclude that the present theories of superconductivity are accurate to a few percent.

The dependence of the energy gap on various parameters has also been widely investigated with the aid of tunnelling. The more important steps were as follows: temperature dependence by Giaever and Megerle [69], magnetic field dependence by Giaever and Megerle [69] and Douglass [70]; dependence on transport current by Levine [71], on film thickness by Wilson [72], on crystal orientation by Zavaritskii [73–76], on pressure by Franck and Keeler [77]. The first experimental proof of the disappearance of the gap (gapless superconductivity) due to the presence of magnetic impurities was given by Reif and Woolf [78].

Tunnelling techniques also proved useful in investigating superimposed films of normal metals and superconductors (proximity effect). Tunnelling into the normal side was first done by Smith, Shapiro, Miles and Nicol [79] who could thus prove the presence of an induced energy gap in the normal metal. Tunnelling into the superconducting side (and thus showing the reduction in the energy gap) was first done by Reif and Woolf [78] and Frerichs and Wilson [80] using magnetic (Fe) and nonmagnetic (Al in the normal state) backing metals respectively.

Conclusions about vortex structure were drawn by Tomasch [81–83], Sutton [84], Nedellec et al. [85] and Donaldson and Brassington [86]. Information about the lifetime of the excited normal electrons was obtained by Ginsberg [87] and Miller and Dayem [88] and about the diffusion constant by Levine and Hsieh [89].

A further interesting application of superconductive tunnelling is spectroscopy. The resonances of molecules trapped in the barrier provide an alternative path of tunnelling and hence appear in the tunnelling characteristics. First indications were obtained by Marcus [90] and Clark [91]; detailed structure by Klein and Leger [92].

Device applications have also been investigated. The negative resistance arising in tunnel junctions consisting of different superconductors was used in a lumped amplifier by Miles et al. [93] and in a distributed amplifier by Yuan and Scott [94]. Detection of photons was analysed by Burstein et al. [95], generation and detection of phonons was shown to be feasible by Eisenmenger and Dayem [96], stimulated emission was observed by Gregory et al. [97]. Applications for thermometry were first envisaged by Giaever et al. [66], for nuclear detection by Wood and White [98].

4. Basic characteristics

4.1 Introduction

This chapter is concerned with the basic phenomena of tunnelling which occur when at least one of the electrodes is a superconductor. In Sections 4.2 to 4.4 we shall try to give a physical picture and then in 4.5 we shall put the argument in mathematical form and derive the relationship between current and voltage. Section 4.6 is concerned with the microscopic theory. An attempt is made there to justify the results of the conceptually simple phenomenological theory but without going into much detail. An alternative diagrammatic representation of superconductive tunnelling is discussed in Section 4.7. A number of deviations from the premises of the theoretical model are mentioned in Section 4.8 and the last Section, 4.9, is devoted to fabrication techniques.

4.2 Normal metal–barrier–superconductor junctions

We shall consider here a junction between a normal metal and a superconductor separated by a thin insulating barrier (usually denoted as N–I–S or simply NS) as shown schematically in Fig. 4.1 in the semiconductor representation (discussed in Section 1.13). At absolute zero temperature all the states are filled up to $E_F - \Delta$ and there are no filled states above the gap. At finite temperature there are electrons above the gap and holes below the gap.

In thermal equilibrium the Fermi energies must match (Fig. 4.1 (a)). When a voltage $V < \Delta/e$ is applied the electrons on the left have no access to empty states and thus no current can flow. At $V = \Delta/e$ there is a sudden rise in current, not only because now electrons may tunnel from left to right, but in addition

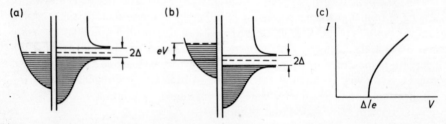

Fig. 4.1 The energy diagram of an NS junction in the semiconductor representation; (a) $V = 0$, (b) $V > \Delta/e$, (c) the I–V characteristic at $T = 0$.

they face a large density of states. For $V > \Delta/e$ further empty states become available for tunnelling (Fig. 4.1 (b)) and the current increases as shown in Fig. 4.1 (c).

For finite temperatures (Fig. 4.2(a)) some of the electrons on the left have energies in excess of $E_F + \Delta$ even at thermal equilibrium and there are also some normal electrons above the gap on the right. Now a very small voltage is

(a)

(b)

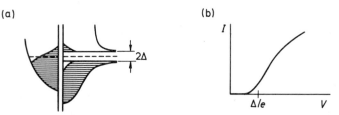

Fig. 4.2 (a) The energy diagram of an NS junction at finite temperature in thermal equilibrium, (b) the I–V characteristic at finite temperature.

sufficient for the current to start to flow but any appreciable rise in current must again occur around $V = \Delta/e$ as shown in the I–V characteristic of Fig. 4.2 (b).

The first experiments were performed by Giaever [43, 45] on aluminium–aluminium oxide–lead junctions. The current–voltage characteristic was found to be linear when both metals were in the normal state (curve 1 in Fig. 4.3) but it turned highly nonlinear when lead became superconducting (curve 2). The first estimate of the superconducting energy gap by tunnelling experiments was made with the aid of the latter curve.

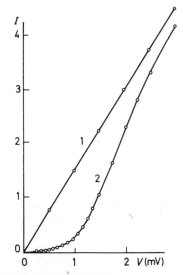

Fig. 4.3 The I–V characteristic of an Al–I–Pb junction (Al normal) after Giaever [44]. Curve 1: Pb normal, Curve 2: Pb superconducting.

4.3 Junctions between identical superconductors

The energy diagram for $T = 0°K$ is shown in Fig. 4.4. All energy levels are filled up to $E_F - \Delta$. In thermal equilibrium (Fig. 4.4 (a)) there is no current flowing. When a voltage $V < 2\Delta/e$ is applied there is still no current flowing because the electrons below the gap on the left have no access to empty states on the right. At $V = 2\Delta/e$ (Fig. 4.4 (b)) there is a sudden rise in current because electrons on the left suddenly gain access to the states above the gap on the right. The corresponding current–voltage characteristic is shown in Fig. 4.4 (c).

(a)   (b)   (c)

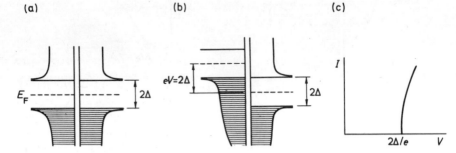

Fig. 4.4 The energy diagram of an SS junction; (a) $V = 0$, (b) $V = 2\Delta/e$, (c) the I–V characteristic at $T = 0$.

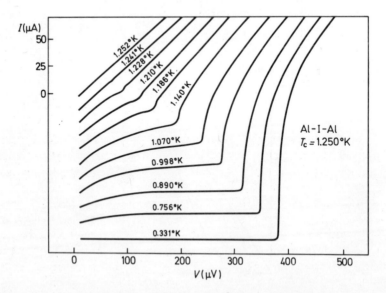

Fig. 4.5 The I–V characteristic of an Al–I–Al junction. Subsequent curves are displaced for clarity. After Blackford and March [99].

For finite temperatures there will be some rounding off* of the sharp features of Fig. 4.4 (c) which, of course, depends on the actual temperature (how near it is to the critical). A very neat set of experimental results (Fig. 4.5) by Blackford and March [99] shows the temperature dependence of the current–voltage characteristic for an aluminium–aluminium oxide–aluminium junction. At $1 \cdot 252°$K aluminium is in the normal state and the characteristic is linear. At $1 \cdot 241°$K (a mere 9 millidegrees below the critical temperature) there is already some sign of the energy gap, and it becomes clearly discernible at $1 \cdot 228°$K. As the temperature decreases the knee in the curves moves to higher and higher voltages (corresponding to higher and higher energy gaps). The characteristic at $T = 0 \cdot 331°$K is practically identical to that at $0°$K.

4.4 Junctions between superconductors of different energy gap

In the same way as the previously discussed case of identical superconductors, at $T = 0°$K no current flows until the applied voltage is sufficiently large to bring the bottom of the gap on the left in line with the top of the gap on the right. This occurs at an applied voltage of $V = (\Delta_1 + \Delta_2)/e$ as shown in Fig. 4.6 (a). The current–voltage characteristic (Fig. 4.6 (b)) is similar to that shown in Fig. 4.4 (c) with the sole difference that the current starts rising at a voltage corresponding to the arithmetical mean of the gap energies.

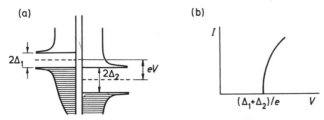

Fig. 4.6 Energy diagram and I–V characteristic of an $S_1 S_2$ junction at $T = 0$.

At finite temperatures we may still assume that the normal electron states above the larger gap are empty but there are some thermally excited normal electrons in the smaller-gap superconductor as shown in Fig. 4.7 (a) for the case of thermal equilibrium.

Applying a voltage the current will start to flow immediately and will increase with increasing voltage (Fig. 4.7 (e)) until $V = (\Delta_2 - \Delta_1)/e$. The energy diagram for this case is shown in Fig. 4.7 (b); at this stage all electrons above the gap on the left can tunnel across into empty states on the right. What happens when the voltage is increased further? The number of electrons capable to tunnel

---

*Taking the BCS density of states, there is theoretically a discontinuity in current at $V = 2\Delta/e$ even for finite temperatures. The experimentally measured rounding off is due to a number of 'nonideal' circumstances as will be discussed in Section 4.7.

across is still the same but they face a smaller density of states, as shown in Fig. 4.7 (c), hence the current decreases. The decrease in current continues until $V = (\Delta_1 + \Delta_2)/e$. At this point (Fig. 4.7 (d)) electrons from below the gap on the left gain access to empty states on the right, and there is a sudden increase in current. Thus the current–voltage characteristic of Fig. 4.7 (e) exhibits a negative resistance in the region

$$\frac{\Delta_2 - \Delta_1}{e} < V < \frac{\Delta_2 + \Delta_1}{e} \tag{3.1}$$

The appearance of a negative resistance was reported simultaneously by Nicol *et al.* [46] and Giaever [45]. A very convincing characteristic presented by the latter author for an Al–Al$_2$O$_3$–Pb junction is shown in Fig. 4.8.

The experimentally found dependence [100] of the negative resistance on temperature is shown in Fig. 4.9 for a Sn–SnO–Pb junction. The current–voltage characteristic turns nonlinear when lead becomes superconducting and the negative resistance appears as soon as tin becomes superconducting as well. The negative resistance may be clearly seen down to 2·39°K but not at 1·16°K. Experimentally the negative resistance always disappears at sufficiently low temperatures but that may be due to insufficient accuracy of measurement and to nonideal circumstances.

The presence of a maximum and minimum in the characteristic gives further help in diagnostic measurements aimed at determining the width of the energy gaps. In addition, the negative resistance may be used in devices which will be discussed in more detail in Section 7.1.

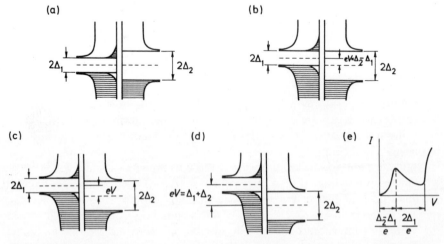

Fig. 4.7 The energy diagram and I–V characteristic of an $S_1 S_2$ junction at finite temperature; (a) $V = 0$, (b) $V = (\Delta_2 - \Delta_1)/e$, (c) $(\Delta_2 - \Delta_1)/e < V < (\Delta_2 + \Delta_1)/e$, (d) $V = (\Delta_1 + \Delta_2)/e$, (e) the I–V characteristic.

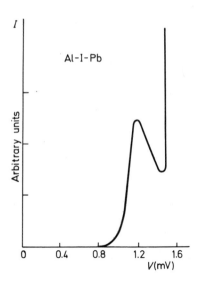

Fig. 4.8 The *I–V* characteristic of an Al–I–Pb junction, both Al and Pb superconducting. After Giaever [45].

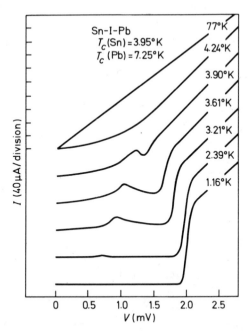

Fig. 4.9 I–V characteristics of an Sn–I–Pb junction.

4.5 Phenomenological theory

In this section we shall extend the treatment of Section 2.2 to tunnelling of normal electrons in superconductors. The essential assumption we have to make is that when the metal turns superconductor the density of states changes but nothing else does. The matrix element remains the same and the factors u_k, v_k, associated with quasiparticles may be ignored.*

We may then use Equation (2.14) for describing the net current flowing across a normal metal–insulator–superconductor junction provided we replace $N_2(E)$ by the density of states in a superconductor (Equation (1.51)):

$$N_S(E) = N_N(E)n_S(E). \tag{4.2}$$

But the energy dependence of $N_N(E)$, the density of states in the normal metal, cancels again and the current may be written as

$$I_{NS} = AN_{1N}(0)N_{2N}(0)\int n_S(E)[f(E-eV)-f(E)]\,dE. \tag{4.3}$$

As may be seen the only difference from Equation (2.16) is the appearance of $n_S(E)$ in the integrand.

At $T = 0$ the Fermi function is unity up to $f(0)$, and takes the value zero above that energy. Hence

$$f(E-eV)-f(E) = \begin{matrix} 1 \\ 0 \end{matrix} \quad \text{for} \quad \begin{matrix} 0 < E < eV \\ E < 0 \quad \text{and} \quad E > eV. \end{matrix} \tag{4.4}$$

Taking account of the above relationship, Equation (4.3) modifies to

$$I_{NS} = AN_{1N}(0)N_{2N}(0)\int_0^{eV} n_S(E)\,dE. \tag{4.5}$$

For the differential conductance we get

$$\frac{dI_{NS}}{dV} = AN_{1N}(0)N_{2N}(0)n_S(eV)e, \tag{4.6}$$

that is the density of states function of the superconductor may be determined by measuring the differential conductance at a temperature very close to absolute zero.

If we assume the BCS density of states as given by Equation (1.51) we get for the current, $I_{NS} = 0$ for $eV < \Delta$, and

$$I_{NS} = AN_{1N}(0)N_{2N}(0)\int_\Delta^{eV} \frac{E\,dE}{(E^2-\Delta^2)^{1/2}}$$

$$= AN_{1N}(0)N_{2N}(0)[(eV)^2-\Delta^2]^{1/2} \tag{4.7}$$

*This is just a lucky coincidence as will be discussed in Section 4.6 where the proper derivation from the microscopic theory is outlined.

for $eV > \Delta$.

When $T \neq 0$ we need to integrate

$$I_{NS} = AN_{1N}(0)N_{2N}(0) \int \frac{|E|}{(E^2 - \Delta^2)^{1/2}} [f(E) - f(E - eV)] \, dE \qquad (4.8)$$

for all the possible energies. The calculations are carried out in Appendix 3 yielding

$$I_{NS} = 2G_{NN}\frac{\Delta}{e} \sum_{m=0}^{\infty} (-1)^{m+1} K_1(m\Delta/kT) \sinh(meV/kT) \qquad (4.9)$$

where

$$G_{NN} = AN_{1N}(0)N_{2N}(0)e \qquad (4.10)$$

is the normal state conductance as may be seen from Equation (2.18), and K_1 is the first-order modified Bessel function of the second kind.

Equation (4.9) was first obtained by Giaever and Megerle [69]; it converges for $eV < \Delta$. When $V \to 0$ it reduces to

$$\lim_{V \to 0} I_{NS} = 2G_{NN}\frac{V\Delta}{kT} \sum_{m=0}^{\infty} (-1)^{m+1} mK_1(m\Delta/kT). \qquad (4.11)$$

When $T \to 0$ we may use the asymptotic form of the modified Bessel function [101]

$$K_1(z) \cong \left(\frac{\pi}{2z}\right)^{1/2} \exp(-z) \qquad (4.12)$$

We may then neglect all the terms $m > 1$ and get*

$$\lim_{\substack{V \to 0 \\ T \to 0}} I_{NS} = I_{NN}(2\pi\Delta/kT)^{1/2} \exp(-\Delta/kT) \qquad (4.13)$$

that is, for sufficiently low voltage at sufficiently low temperature, Δ may be determined by measuring I_{NS}/I_{NN}.

When both materials are superconducting the density of states functions of the normal metals cancel again and Equation (2.14) modifies to

$$I_{SS} = AN_{1N}(0)N_{2N}(0) \int \frac{|E - eV|}{[(E - eV)^2 - \Delta_1^2]^{1/2}} \frac{|E|}{(E^2 - \Delta_2^2)^{1/2}} [f(E - eV) - f(E)] \, dE$$

$$(4.14)$$

When $T = 0$ and the gaps are identical, $\Delta_1 = \Delta_2$, the above integral may be expressed in the form (see Appendix 4), $I_{SS} = 0$ for $V < 2\Delta/e$ and

$$I_{SS} = \frac{G_{NN}}{e}\left[(2\Delta + eV)E(\alpha) - 4\frac{\Delta(\Delta + eV)}{2\Delta + eV}K(\alpha)\right], \qquad V \geqslant 2\Delta/e \qquad (4.15)$$

* Note that the specific heat has the same dependence on Δ and T in the vicinity of absolute zero.

where

$$\alpha = \frac{eV - 2\Delta}{eV + 2\Delta} \qquad (4.16)$$

and $K(\alpha)$ and $E(\alpha)$ are complete elliptic integrals [101].

It is interesting to note that Equation (4.15) tends to a finite value as $eV \rightarrow 2\Delta$, that is, owing to the infinitely large density of states there is a discontinuity in current. Since [101]

$$\lim_{\alpha \to 0} K(\alpha) = \lim_{\alpha \to 0} E(\alpha) = \frac{\pi}{2}. \qquad (4.17)$$

this value of current comes to

$$I_{SS}(T = 0, eV = 2\Delta) = G_{NN}\frac{\pi}{2}\frac{\Delta}{e} = I_{NN}\frac{\pi}{4}. \qquad (4.18)$$

It follows from our previous qualitative approach (and may be shown from Equation (4.14)) that for unequal gaps at $T = 0$

$$I_{SS}(T = 0) = 0 \qquad \text{for} \qquad 0 < eV < \Delta_1 + \Delta_2 \qquad (4.19)$$

whereas for $eV \geqslant \Delta_1 + \Delta_2$ it is possible to express again the current in terms of complete elliptic integrals [33]

$$I_{SS}(T = 0) = \frac{G_{NN}}{e}[-2\Delta_1\Delta_2\beta K(\gamma) + \beta^{-1}E(\gamma)] \qquad (4.20)$$

where

$$\beta = [(eV)^2 - (\Delta_2 - \Delta_1)^2]^{1/2} \quad \text{and} \quad \gamma = \beta[(eV)^2 - (\Delta_1 + \Delta_2)^2]^{1/2}. \quad (4.21)$$

When $eV \rightarrow \Delta_1 + \Delta_2$ there is a discontinuity in current

$$I_{SS}(T = 0, eV = \Delta_1 + \Delta_2) = \frac{G_{NN}}{e}\frac{\pi}{2}\sqrt{(\Delta_1\Delta_2)}. \qquad (4.22)$$

When $T \neq 0$ (which is the more interesting case for junctions between superconductors of unequal gaps) the current cannot be expressed in terms of tabulated functions; Equation (4.14) must be solved numerically. This was first done by Nicol et al. [46] and Shapiro et al. [102]; the latter authors showed that there is a logarithmic singularity in current at $eV = \Delta_2 - \Delta_1$, a negative resistance region for $\Delta_2 - \Delta_1 < eV < \Delta_1 + \Delta_2$ and a discontinuity at $eV = \Delta_1 + \Delta_2$ of magnitude

$$\Delta I_{SS} = \frac{G_{NN}}{e}\frac{\pi}{4}\sqrt{(\Delta_1\Delta_2)}\frac{1 - \exp[-(\Delta_1 + \Delta_2)/kT]}{[1 + \exp(-\Delta_1/kT)][1 + \exp(-\Delta_2/kT)]}. \qquad (4.23)$$

It should be noted that in the numerical solution of Taylor *et al.* [103] a negative resistance region was shown to exist even for the case of equal gaps, appearing at $T < 0.3\Delta/k$.

4.6 Microscopic theory

An expression for I_{NS} starting from microscopic theory was first given by Cohen *et al.* [50]. They used the Bogoliubov transformation (expressing the partial creation and partial destruction of an electron mentioned in Section 1.12) to obtain the tunnelling Hamiltonian in terms of the quasiparticle* operators and computed the current from the formula

$$I_{NS} = e\langle\frac{d\mathcal{N}_S}{dt}\rangle = -e\langle\frac{d\mathcal{N}_N}{dt}\rangle = -\frac{ie}{\hbar}\langle\mathcal{N}_N H_T - H_T \mathcal{N}_N\rangle \qquad (4.24)$$

where $\mathcal{N}_N$ is the total number of electrons in the normal metal, $\mathcal{N}_S$ the total number of electrons in the superconductor, and the brackets indicate expectation values at a given temperature.

An alternative derivation in terms of particle conserving operators (a slightly modified form of the Bogoliubov operators) was given by Bardeen [104]. We shall follow here Schrieffer [105, 106] who writes the matrix element in the form

$$|M_{lr}|^2 = |T_{lr}|^2 u_k^2 \qquad (4.25)$$

where u_k is specified in Equations (1.43) and (1.44). Thus the matrix element is *not* identical to that in the normal state; there is an additional factor u_k^2. The other important difference is that there are two quasiparticle states for a value

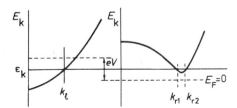

Fig. 4.10 The energy–momentum diagram of an NS junction at an applied voltage, V, in the quasiparticle representation.

of energy above the gap as was shown in Fig. 1.3 (*b*) and redrawn in more detail in Fig. 4.10. Thus an electron of momentum k_1 on the left-hand side may tunnel both into k_{r1} and k_{r2}. Hence the current will be proportional to the sum

$$|T_{lr}|^2(u_{k_{r1}}^2 + u_{k_{r2}}^2) = |T_{lr}|^2\frac{1}{2}\left(2 + \frac{\varepsilon_{k_{r1}}}{E_k} + \frac{\varepsilon_{k_{r2}}}{E_k}\right) \qquad (4.26)$$

---

* We are using the term quasiparticles in this Section but merely for showing that as far as tunnelling is concerned the 'normal electron' description is perfectly adequate.

But according to Equation (1.46)

$$\varepsilon_{k_{r1}} = -\varepsilon_{k_{r2}} \tag{4.28}$$

and we get the result that the current is proportional to $|T_{lr}|^2$. This rather accidental simplification is the reason for the correctness of the simple tunnelling picture (and of the phenomenological theory) discussed in the preceding sections.

The cancellation holds also for junctions between superconductors as proved by Josephson [47].

4.7 The Adkins model

In the previous section we presented N–I–S tunnelling in terms of the E–k diagrams. For S_1–I–S_2 tunnelling (just taking the simple case when tunnelling is between maximum density of states) the diagrams take the form shown in Fig. 4.11. In case of Fig. 4.11 (a) energy may be conserved by simply tunnelling

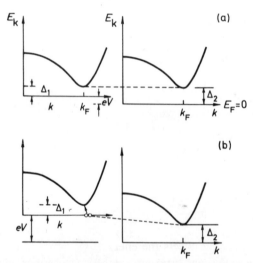

Fig. 4.11 The energy–momentum diagram of an S_1S_2 junction for (a) $V = (\Delta_2 - \Delta_1)/e$, (b) $V = (\Delta_2 + \Delta_1)/e$.

from S_1 to S_2. But for an applied voltage $V = (\Delta_1 + \Delta_2)/e$ the tunnelling process will occur by first breaking a Cooper-pair (located at zero excitation energy); one of the electrons becomes part of the normal fluid in S_1 and the other electron tunnels across. Note that energy is conserved in the process.

The representation in terms of the E–k diagrams certainly depicts better the real physical situation but it is very complicated especially when both branches of the diagram need to be taken into account. The semiconductor model is perfectly adequate in the sense that it gives the correct answer* but perhaps it takes away too much of the physics.

An intermediate picture was proposed by Adkins [62, 107]. Tunnelling for $V = (\Delta_1 + \Delta_2)/e$ is given in that picture by Fig. 4.12. There is strong similarity to Fig. 4.11 (b); the concept of excitation is retained but the E–k diagram is eliminated. In the following chapters we shall mostly rely on Adkins' model.

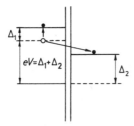

Fig. 4.12 Tunnelling in an S_1S_2 junction in the Adkins representation.

4.8 Non-ideal behaviour

One tends to call a behaviour 'non-ideal' whenever the sample behaves differently from theoretical expectations. Some of these differences may be later explained by modifications of the theory and then they are no longer called 'non-ideal'. Some other discrepancies will never be resolved because the real material is bound to differ from the theoretical model. We shall in this section mention a number of factors which may give rise to the observed differences.

(i) *Lifetime effects.* The normal electrons which can exist in a superconductor at non-zero temperatures have finite lifetime. Hence the density of states is smeared out and there is no singularity at Δ.

(ii) *Gap anisotropy.* The energy gap is known to depend somewhat on crystal direction. Hence in a polycrystalline material one measures a certain average gap, reducing again the sharpness of the density of states function.

(iii) *Strain.* Since strain effects the energy gap, a non-uniform strain in the sample will again lead to some average gap.

(iv) *Leakage currents.* Besides the tunnel current there are some other sources of current as well which in some samples may be dominant. Since the tunnel current depends exponentially on temperature, the leakage current can always be detected at sufficiently low temperatures. Indeed the surest way of finding out which fraction of the current is due to tunnelling is to measure the current when both electrodes are superconducting.

---

*A notable exception is the Tomasch effect which needs to be explained in terms of the $E-k$ diagram.

Fig. 4.13 shows the current–voltage characteristic of a relatively good sample. It may be seen that there is a temperature-independent (which

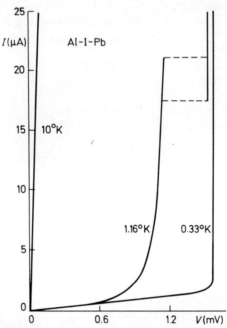

Fig. 4.13 I–V characteristics of an Al–I–Pb junction measured with a constant current generator. For low voltages the leakage current dominates. After Giaever *et al.* [66].

is not always the case) leakage current which suppresses the negative resistance at 0·33°K.

(v) *Trapped magnetic flux.* This effect is shown in Fig. 4.14. Curve 1 is a virgin curve showing the negative resistance. Curve 3 is taken with a magnetic field applied and curve 2 with the magnetic field removed. Curve 2 shows no negative resistance because some trapped magnetic flux makes some of the regions normal. Hence one measures an average of various mechanisms of tunnelling and the fine details are smeared out.

The earth's magnetic field trapped by the superconducting shield may also cause excess current as pointed out by Donaldson [108].

(vi) *Measuring current.* The tunnelling current flowing through the junction may itself affect the *I–V* characteristic. Its magnetic field may be trapped, it may cause some local heating or more significantly it may induce transition into the normal or intermediate state when one of the super-conductors is near to its critical temperature [109].

(vii) *Edge effects.* An evaporated film usually has small crystals at the edge which are imperfectly connected to the main body of the film [109].

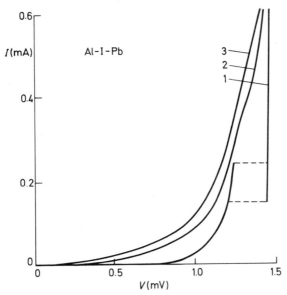

Fig. 4.14 The effect of trapped flux on the *I–V* characteristic of an Al–I–Pb junction. After Giaever and Megerle [69].

These may have higher critical temperatures and higher energy gaps, hence the gap may appear smeared or one might even measure a multiplicity of gaps [110]. The remedy often used against edge effects is to put some material (by evaporating a thicker insulator [110] or painting [111]) before the second metal is evaporated.

4.9 Fabrication

Oxide layers. The most widely used technique of preparing tunnel junctions is vacuum deposition of the metals, and oxidation in air for producing the insulating layer. This method is indeed the same as used for the investigation of tunnelling between normal metals. The circumstance that one or both metals become superconductive below a certain temperature does not affect the fabrication.

It is usual to realise the junctions in the form of cross-strips and to prepare simultaneously several junctions. In an experiment Fisher and Giaever [112] evaporated 99·999% pure aluminium film $\frac{6}{64}$ inch wide and several tens of nm thick (Fig. 4.15 (a)), oxidised the film in air or oxygen atmosphere (obtaining an oxide layer of several nm) and finally deposited across it five aluminium strips from $\frac{1}{64}$ to $\frac{5}{64}$ inch wide as shown in Fig. 4.15 (c). The current voltage characteristics were found linear at low voltages (Fig. 4.16) and exponential at higher voltages; the currents were roughly proportional to the relative areas of the oxide films as shown in Fig. 4.16. The increase in resistance with thickness followed exponential curves though a fair amount of scattering was obviously present.

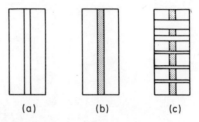

Fig. 4.15 Junction preparation. (a) An Al film is deposited on a glass slide, (b) the surface is oxidised, (c) Al films are deposited across the oxide.

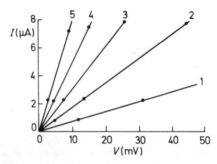

Fig. 4.16 The I–V characteristics at low voltages at room temperature. Junction areas are in the proportions 5:4:3:2:1. After Fisher and Giaever [112].

The finished product for an aluminium*–insulator–lead junction made by a similar technique by Rowell and Kopf [113] is shown in Fig. 4.17. Electrical contact to the films was made using 0·002 inch thick aluminium foil leads attached with silver paste, both ends of all films being used to give a four-terminal connection to each junction.

It should be noted that the junctions made by this technique should better be kept at liquid nitrogen temperatures. The structures are usually not stable enough to survive frequent cycling to room temperature, most of the trouble coming from the large number of O_2 molecules which remain adsorbed on the surface. These molecules diffuse into and react with the metals deposited afterwards provided sufficient energy is available. This energy is provided at room temperature by the thermal energy kT. Thus quite apart from atmospheric moisture there is a clear mechanism causing the deterioration of the oxide layer

*Aluminium is often preferred as the base electrode in this technique because it readily forms a thin and coherent oxide layer.

during any prolonged storage at room temperature. A similar deterioration may also occur during operation when the electric fields in the junction may be fairly large, initiating ion migration or dipole flipping.

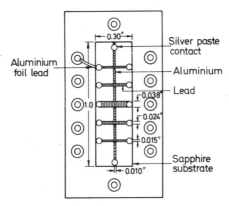

Fig. 4.17 Five Al–I–Pb junctions with electric contact arrangement. After Rowell and Kopf [113].

Many good junctions were made by thermal oxidation and long life achieved by storing the junctions at liquid nitrogen temperature. There is, however, a more reliable method by means of gaseous anodisation (also known as plasma or glow discharge method) introduced by Miles and Smith [114]. The basic set-up is shown in Fig. 4.18 in which an oil diffusion pump can achieve a pressure

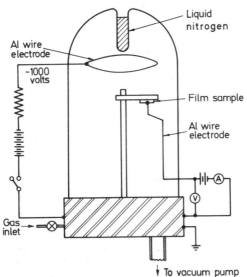

Fig. 4.18 Schematic of glow discharge system. After Miles and Smith [114].

of about 10^{-6} torr. First a pure material is evaporated (most of their work involved Al but the technique was proved successful for a large number of other metals) through a mask on an ordinary glass microscope slide. After the evaporation is completed oxygen gas is admitted to bring the system to a pressure of about $5 \cdot 10^{-2}$ torr. A glow discharge is then initiated by applying several hundred to one thousand volts negative potential to a pure aluminium wire ring mounted in the bell jar. The metal base plate served as the ground electrode.

If a relatively thin oxide is desired the glow is maintained for a few seconds to some tens of minutes. The oxide thickness as a function of time is shown in Fig. 4.19. If a thicker oxide is desired a direct electrical connection is made to the

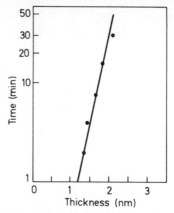

Fig. 4.19 Oxide thickness as a function of time obtained by the glow discharge technique (without external potential). After Miles and Smith [114].

sample by means of a probe formed of pure aluminium wire and a small positive d.c. voltage is applied. The current flowing in the circuit decays with time as in an ordinary aqueous anodising bath. When the current has decayed, the discharge is shut off and the system pumped down for the evaporation of the second metal. The oxide thickness in this case was found to be proportional to the d.c. voltage applied.

Another method used by Miles and Smith [114], which actually did not gain acceptance, is the so-called solid state anodisation. In that method a very thin oxide layer is formed first (say by a few seconds of glow discharge) and the top metal is slowly (1–1·5 nm/sec) evaporated in the presence of oxygen, the pressure being kept at 10^{-4} torr. The metal chosen for the top electrode (both tin and lead were found satisfactory) must be able to dissolve relatively large amounts of oxygen. After evaporation the sample is removed from the vacuum system and a small d.c. voltage is applied to it, the first electrode, aluminium, being maintained positive. Again a current is observed to flow; it decays with time indicating the growth of oxide.

Lead–lead oxide–lead junctions were made by Schroen [115] by the glow discharge method with the aim of using them for Josephson tunnelling (the oxide layer needs to be thin, see Section 15.2). Schroen claims that the essential difference between thermal and glow discharge oxidation is that in the latter case the oxygen molecules arrive at the surface of the metal with some energy gained from the discharge and are therefore capable to form a more densely packed structure. The main processes likely to take place are as follows:

(i) Some electrons from the discharge move to the sample and charge it up negatively. This negative potential helps to attract positively-charged oxygen molecules.

(ii) The energy of the impinging oxygen molecule is used partly for dissociation (≈ 5 eV) and partly for moving into the layer of lead and forming lead oxides of various compositions.

(iii) When the discharge is extinguished and while the oxygen gas is pumped off, at least one monolayer of O_2 molecules is adsorbed on the surface.

(iv) The O_2 molecules react with the lead layer deposited afterwards and may tend to diffuse further under the effect of thermal energy (at room temperature) or electric fields.

This last problem is the same when the sample is exposed to ambient oxygen but with the glow discharge deposition there is already a tightly packed oxide structure and that apparently is sufficient for long term stability. Life tests showed that the electric properties of the junctions showed no change after eight months of room temperature storage.

It is difficult to know what is exactly happening in a glow discharge. As Schroen [115] pointed out, another type of oxidation may exist simultaneously in which

(i) lead atoms are sputtered by energetic oxygen atoms;

(ii) lead atoms leaving the metal surface react with O_2 molecules to form PbO;

(iii) PbO is adsorbed on the surface by unsaturated metal bonds.

All the oxides mentioned so far were those of the superconducting electrodes to be measured. Another possible solution was used by Adkins [62] whose insulator was a layer of aluminium oxide in Pb–I–Pb and Sn–I–Pb junctions. About 2 nm aluminium was evaporated on the top of the first superconductor and then oxidised.

Other thin film barriers. A monomolecular layer of barium stearate was used as the barrier by Miles and McMahon [116] and Shapiro *et al.* [102] applying the technique developed by Blodgett and Langmuir [117] well before them. An advantage of such layer is the constant thickness but stability and reproducibility of resistance were poor.

Semiconductor barriers have also been used. The advantage is that the incident electrons encounter a smaller energy barrier thus the films can be made thicker; and it is easier to attain continuity and uniformity with thicker films. CdS and ZnS was used by Giaever [118] and Giaever and Zeller [119, 120], and a number of compounds and elemental semiconductors were tried by McVicar *et al.* [121]. Their conclusion is that the only semiconductor which can readily form reliable barriers is carbon. The films (in the range of 9 to 27 nm thickness) were prepared by sublimation from a point contact between two carbon rods, the heating produced by passing a large current between the rods.

Layers of Ge, CdS and ZnS exhibited pinholes which resulted in metallic shorts after the evaporation of the top metal. They can, however, be used if after the evaporation of the semiconductor the vacuum is broken for oxidising the bottom metal. The result is that tunnelling can proceed in two distinct paths, either through the oxide or through the semiconductor. Giaever [118] could estimate the relative magnitudes of these contributions by shining light upon the CdS barrier. The *I–V* characteristics for the unexposed and exposed cases are shown in Fig. 4.20. The effect of light is to generally increase the current.

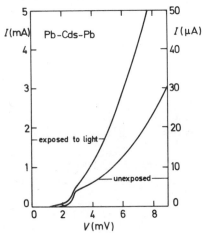

Fig. 4.20 I–V characteristics of Pb–CdS–Pb junctions. After Giaever [118].

Since the oxide is unaffected by light one may conclude that the main contribution to the tunnelling current must come through the semiconductor.

Point contact junctions. These were developed by Levinstein and Kunzler [122, 123] in the form shown in Fig. 4.21. The barrier is prepared by heavily anodising a freshly etched tip of Al, Nb, Ta, etc. The diameter of the junction at the point of contact was estimated to be less than 10 μm. Tunnelling characteristics were observed in a large resistance range from 10^2 to 10^5 ohm.

The advantage of point contacts is that tunnelling measurements can be made on materials not accessible in thin film form. Furthermore, the tunnelling is generally from one single crystal to another since the grain size of the material

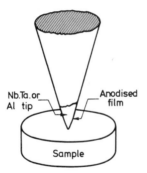

Nb.Ta.or Al tip

Anodised film

Sample

Fig. 4.21 Point contact junction. After Levinstein and Kunzler [122].

both in the tip of the point contact and in the bulk is considerably larger than the contact area. Notable success of the point contact technique was to obtain the correct value for the energy gap of Nb_3Sn where thin film measurements consistently gave the wrong value.

Pressure contacts. We shall further mention here a fabrication method reported by Sullivan and Roos [314] whose junctions were formed by inserting super-conducting wire into a small hole in the normal metal and subjecting the normal metal to pressures up to several kbars. Note that the resulting current densities are greater by about four orders of magnitude than those obtained for other types of junctions.

5. Special tunnelling effects

5.1 Two-particle tunnelling

Apart from the usual (single-particle) tunnel current the following three types of excess currents were observed in a set of measurements by Taylor and Burstein [61].

  (*i*) An excess current characterised by a sharp temperature-independent jump at $V = \Delta/e$ for identical superconductors and at $V = \Delta_1/e$ and $V = \Delta_2/e$ for two different superconductors, as may be seen in Fig. 5.1.

  (*ii*) A temperature-independent excess current which has an exponential dependence on applied voltage.

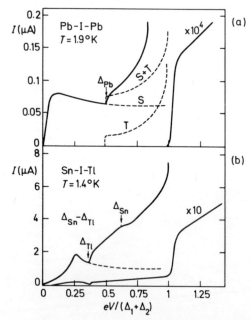

Fig. 5.1 I–V characteristics for Pb–I–Pb and Sn–I–Tl junctions. The solid lines represent the experimental data. S: theoretical single-particle tunnelling curves; T: theoretical two-particle tunnelling curves; S+T: sum of the single-particle and two-particle curves. There is a change in slope at voltages corresponding to the half-gap of either superconductor comprising the junction. After Taylor and Burstein [61].

(*iii*) An excess current which has a strong dependence on temperature as well as an approximately exponential dependence on applied voltage (observed only for Pb–I–Pb junctions).

Similar results obtained by Adkins [62, 107] are shown in Fig. 5.2 where the temperature independence of the excess current can be clearly seen.

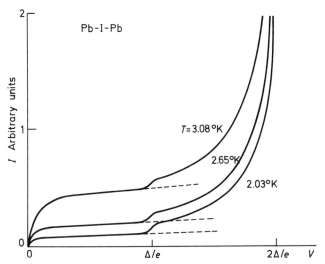

Fig. 5.2 I–V characteristics of a Pb–I–Pb junction showing the temperature independence of the two-particle contribution. After Adkins [62].

A theoretical explanation for (*i*) and (*ii*) was given by Schrieffer and Wilkins [124] (for a more detailed analysis see Wilkins [125]) in terms of two-particle

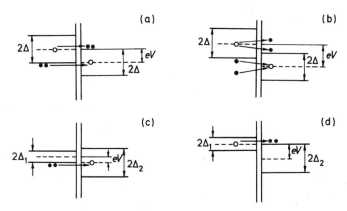

Fig. 5.3 Energy diagrams of two-particle processes in the semiconductor representation; (*a*) $V = \Delta/e$, (*b*) $V > \Delta/e$, (*c*) $V = \Delta_1/e$, (*d*) $V = \Delta_2/e$.

tunnelling which we shall illustrate here both in the semiconductor and in the Adkins representation.

First of all we have to modify slightly the semiconductor-type energy diagram. In addition to electrons above and holes below the gap we must also take into account the Cooper-pairs all condensed at the Fermi energy. For identical superconductors at $V = \Delta/e$ the bottom of the gap on the left is in line with the Fermi energy on the right (Fig. 5.3 (a)). Hence at this value of the applied voltage two electrons residing just below the gap on the left may tunnel to the right and turn into a Cooper-pair. Similarly, a Cooper-pair residing at the Fermi energy on the left may tunnel into the superconductor on the right and dissociate into two electrons. At higher applied voltages both processes are still possible but in addition there will be transitions involving states of different energy (to satisfy the conservation laws) as shown in Fig. 5.3 (b).

For junctions consisting of two different superconductors there are two (in addition to the onset of negative resistance occurring at $V = (\Delta_2 - \Delta_1)/e$) salient points in the current–voltage characteristic as shown in Fig. 5.1 (b).

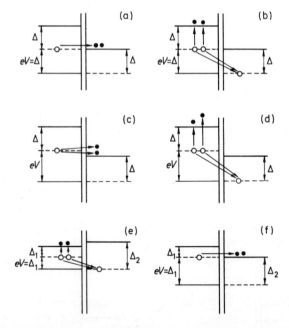

Fig. 5.4 Energy diagrams of two-particle processes in the Adkins representation; (a) $V = \Delta/e$, one Cooper-pair breaks up, two electrons tunnel across; (b) $V = \Delta/e$, two Cooper-pairs break up, two electrons move up into the continuum and one Cooper-pair tunnels across; (c) $V > \Delta/e$, one Cooper-pair breaks up, two electrons tunnel across into states of different energies; (d) $V > \Delta/e$, two Cooper-pairs break up, two electrons move up into the continuum into states of different energies and one Cooper-pair tunnels across; (e) $V = \Delta_1/e$, equivalent to (b) for unequal gaps; (f) $V = \Delta_2/e$, equivalent to (a) for unequal gaps.

The difference from the previous case is that the processes shown in Fig. 5.3 (a) occur now at separate threshold voltages $V = \Delta_1/e$ and $V = \Delta_2/e$ respectively, as shown in Figs. 5.3 (c) and 5.3 (d).

The same processes in Adkins' representation are shown in Fig. 5.4. The tunnelling of the Cooper-pair in Fig. 5.3 (a) from left to right looks nearly the same in Fig. 5.4 (a) but since electrons below the gap (unexcited electrons) play no role in the Adkins model, a Cooper-pair on the right must be created by the rather different process shown in Fig. 5.4 (b). Two Cooper-pairs break up, two electrons move up into the continuum and one Cooper-pair moves across the barrier to the right. The process shown in the upper part of Fig. 5.3 (b) has a similar representation in Fig. 5.4 (c) but the process in the lower part (the creation of a Cooper-pair on the right) must again proceed by the break-up of Cooper-pairs as shown in Fig. 5.4 (d). The equivalents of Fig. 5.3 (c and d) may now be easily constructed; they are shown in Figs. 5.4 (e and f).

It should be noted that since two-particle tunnelling involves the simultaneous tunnelling of two electrons, the matrix element for the process will depend on the square of the matrix element involved in the single-particle tunnelling process.* Hence the contribution from two-particle tunnelling is small in comparison with single-particle tunnelling. However, the magnitude of the single-particle tunnelling current decreases exponentially with temperature whereas the two-particle tunnelling process is relatively independent of temperature (since Cooper-pairs from an infinite reservoir are involved). Hence it follows that the chances of observing two-particle tunnelling are better at lower temperature. Similarly, by reducing the barrier thickness (that is increasing the tunnelling probability) the proportion of two-particle to single-particle tunnelling should increase.

Two-particle tunnelling is a beautiful illustration of a special superconductive effect which does not and cannot occur in tunnelling between normal materials. It is, however, doubtful whether it was ever observed. The experimental results discussed here (and interpreted at the time as two-particle tunnelling) may be special cases of another effect called subharmonic tunnelling (see Section 5.4) which produces current jumps at submultiples of the gap voltage.

5.2 Photon-assisted tunnelling

The tunnelling current may be modified by illuminating the junction with electromagnetic waves. It is easy to see that if the energy of the incident photons is in excess of 2Δ they will break up Cooper-pairs and create two electrons above the gap as shown in Fig. 5.5 (a). Since the number of electrons above the gap increases this way above its equilibrium value, some of these extra electrons will

---

* In the independent particle model the second order perturbation vanishes. It is finite for superconductive tunnelling, and provides the two-particle current, because the particles are *not* independent.

tunnel across the barrier (Fig. 5.5 (b)) creating thereby an extra current. We shall return to this problem in Section 7.2, for the moment we shall concentrate on the case when the energy of the incident photon is insufficient to break up a Cooper-pair. Influence on the tunnelling characteristics is still possible then if the photons act jointly with the applied voltage.

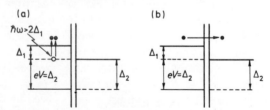

Fig. 5.5 Effect of incident photons on a tunnel junction; (a) a photon creates two electrons by breaking up a Cooper-pair, (b) one of the electrons created tunnels across.

Let us take $T = 0°K$ again and recall the case when $V = (\Delta_1 + \Delta_2)/e$. Then a Cooper-pair may break up into two electrons, one of them tunnelling across the barrier as has been shown in Fig. 4.12. If $V < (\Delta_1 + \Delta_2)/e$ no current flows. A Cooper-pair breaking up could not cause a current because the transition shown in Fig. 5.6 (a) with dotted lines is not permissible. However, if a photon

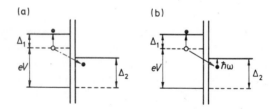

Fig. 5.6 (a) Tunnelling not allowed. (b) Tunnelling allowed if assisted by a photon.

of the right energy is available the liberated electron may follow the path shown in Fig. 5.6 (b) and get into an allowed state just above the gap. We may say that the electron tunnelled across the barrier by absorbing a photon, and refer to the phenomenon as photon-assisted tunnelling. The mathematical condition for the onset of tunnelling current is

$$\hbar\omega = \Delta_1 + \Delta_2 - eV. \tag{5.1}$$

If the energy of the photon is above this value tunnelling is still possible, though with a reduced probability because of a less favourable density of states. If the energy of the incident photon is below the value given by Equation (5.1) tunnelling may still be possible with the aid of a multi-photon process. An electron absorbing for example three photons simultaneously may tunnel across the (a)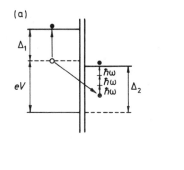

(b)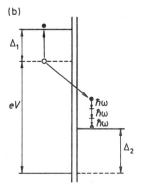

Fig. 5.7 Tunnelling assisted (*a*) by absorption of three photons, (*b*) by emission of three photons.

barrier in the way shown in Fig. 5.7 (*a*). Hence we may expect sudden rises in the tunnelling characteristics whenever the condition

$$n\hbar\omega = \Delta_1 + \Delta_2 - eV \qquad (5.2)$$

is satisfied, that is for a series of voltages in the range $0 < V < (\Delta_1 + \Delta_2)/e$.

When $V > (\Delta_1 + \Delta_2)/e$ we know that a tunnelling current will flow even in the absence of an incident electromagnetic wave. However, if photons of the right energy are available they can assist the tunnelling in this case as well, as shown in Fig. 5.7 (*b*) for a three-photon process. A Cooper-pair breaks up; one of the electrons goes into a state just above the gap on the left, and the other electron tunnels across into the superconductor on the right at an energy demanded by energy conservation (the sum of electron energies must equal the energy of the Cooper-pair). This process would occur with much higher probability if the electron could tunnel into the high density states lying just above the gap on the right. In Fig. 5.7 (*b*) this becomes energetically possible when three photons are *emitted* at the same time. Thus the mechanism of current rise is photon emission *stimulated* by input photons. For an *n*-photon emission process the current rises occur when

$$V_n = \frac{1}{e}(\Delta_1 + \Delta_2 + n\hbar\omega). \qquad (5.3)$$

For finite temperatures there is one more instance where electrons tunnel between maximum density states and that occurs at $V = (\Delta_2 - \Delta_1)/e$, as shown in Fig. 5.8 (*a*). Tunnelling between those states may also be assisted by photons as shown in Figs. 5.8 (*b* and *c*) for photon absorption and emission respectively. In general, multi-photon absorptions and emissions are possible again, and thus for finite temperatures there is another set of voltages,

$$V_m = \frac{1}{e}(\Delta_2 - \Delta_1 + m\hbar\omega), \qquad m = \pm 1, \pm 2, \pm 3 \qquad (5.4)$$

at which current rises can be expected.

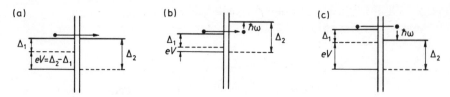

Fig. 5.8 Tunnelling between maximum density states at finite temperature (a) directly, (b) by photon absorption, (c) by photon emission.

The first experiments on tunnel junctions in the presence of electromagnetic waves were performed by Dayem and Martin [57] using junctions between Al and Pb, In or Sn. The frequency of the electromagnetic wave employed was 38·83 GHz so the experimental solution was to place the sample inside a cavity. The current–voltage characteristic was measured and rises in current were indeed found as may be seen in Fig. 5.9 (a) where the solid and dotted lines show the characteristic in the absence and presence of microwaves respectively.

Quantitative explanations were given nearly simultaneously by Tien and Gordon [58] and Cohen, Falicov and Phillips [126]. The methods in their papers were different but obtained essentially the same results. Cohen, Falicov and Phillips assumed that the magnetic field of the microwaves modulates the energy gap, whereas Tien and Gordon added an electrostatic perturbation term to the Hamiltonian. We shall follow here the latter derivation.

The simplest assumption one can make is to regard the junction as a capacitance with a time-varying but spatially constant electric field between the

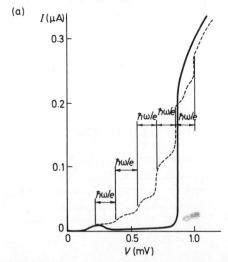

Fig. 5.9 (a) I–V characteristic of an Al–I–In junction in the absence (solid lines) and presence (dotted lines) of microwaves of frequency 38·83 GHz. Measurements by Dayem and Martin, quoted by Tien and Gordon [58].

plates. Regarding the potential of one of the superconductors (2) as the reference we may argue that the only effect of the microwave field is to add an electrostatic potential of the form

$$V_{rf} \cos \omega t \tag{5.5}$$

to the energy of the electrons in the other superconductor (1). Hence, for electrons in superconductor (1) we may use the new Hamiltonian

$$H = H_0 + e V_{rf} \cos \omega t \tag{5.6}$$

where the first term is the unperturbed Hamiltonian in the absence of microwaves.

If the unperturbed wavefunction was

$$\Psi_0(x, y, z, t) = f(x, y, z) \exp(-iEt/\hbar) \tag{5.7}$$

then the solution for the new wavefunction may be sought in the form

$$\Psi(x, y, z, t) = \Psi_0(x, y, z, t) \sum_{n=-\infty}^{\infty} B_n \exp(-in\omega t). \tag{5.8}$$

Substituting Equation (5.8) into Schrödinger's equation

$$H\Psi = i\hbar \frac{\partial \Psi}{\partial t} \tag{5.9}$$

we find

$$2nB_n = \frac{eV_{rf}}{\hbar \omega}(B_{n+1} + B_{n-1}) \tag{5.10}$$

which is satisfied by [101]

$$B_n = J_n(eV_{rf}/\hbar\omega) \tag{5.11}$$

where J_n is the n_{th} order Bessel function of the first kind. The new wavefunction is then

$$\Psi(x, y, z, t) = f(x, y, z, t) \exp(-iE\hbar/t) \sum_{n=-\infty}^{\infty} J_n(\alpha) \exp(-in\omega t), \tag{5.12}$$

where

$$\alpha = \frac{eV_{rf}}{\hbar\omega}. \tag{5.13}$$

It may be seen that in the presence of microwaves the wavefunction contains components with energies

$$E, E \pm \hbar\omega, E \pm 2\hbar\omega, \ldots \tag{5.14}$$

respectively. Without the electric field, an electron of energy E in superconductor (1) can only tunnel to the states in superconductor (2) of the same energy. In the presence of the electric field, the electron may tunnel to the states in superconductor (2) of energies E, $E \pm \hbar\omega$, $E \pm 2\hbar\omega$, etc. Let $N_{20}(E)$ be the unperturbed density of states of the superconductor (2). In the presence of microwaves we then have an effective density of states given by

$$N_2(E) = \sum_{n=-\infty}^{\infty} N_{20}(E+n\hbar\omega)J_n^2(\alpha). \qquad (5.15)$$

We may now obtain the tunnelling current by substituting Equation (5.15) into the general expression Equation (2.14), yielding*

$$I = A \sum_{n=-\infty}^{\infty} J_n^2(\alpha) \int_{-\infty}^{\infty} N_1(E-eV)N_{20}(E+n\hbar\omega)[f(E-eV)-f(E+n\hbar\omega)]\,dE$$

$$= A \sum_{n=-\infty}^{\infty} J_n^2(\alpha)I_0(eV+n\hbar\omega) \qquad (5.16)$$

where $I_0(eV)$ is the tunnelling current in the absence of microwaves.

In the limit $\hbar\omega \to 0$ it may be shown (see Appendix 5) that the above expression reduces to the classical value

$$I = \frac{1}{\pi} \int_{-\pi/2}^{\pi/2} I_0(V + V_{\text{rf}} \sin \omega t)\,d(\omega t). \qquad (5.17)$$

The comparison between theory and experiments has a long and tangled story. The first attempt was made by Tien and Gordon [58] who could repro-

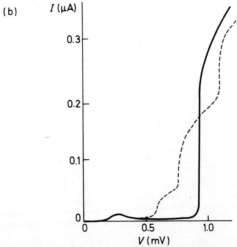

Fig. 5.9 (b) Theoretical curves by Tien and Gordon [58] for $\alpha = 2$.

*The same formula was also derived by Riedel [127] and Werthamer [128] from microscopic theory.

duce the experimental results of Dayem and Martin [57] by taking $\alpha = 2$ as shown in Fig. 5.9(b). The experimental value of α (that is the voltage in the junction) was, however, not known. Estimates by Tien and Gordon indicated a discrepancy as large as an order of magnitude.

The experiments of Dayem and Martin were repeated by Cook and Everett [129] and extended to Sn–I–Pb junctions as well. For a junction between different superconductors there should be another set of current rises centred on $V = (\Delta_2 - \Delta_1)/e$ as predicted by Equation (5.4). This was indeed found experimentally as shown in Fig. 5.10 where dI/dV is plotted against voltage

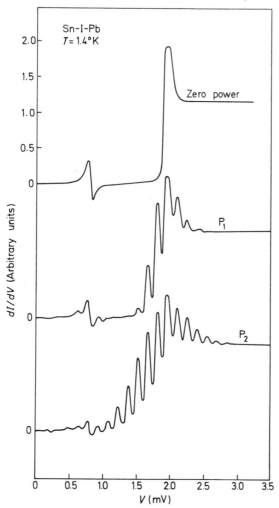

Fig. 5.10 The differential conductance as a function of voltage for microwave input powers of 0, P_1 and P_2. After Cook and Everett [129].

for a microwave power level P_1 (adjusted so that the peaks spreading out from the two different centres should not overlap). The peaks in the curves correspond to points of maximum current rise. It may be seen clearly that the structure in the characteristic is caused by the microwaves, that the distance between the peaks is an integral multiple of $\hbar\omega/e$, that they are centred upon the voltages $V = (\Delta_1 + \Delta_2)/e$ and $V = (\Delta_2 - \Delta_1)/e$ and that the peaks vanish for higher values of n. Increasing the microwave power by 7 db to P_2 the peaks may be seen for higher values of n but there is too much overlapping so the centre at $V = (\Delta_2 - \Delta_1)/e$ is no longer conspicuous. Cook and Everett also investigated the dependence of given peaks ($n = 2$ and 3) on microwave power. They found that the peaks tended to a constant value in contrast to the Bessel function dependence predicted by Equation (5.16).

Cook and Everett [129] proposed a new theory to explain their experimental results. They obtained good agreement with the experiments by rewriting the Tien–Gordon Hamiltonian in a symmetric form and thus arriving at a different density of states function.

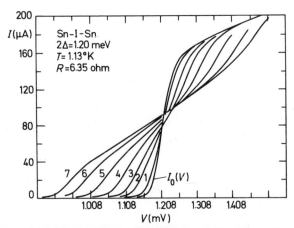

Fig. 5.11 I–V characteristics of a 6·3 ohm normal resistance Sn–I–Sn junction. $I_0(V)$ in the absence of microwave power, 1–7 increasing microwave power. After Sweet and Rochlin [132].

The next set of experimental results were reported by Bonnet and Rabenhorst [130] on Nb–I–Sn junctions. They found current jumps at 37 GHz but only a general increase in current at 10 GHz. Similar results were reported by Teller and Kofoed [131] for Sn–I–Pb junctions. As many as fifteen peaks were found in the dI/dV against V function at 35 GHz but none at 10 GHz. This is in agreement with expectations; as the classical limit is approached it is difficult to resolve the peaks. Teller and Kofoed [131] also compared their experimental results with the Tien–Gordon and Cook–Everett theories and found good agreement with the latter one. The experimental results of Sweet and Rochlin

[132] did, however, point to the correctness of the Tien–Gordon theory. A set of I–V characteristics obtained by them is shown in Fig. 5.11. With increasing microwave power (1 to 7) the current increased below the gap and decreased above the gap. The excess current

$$\Delta I = I - I_0 \tag{5.18}$$

is plotted in Fig. 5.12 and compared with the Tien–Gordon theory. The agreement is remarkable. Sweet and Rochlin came to the conclusion that for high

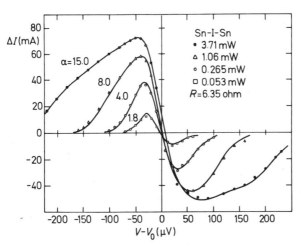

Fig. 5.12 $\Delta I(V)$ derived from I–V characteristics of Fig. 5.11. Solid lines represent Tien–Gordon theory. V_0 is an arbitrary voltage near $V = 2\Delta$ chosen for convenience in data reduction. Correspondence between α and microwave power was determined by fitting curve 7 of Fig. 5.11 at one point. After Sweet and Rochlin [132].

resistance junctions (several ohms are already regarded as high) and at least for the low frequency used (3·93 GHz) there is excellent agreement with the Tien–Gordon theory but not for lower values of resistance as may be seen in Fig. 5.13 for a junction resistance of 0·352 ohm.

The Cook–Everett theory came under attack from Buttner and Gerlach [133] who showed that proper symmetrisation of the Tien–Gordon Hamiltonian should reproduce the Tien–Gordon results. Somewhat later Sweet and Rochlin [134] showed that the Cook–Everett theory does not reduce to the correct classical limit. So it looks as if the agreements with the Cook–Everett theory were purely fortuitous. The experimental facts, however, did remain; the height of the current jumps in the high-frequency experiments did *not* follow the Tien–Gordon theory.

The problem has now been resolved by Hamilton and Shapiro [135] who proved convincingly that the geometry of the thin film junction is responsible for the discrepancy. Although the junction dimensions are small (of the order

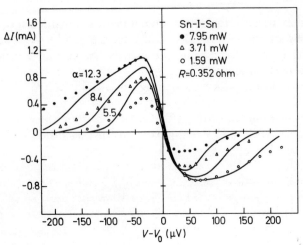

Fig. 5.13 $\Delta I(V)$ against $V-V_0$ for a junction of lower resistance. Correspondence between α and microwave power was determined by fitting the 7·95 mW curve at one point. After Sweet and Rochlin [132].

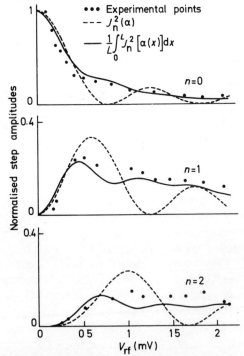

Fig. 5.14 Photon-assisted step amplitudes at a frequency of 80 GHz as a function of rf voltage for $n = 0, 1, 2$. After Hamilton and Shapiro [135].

of 0·1 mm) in comparison with the free space wavelength that is not the relevant factor. It was shown by Swihart [136] that since the penetration depth in the superconductors is much larger than the barrier thickness the velocity of a propagating electromagnetic wave is drastically reduced and of course the same applies to the wavelength which is calculated by Hamilton and Shapiro to be 0·17 mm at 80 GHz. Hence a standing wave may exist in the junction meaning both a spatially varying and a higher effective V_{rf}. By modifying the Tien–Gordon equation (replacing $J_n^2(\alpha)$ by its spatial average over the junction) it does follow that the zeros of the Bessel function are smeared out. The experimental points of Hamilton and Shapiro in conjunction with the original (dotted lines) and modified (full lines) theory are shown in Fig. 5.14 for $n = 0$, 1 and 2. This certainly proves that the modified theory is on the right track. The difficulty in making any quantitative comparison is that one still does not know the spatial variation of V_{rf} and can only make inspired guesses (Hamilton and Shapiro assume a reflection coefficient of 2/3 at the boundary; the number of wavelengths in the junction is apparently not a critical variable).

To prove the point that it is the spatial variation which is responsible for the discrepancy, Hamilton and Shapiro [135] conducted another series of experi-

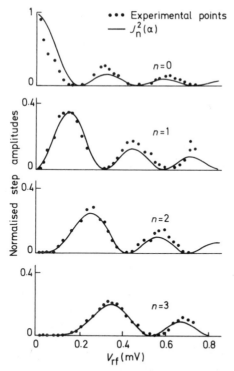

Fig. 5.15 Same as Fig. 5.14 for a very small junction. After Hamilton and Shapiro [135].

ments on a very small (hardly overlapping in an in-line geometry) junction. The results then did agree with the Tien–Gordon theory as shown in Fig. 5.15.

Two more proofs in favour of the Tien–Gordon theory are the measurements of Hamilton and Shapiro [135] at 200 Hz where V_{rf} could be easily measured and the microwave experiments of Longacre and Shapiro [137] conducted on point contact (that is, very small) junctions.

5.3 Phonon-assisted tunnelling

The energy of a phonon is given by $\hbar\omega$, that is, by exactly the same formula as for photons. Thus one may expect that whatever was said in the last section about the effect of photons upon the tunnelling characteristics, would also apply to phonons. Accordingly, Equation (5.16) is still valid if we reinterpret α. As may be seen in Equation (5.13), the value of α depends on the voltage difference across the junction. This voltage difference can be caused by coherent phonons incident upon the junction.

An acoustic wave will influence the distance of the lattice atoms from each other, thereby changing the local energy band structure and leading to the appearance of a potential, usually referred to as deformation potential [138], and expressed mathematically (for small deformations) as

$$eV = Cs \tag{5.19}$$

where C is the deformation potential constant and s the acoustic strain in the material. Hence α in Equation (5.16) may be replaced by*

$$\alpha_p = \frac{C_1 s_1 - C_2 s_2}{\hbar\omega}. \tag{5.20}$$

Taking the first three terms ($n = 0, \pm 1$) in Equation (5.16) and using the Bessel function expansions [101] for small α

$$J_0^2(\alpha) \cong 1 - \frac{\alpha^2}{2} \quad \text{and} \quad J_1^2(\alpha) = J_{-1}^2(\alpha) \cong \frac{\alpha^2}{4} \tag{5.21}$$

we get for the excess current in the presence of phonons

$$\Delta I = I - I_0$$
$$= \frac{\alpha_p^2}{4} \left[I_0(eV + \hbar\omega) + I_0(eV - \hbar\omega) - 2I_0(eV) \right]. \tag{5.22}$$

Expanding again in $\hbar\omega$, the odd power terms cancel and we obtain

$$\Delta I \cong \frac{(\alpha_p \hbar\omega)^2}{4} \left[\frac{d^2 I_0}{dV^2} + \frac{(\hbar\omega)^2}{12} \frac{d^4 I_0}{dV^4} \right]. \tag{5.23}$$

*The same relationship is derived, in order of increasing rigour, by Lax and Vernon [60] and Goldstein et al. [139].

Experimental results [60] on Al–Al$_2$O$_3$–Pb junctions obtained by 8·63 GHz longitudinal phonons are in good agreement with this theory (taking only the first term in Equation (5.23)) as shown in Fig. 5.16. The agreement is somewhat worse for Pb–Pb junctions [139] as shown in Fig. 5.17 where the theoretical curve is calculated from Equation (5.16) using the measured characteristic for I_0.

Apparently, phonons assist tunnelling in the same way as photons. Hence one may expect that future experimental curves taken at higher phonon frequencies will display the equispaced peaks in dI/dV just as it was found for photons.

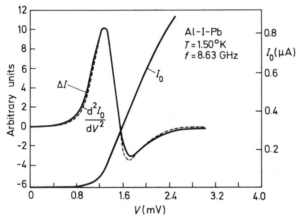

Fig. 5.16 Right scale: the *I–V* characteristic of an Al–I–Pb junction in the absence of input phonons. Left scale: the measured value of $\Delta I(V)$ compared with the normalised value of the second derivative of the $I_0(V)$ curve. After Lax and Vernon [60].

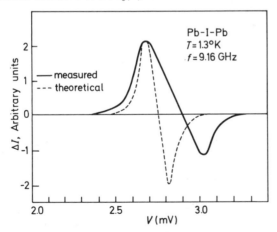

Fig. 5.17 $\Delta I(V)$ for a Pb–I–Pb junction using 9·16 GHz longitudinal phonons. After Goldstein, Abeles and Cohen [139].

We wish to note here that the temperature-dependent excess current found by Taylor and Burstein [61] in Pb–I–Pb junctions (mentioned in Section 5.1) was explained by Kleinman [140] and Kleinman *et al.* [141] as tunnelling assisted by thermal phonons. The reason that this type of excess current was found only in Pb–I–Pb junctions is probably due to the fact that the electron–phonon coupling is stronger in Pb than in the other superconductors.

5.4 Subharmonic structure

Changes in current at voltages corresponding to a submultiple of the energy gap ($eV = 2\Delta/n$) came to be called subharmonic structure (not to be confused with subharmonic steps to be discussed in Section 11.6 which bears no relation to the energy gap). The first experimental evidence of this structure was provided by Taylor and Burstein [61] who discovered besides the $n = 2$ case, at least for one of the junctions, structure at $n = 3$ and $n = 4$ as well. As mentioned before in Section 5.1 the $n = 2$ case was explained as two-particle tunnelling so a natural extension was to regard the values of $n = 3$ and 4 as due to multi-particle tunnelling. For $n = 3$ this process is illustrated in Fig. 5.18. The difficulty

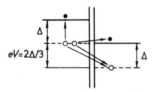

Fig. 5.18 The three-particle tunnelling process. Two Cooper-pairs break up, one electron moves into the continuum; one Cooper-pair and one electron tunnel across.

with this explanation is that higher order processes are proportional to higher and higher powers of the matrix element so they become less and less probable. Perhaps $n = 3$ might still be regarded feasible for very thin (at least at certain spots) barriers but when Rowell [63] reported that such structure was observable up to $n = 12$ it became clear that high values of n call for a different explanation. Of course the question immediately arose whether the whole $2\Delta/n$ structure could be explained by some other effect.

The experimental results were reviewed by Rowell and Feldman [142], in the light of their own experiments coming to the conclusion that there are two types of phenomena, the multi-particle tunnelling already discussed and subharmonic tunnelling (characterised mainly by a more gradual rise of current in the I–V characteristic) caused by metallic shorts.

Theoretical explanations starting from microscopic theory were given by Werthamer [128] and Ivanchenko [143] both involving the equations of Josephson tunnelling as well.

In a more recent study by Giaever and Zeller [144] it is claimed that the answer is contained in Werthamer's equations [128] which have not yet been solved numerically (not in sufficient generality). The experimental results of Giaever and Zeller [144] support this claim and are strongly against the explanation based on multi-particle tunnelling. The main evidence is obtained on Sn–CdS–Sn junctions where the matrix element could be varied by applying light to the photosensitive semiconductor barrier. It was shown that the structure at $V = \Delta/e$ increases faster than M^2 for a weakly coupled system and proportional to M^2 for a strongly coupled system whereas the increase should be proportional to M^4 if the mechanism was two-particle tunnelling. Further results pointing to the validity of the Werthamer theory are sensitivity to external microwave radiation and the observed dependence of structure on whether n is even or odd.

We shall not discuss any more aspects of this problem as it is obviously still in a state of flux. We shall conclude this section by a qualitative description of the Werthamer mechanism of subharmonic structure.

First we have to anticipate one of the results coming from the study of Josephson tunnelling, namely that an applied d.c. voltage, V, generates electromagnetic waves at the frequencies

$$\omega = \frac{2eV}{\hbar}n. \tag{5.24}$$

If $\hbar\omega \geq 2\Delta$, the radiation is absorbed by the superconductors comprising the junction. So the threshold for this new type of loss mechanism occurs at

$$eV = \frac{2\Delta}{2n}. \tag{5.25}$$

The mechanism for odd submultiples is somewhat different; we have to evoke the possibility of photon-assisted tunnelling in addition to our previous argument. Then it follows from Equations (5.2) and (5.24) that structure arises when

$$eV = 2\Delta - n\hbar\omega = 2\Delta - 2eVn \tag{5.26}$$

that is

$$V = \frac{2\Delta}{2n+1}. \tag{5.27}$$

For junctions of different superconductors the corresponding results are

$$eV = 2\Delta_1/2n, \quad 2\Delta_2/2n, \quad (\Delta_1 + \Delta_2)/(2n+1). \tag{5.28}$$

Note that the $(\Delta_1 + \Delta_2)/2n$ series is missing in accordance with the experimental results of Giaever and Zeller [144]. In contrast, Rowell and Feldman [142] claim to have detected structure at $(\Delta_1 + \Delta_2)/2$ and $(\Delta_1 + \Delta_2)/4$.

5.5 Geometrical resonances; the Tomasch effect

The Tomasch effect was discovered, as the name implies, by Tomasch [65, 145–147]. He measured the current–voltage characteristics of Al–I–S sandwiches (S stands for Pb, In or Sn) and found a number of roughly equidistant peaks when d^2V/dI^2 was plotted against V (Fig. 5.19). The same effect was

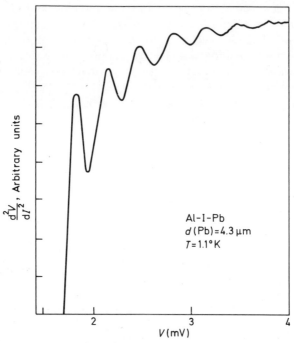

Fig. 5.19 Structure in the d^2V/dI^2 characteristic found by Tomasch [65] for thick films.

obtained on a number of diodes and was apparently independent of junction area and specific junction resistance. The essential clue was given by the fact that the effect was displayed by thick films (about 2·5–30 μm) of S but not by thin films ($\sim$ 0·1 μm), and only by those superconductors where the mean free path was in the micron range. So Tomasch suspected that the geometry of the S film is responsible for the effect and indeed further experimental work showed that the distance between the peaks was a nearly linear function of inverse film thickness (d) as may be seen in Fig. 5.20. So it was quite natural to assume the existence of a standing wave pattern where for a given length higher resonances (more de Broglie wavelengths) lead to higher energies. A further indication of the correctness of this assumption was given by measurements on diodes which had an additional layer of Ag evaporated upon the surface of the thick film S. The layer of Ag apparently gave a better boundary for the standing waves and the effect was considerably enhanced.

A theoretical explanation of the Tomasch effect was first given by McMillan and Anderson [148]; the enhancement of the structure by the Ag overlay was explained by Schattke [149] who extended the McMillan–Anderson theory to include the effect of surface conditions. We shall follow here a much simplified account given by Tomasch [150].

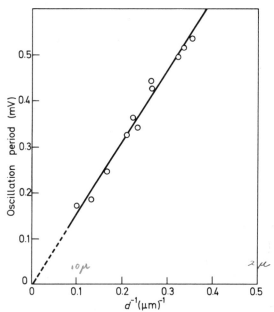

Fig. 5.20 Variation of oscillation period with inverse thickness of Pb film. After Tomasch [65].

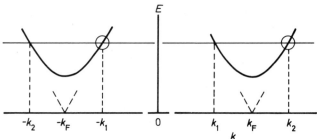

Fig. 5.21 Excitation spectrum for quasiparticles moving normal to the plane of the thick film. After Tomasch [125].

Let us consider the one-dimensional $E-k$ plot of the excitation spectrum of a superconductor as discussed in Section 1.11. As shown in Fig. 5.21 there are four values of k for a given energy, namely

$$k_1 = k_F - q, \quad k_2 = k_F + q, \quad -k_1 = -k_F + q, \quad -k_2 = -k_F - q. \quad (5.29)$$

One may expect that a suitably chosen linear combination of these four waves will give the required standing wave pattern

$$\psi = \exp i[k_1 x - (E/\hbar)]t + \exp i[-k_2 x - (E/\hbar)t]$$
$$- [\exp i[k_2 x - (E/\hbar)t] - \exp i[-k_1 x - (E/\hbar)t]$$
$$= -4i \exp[-i(E/\hbar)t] \cos k_F\, x \sin qx. \qquad (5.30)$$

In fact, we have not one but two standing wave patterns: one in terms of $\lambda_F = 2\pi/k_F$ and the other for $\lambda = 2\pi/q$. Since $k_F \gg q$ we get $\lambda_F \ll \lambda$. Now λ_F is too small to cause any resonances in the thick film but λ is of the right order.

Using the analogy of an organ pipe, resonances will occur when the length (film thickness in this case) is an integral multiple of quarter wavelength, that is, when

$$d = l\frac{\lambda}{4}. \qquad (l = 1, 2, \ldots) \qquad (5.31)$$

The corresponding energy may be calculated in the free electron approximation as follows

$$E_F + \varepsilon = \hbar^2 (k_F \pm q)^2 / 2m^* \qquad (5.32)$$

whence neglecting q^2

$$\varepsilon = \frac{\hbar^2 k_F q}{m^*} = \hbar v_F q = l\frac{h v_F}{4d} = \varepsilon_l \qquad (5.33)$$

and the excitation energy from Equation (1.46)

$$E_l = (\Delta_s^2 + \varepsilon_l^2)^{1/2}. \qquad (5.34)$$

One may expect that due to the resonance the density of states has peaks at values of energy corresponding to Equation (5.34). So there will be an increase in current whenever the applied voltage satisfies the condition

$$eV_l = E_l - \Delta_A \qquad (5.35)$$

as shown in Fig. 5.22. Thus the linear relationship against $1/d$ applies to ε and not to the experimentally obtained V_l. This may be seen particularly well on

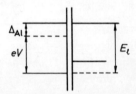

Fig. 5.22 Energy relations at geometrical resonance.

thick samples which led to an ε_l small in comparison with Δ_s. The experimental results [151] for a sample with a 30·0 μm thick indium electrode are shown in Fig. 5.23 and the values of ε_l and E_l derived from them are plotted in Fig. 5.24. The agreement between theory and experiment is remarkable.

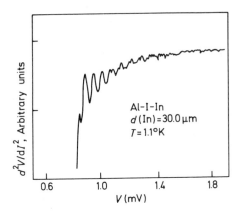

Fig. 5.23 d^2V/dI^2 as a function of V for a 30·0 μm thick indium film. After Tomasch and Wolfram [151].

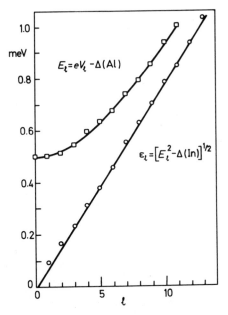

Fig. 5.24 Plots of E_l and ε_l as a function of l obtained from the data of Fig. 5.23. After Tomasch and Wolfram [151]

A further test of Equation (5.33) is to determine $d\varepsilon_l/dl$ from the experimental results and compare the obtained Fermi velocity

$$v_F = \frac{2d}{h}\frac{d\varepsilon_l}{dl} \tag{5.29}$$

with those from other sources. The Fermi velocity obtained for the In film from Fig. 5.24 is in close agreement with that obtained from cyclotron resonance measurements [152], there is though a discrepancy of about a factor 2 for the lead sample shown in Fig. 5.19. The technique can be extended to single crystal materials in which case the variation of Fermi velocity with crystal orientation may be determined [153].

It may be further noted that in view of Equation (5.35) the Tomasch effect may be used for the experimental determination of the $E(k)$ curve as discussed by Maki and Griffin [154].

Finally, we wish to mention some related experiments by Rowell and McMillan [155] who noticed similar (though considerably smaller) interference effects in Al–I–Ag–Pb sandwiches. The geometrical resonance is now in the silver film which remains normal but the presence of the backing superconductor (Pb) is essential for producing the right kind of electron wave. This is a manifestation of the proximity effect which will be discussed in more detail in Section 6.13.

6.   Diagnostic applications

6.1 Introduction

There are a number of interesting effects associated with superconductive tunnelling and as we shall see in the next chapter there are a number of potential applications in the device field as well. However, far the most important application of superconductive tunnelling is diagnostics. We use superconductive tunnelling to reveal the properties of the superconducting state. It is true to say that the spectacular development of superconductivity in the last decade owes a lot to the simple technique of tunnelling; in fact superconductive tunnelling is the most sensitive probe of the superconducting state.

A detailed description of all the diagnostic applications with their successes and failures would mean covering most of the theories of superconductivity. There is plenty of material here for a treatise and the difficulty is how to restrict it to the size of a chapter, and how to include it in a book not concerned with microscopic theory. There is obviously no room here to review in any detail the theories concerned with various aspects of superconductivity, e.g. magnetic impurities, gap anisotropy or strong-coupling effects, as they are in no way related to tunnelling. The solution chosen is to give references to the major theories, outline (just in a few cases) the main steps in the historical development and, mainly, emphasise the role of tunnelling experiments in testing the validity of the theories.

We shall briefly mention the methods used in diagnostic measurements and then discuss in more detail the first, and perhaps greatest, success of the technique of superconductive tunnelling, namely the determination of the energy gap. We shall go on from there to investigate the dependence of the energy gap on temperature, magnetic field, crystal orientation, transport current, magnetic impurities, film thickness and pressure. In addition we shall discuss how the technique of tunnelling was instrumental in setting up a strong-coupling theory of superconductivity, and how useful information may be obtained about such diverse subjects as the proximity effect, vortex structure, normal electron lifetime, spectra of trapped molecules and superconductivity of small particles.

6.2 Measurements methods

In principle, tunnelling measurements may be made with no more than a power supply and a voltmeter which certainly shows the basic simplicity of the technique. The practical measurement of I–V characteristics is, of course, done by more sophisticated circuitry using current sweep, an X–Y detector, stabilised d.c. amplifiers, etc. For tracing the negative resistance region, for example, a low impedance source is required and facilities are needed for suppressing unwanted high-frequency oscillations. All this is, however, conventional instrumentation; the need for a new type of measuring equipment arose only later when finer structure in the tunnelling characteristics was to be explored.

It was shown in Section 4.4 that the density of states may be determined by measuring the derivative dI/dV at a sufficiently low temperature. If for some reason there is a step in the density of states above the gap, then at a voltage V_s the I versus V characteristic changes slope, there is a somewhat smeared step in dI/dV, and d^2I/dV^2 has a relatively sharp minimum as shown in Fig. 6.1.

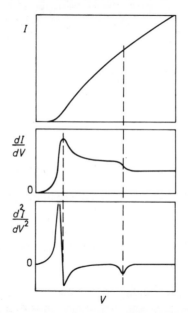

Fig. 6.1 The I–V characteristic for an N–I–S junction with corresponding first and second derivative plots. After Rowell and Kopf [113].

Thus features hardly recognisable on the I–V characteristic may be spotted with great accuracy from the second derivative plot.*

The usual technique of measuring the first derivative is to superimpose a small a.c. modulating signal ($\delta \cos \omega t$) upon the d.c. bias current and measure the a.c. voltage output. Similarly, a signal proportional to the second derivative

---

* We have already come across one example, the Tomasch effect, where the discontinuity in the density of states is reflected in the second derivative plot.

at the point I_0 may be obtained by measuring the second harmonic output of the same a.c. modulating signal. Both follow from the Taylor expansion

$$V(I) = V(I_0) + \left(\frac{dV}{dI}\right)_{I_0} \delta \cos \omega t + \frac{1}{2}\left(\frac{d^2V}{dI^2}\right)_{I_0} (\delta \cos \omega t)^2 \qquad (6.1)$$

which gives for the amplitudes of the first and second harmonic

$$A_1 = \delta\left(\frac{dV}{dI}\right)_{I_0} \quad \text{and} \quad A_2 = \frac{1}{4}\left(\frac{d^2V}{dI^2}\right)_{I_0} \delta^2. \qquad (6.2)$$

The output voltages are very small since the modulation voltage amplitude must be kept below the thermal energy kT (at $1°K$ about $50~\mu V$ rms or less is needed) so that the resolution be limited by thermal rather than instrumental smearing. There are several problems in obtaining this resolution (e.g. stability at high gain, suppression of the harmonics of the modulating signal, avoiding the creation of second harmonics while amplifying the first harmonic, impedance transformation of the sample for avoiding noise) aggravated by the fact that in most applications the normal state conductance is required as well. The additional problems are to eliminate the effect of lead resistance and to keep the properties of the measuring circuit constant until all the normal state measurements are finished. It is the normalised conductance

$$\sigma(V) = \frac{(dI/dV)_{NS}}{(dI/dV)_{NN}} \qquad (6.3)$$

that is usually required.

The problem was solved by Thomas and Rowell [156] and Adler and Jackson [157] by using improved phase detection techniques. The former workers employed a selection–rejection network whereas the latter ones used a sophisticated bridge. There are a number of other solutions reported [158–162] some of them concerned with first derivative measurements only.

6.3 Determination of the energy gap; temperature dependence

A straightforward determination of the energy gap was the first claim to fame of the technique of superconductive tunnelling. A number of methods have been used which we shall briefly review here.

(i) From the I–V characteristics of S–I–S junctions. This is easy enough to do when the gap is clearly defined, as for example in the lower temperature curves of Fig. 4.5. For higher temperatures the usual practice is to take the extrapolations of the characteristic from above and below the kink as may be seen in Fig. 6.2 for a Pb–I–Pb junction [163].

The agreement with the BCS theory is usually good, sometimes spectacular.

Fig. 1.5, which showed the temperature dependence of the energy gap, was in fact obtained by this method from the curves (Fig. 4.5) of Blackford and March [99].

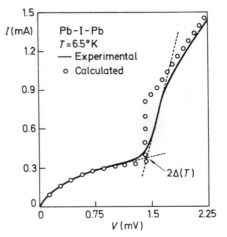

Fig. 6.2 Geometrical construction (dotted line) used for determining the temperature-dependent energy gap, $2\Delta(T)$, from experimental *I–V* characteristics. The open circles are numerically calculated points using the BCS density of states. After Gasparovic *et al.* [163].

(*ii*) *From the I–V characteristics of S_1–I–S_2 junctions.* As discussed in Section 4.2 there is a peak in the *I–V* characteristic at $V = (\Delta_2 - \Delta_1)/e$ and a sudden jump at $V = (\Delta_2 + \Delta_1)/e$, but this is true only when the junction is fed by a constant voltage generator. For a constant current generator the current flowing through the junction must be a monotonically increasing (or decreasing) function of voltage so the negative resistance region is replaced by a hysteresis curve as shown in Fig. 6.3 for Pb–I–Al junctions [164]. By measuring the beginning and the end of the hysteresis region the values of both energy gaps can be deduced. The actual method of determining [164] the Al gap (the smaller of the two) is illustrated in Fig. 6.4. Line (a) is tangent to the curve. The gap is defined as the horizontal distance between (a) and (b). The advantage of this definition is that it is still applicable for small gaps when the negative resistance region disappears but the influence of the gap may still be discerned (see $T_c = 1.328°$K curve) on the *I–V* characteristics.

At higher temperatures the current jump at $V = (\Delta_2 + \Delta_1)/e$ is not so well defined (see, for example, the experimental curves of Shapiro *et al.* [102] on Sn–I–Pb junctions). In that case one may calculate the theoretical curve from Equation (4.14) and try to get a 'best fit' to the experimental points.

For a number of superconductors the BCS temperature variation may be regarded a sufficiently good approximation. In that case we need only one point on the temperature scale where the energy gap can be accurately deter-

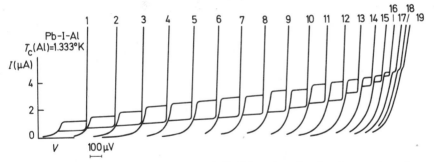

Fig. 6.3 Partial I–V characteristics for a Pb–I–Al junction showing temperature dependence of the energy gap of Al. 1: 0·872°K, 2: 0·969, 3: 1·021, 4: 1·060, 5: 1·088, 6: 1·112, 7: 1·139, 8: 1·170, 9: 1·199, 10: 1·219, 11: 1·243, 12: 1·267, 13: 1·286, 14: 1·300, 15: 1·312, 16: 1·320, 17: 1·325, 18: 1·328, 19: 1·330. After Douglass and Meservey [164].

mined. This point was chosen by McMillan and Rowell [165] in the following manner: The I–V characteristic of an Al–I–S_2 junction was observed as the temperature slowly decreased. When the critical temperature of Al is reached there is a sudden change from the N–I–S_2 to the S_1–I–S_2 characteristic and accordingly a small sharp break appears in the curve at Δ_2. If the critical temperature of Al is measured at the same time we know accurately one point on the $\Delta(T/T_c)/\Delta(0)$ curve from which all other points can be obtained.

(iii) *From the differential conductance of S–I–S and S_1–I–S_2 junctions*. A sudden change in the I–V characteristic appears as a peak in the curve of dI/dV against V. Hence an alternative way of measuring 2Δ or $\Delta_1 + \Delta_2$ is to choose the point where dI/dV is maximum [166, 167].

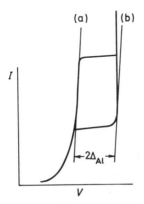

Fig. 6.4 Geometrical construction for determining the temperature-dependent energy gap. Line (a) is tangent to the curve at the first point of inflection; line (b) is parallel to (a) and is tangent to the curve. After Douglass and Meservey [164].

(iv) From the relative differential conductance of N–I–S junctions. This is one of the first methods used; $\sigma(V)$ as defined by Equation (6.1) is measured. Giaever [43, 44] (and more recently some other authors [168]) used the criterion that the half energy gap is approximately equal to the voltage where $\sigma(V) = 1$. This is not far from the truth at low temperatures but is in serious error at higher temperatures. From Bermon's tabulation [169] (applied to an Al–I–Pb junction) the above criterion corresponds to 0.92Δ at $1°K$ but to 1.4Δ at $6°K$. A better criterion (used by Giaever *et al.* [66]) is to compare the calculated and experimental curves of $\sigma(V)$ and choose the value of Δ which gives the best fit. Since the curves have a steeply rising portion (see Fig. 6.5) it is apparent that only one value of Δ will give a good fit.

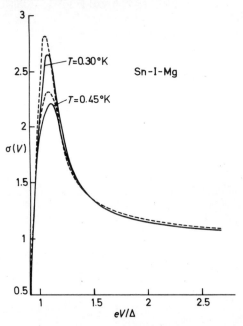

Fig. 6.5 Comparison of experimental and calculated values of the normalised conductance, $\sigma(V)$. After Giaever *et al.* [66].

A more widely used method is to determine the energy gap from the zero bias* value of σ. The relationship was tabulated by Bermon [169] and plotted by Harden and Collier [170] and Douglass and Falicov [33] as shown in Fig. 6.6 (*a*) and (*b*).

It should be noted that by relying on the BCS relationship between Δ and $\sigma(0)$ we explicitly exclude all cases (like proximity effect, magnetic field effects,

---

*Note that I_{NS}/I_{SS} for small d.c. bias (as given by Equations (4.11) and (4.13)) is equal to $\sigma(0)$.

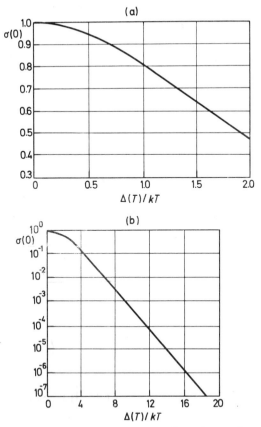

Fig. 6.6 The zero bias value of the normalised differential conductance as a function of $\Delta(T)/kT$, (a) for values of $\sigma(0)$ close to unity, after Harden and Collier [170]. (b) For seven decades on a logarithmic scale, after Douglass and Falicov [33].

paramagnetic impurities, etc.) where the density of states of the superconducting electrode is given by a different function.

To date a large number of energy gaps have been determined by tunnelling measurements. It is a simple method that has already reached undergraduate laboratories [171] but not entirely free of difficulties. For example La was first reported [172] to have $2\Delta/kT_c \cong 1.65$ less than half the BCS value but this was repudiated by later measurements [173–176] showing no major deviation. The erroneous result was probably due to the condition of the film. If several layers of different critical temperature are present, tunnelling measures the gap of the first deposited layer (which may have a smaller critical temperature and thus a smaller gap) while resistance measurements give the highest critical temperature. For Nb_3Sn there are three reports of low values (1.8 [177], 1.3 [178], and 2.1 average measured on single crystal specimen [179]) of $2\Delta/kT_c$

and only one [122] in agreement with the BCS value. Since thermal conductivity measurements also give the BCS value, the failure of the three tunnelling measurements must be explained away by surface conditions. Cohen *et al.* [180] get reasonable agreement with the BCS theory by assuming that there is a drastic drop of the gap, right down to zero, in a distance corresponding to the mean-free-path, causing the observed reduction in the 'effective' gap.

6.4 Effect of magnetic impurities

The critical temperature of superconductors is very sensitive to the presence of paramagnetic impurities as it was found experimentally by Matthias *et al.* [181, 182] and Schwidtal [183]. In contrast to non-magnetic impurities a magnetic impurity concentration of as little as 1% causes an appreciable depression of the critical temperature. The explanation was given by Abrikosov and Gorkov [184] in terms of an exchange interaction coupling the spin of the conduction electrons to the impurity spin. In the BCS model there is correlation between electron pairs of opposite momenta and spins which are often referred to as time-reversed states. Since a magnetic impurity may cause a spin flip, the time reversal invariance is broken and the pairs have a finite lifetime τ which causes an energy spread $\Gamma = \hbar/\tau$ introducing states into the gap. The remarkable prediction coming out of the Abrikosov–Gorkov theory is that the energy gap decreases faster than the critical temperature. Thus at a certain concentration (91% of the concentration which reduces T_c to zero) the energy gap vanishes but the critical temperature remains finite; in other words we have a *gapless super-conductor*. $\Delta(T)$ which used to denote the energy gap appears now as an order parameter. Whenever $\Gamma = \Delta$ the energy gap vanishes. This was calculated by Skalski *et al.* [185] and is shown in Fig. 6.7 where $\Delta(T)/\Delta^P(0)$ is plotted against

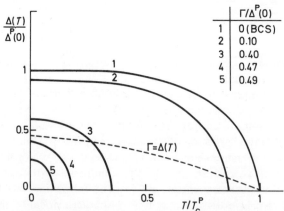

Fig. 6.7 The order parameter as a function of temperature (normalised to the critical temperature of the impurity-free material, T_c^P). The dashed curve is a boundary curve separating the gapless region from the region with a gap. After Skalski *et al.* [185].

T/T_c^p. The superscript refers to the value in the absence of magnetic impurities ($\Gamma = 0$).

The prediction of gapless superconductivity presented another challenge to the technique of tunnelling. The experiments were first performed by Reif and Woolf [78, 186, 187] using Pb as base metal in order to reach the required low reduced temperatures. Since it is difficult to dissolve magnetic impurities in lead the following technique was used.

The desired amount of impurity is added to the pure metal and the resulting mixture is repeatedly (as many as 75 times) folded and rolled until a foil is obtained in which the two components are homogeneously distributed. The foil is then cut into pellets of about 0·5 mm in size which are successively evaporated upon oxidised Al held at about 1°K. Since diffusion at this temperature is negligible and since each pellet contributes only about 1 nm to the alloy film thickness, the resulting film is a microscopically homogeneous alloy.

A comparison between the experimental results and theory is shown in Fig. 6.8. The circles and triangles are the experimental points for the critical tempera-

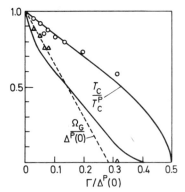

Fig. 6.8 Critical temperature and half-energy gap at T_c/T_c^p as a function of $\Gamma/\Delta^p(0)$. The open circles and triangles are the experimental points for the critical temperature and half-energy gap respectively as measured by Reif and Woolf [78]. The dashed curve is the linear extrapolation of the low-impurity-concentration points. The triangle at $\Gamma/\Delta^p(0) = 0.3$ is roughly the concentration at which no gap is observed. After Skalski *et al.* [185].

ture (determined resistively) and the energy gap (denoted by Ω_G, determined by tunnelling) respectively. The dashed curve is the linear extrapolation of the low impurity concentration points. The full lines are calculated by Skalski *et al.* [185]. It may be clearly seen that the energy gap decreases faster than the critical temperature. The agreement between theory and experiments may be regarded satisfactory. The magnetic impurity was Gd in this case; it should be noted that experimental results [187] for Fe (in In) and Mn (in Pb) were less in agreement with the Abrikosov–Gorkov theory.

Large deviation from the Abrikosov–Gorkov theory was found by Edelstein

[188, 189] and Tsuda [190] for LaCe alloys. The effect of the Ce impurity on both energy gap and critical temperature was considerably larger than predicted by the theory. The apparent reason is that the approximations used by Abrikosov and Gorkov are not valid for LaCe because it exhibits the Kondo effect (a resistivity minimum in the normal state). The additional states found in the gap are explained by Edelstein [189] as resulting from electron scattering from the impurities.

6.5 Magnetic field dependence of the energy gap

Tunnelling played a major role in finding the effect of magnetic field on the energy gap. The first measurements were performed by Giaever and Megerle [69] who showed that a magnetic field applied parallel to the plane of the junction decreased the energy gap monotonically to zero. More detailed measurements were done later by Douglass [191, 70] who found that for large values of d/λ (d, film thickness) the energy gap does not reduce smoothly to zero but drops suddenly at the critical magnetic field, H_c. The measured results for two different film thicknesses are shown in Fig. 6.9 where $[\Delta(H)/\Delta(0)]^2$ is plotted against $(H/H_c)^2$. Douglass [70, 191] worked out the theoretical value of the energy gap as well on the basis of the Ginzburg–Landau equations. He assumed $\kappa = 0$ (or alternatively that the order parameter is independent of position), solved the

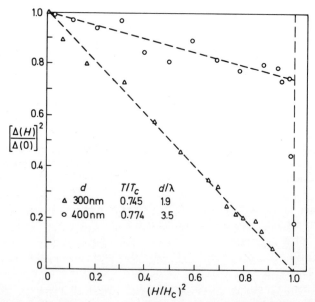

Fig. 6.9 Energy gap of Al versus magnetic field for films of thicknesses 300 nm and 400 nm measured by electron tunnelling. The dashed curves are the best straight lines through the data points. After Douglass [70].

boundary condition problem for the relevant thin-film geometry and obtained the gap by simply identifying the order parameter with the energy gap. The calculated curves are shown in Fig. 6.10. There is no quantitative agreement between theory and experiments but the main features are identical. Investigations by Meservey and Douglass [192] showed further discrepancies for $d/\lambda < 1$.

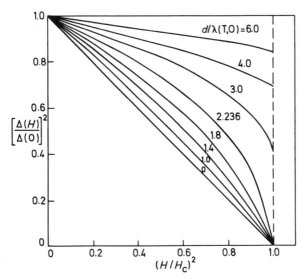

Fig. 6.10 Energy gap versus magnetic field for various ratios of thickness to penetration depth. After Douglass [70].

Collier and Kamper [193] recalculated the curves of Douglass by assuming a space-dependent order parameter. This modified theory was in good agreement with their experimental results on 'clean' tin films. A common deficiency of all the above attempts is that they rely on the BCS density of states for interpreting their results. In particular Collier and Kamper [193] determined the gap from the measured value of $\sigma(0)$.

Conceptually different theories were constructed by de Gennes [194], Maki [195–198] and Maki and Fulde [199]. They are essentially extensions of the Abrikosov–Gorkov theory to the case of magnetic fields. Interestingly, the role of magnetic impurities is taken over by $(H/H_c)^2$, hence gaplessness occurs at a magnetic field $H = \sqrt{(0.91)}H_c = 0.955\ H_c$. The density of states is also radically altered which makes it a suitable* quantity for comparison with experiments. It was shown by Guyon et al. [200] that their experimental

*Another quantity also used for comparison is [200]
$$G(V) = V(I_{NN} - I_{SN})/I_{NN}.$$

results on Al–I–SnIn and Al–I–PbBi junctions (alloys are chosen to reach the 'dirty' limit) in the vicinity of the upper critical field cannot be explained with the BCS density of states but the agreement with de Gennes' theory [194] is excellent. Similar conclusions were reached by Levine [201] who measured $\sigma(V)$ for a series of magnetic fields. The comparison with Maki's theory (a more formal approach which otherwise agrees in nearly all its conclusions with the theory of de Gennes) is shown in Fig. 6.11. As may be seen the agreement deteriorates as the magnetic field increases. This may be due to the effect of the non-zero mean-free-path. Millstein and Tinkham [202] compared their experimental results both with Maki's theory and with the Strassler–Wyder theory [203] (taking into account mean-free-path) (Fig. 6.12) finding spectacular agreement with the latter.

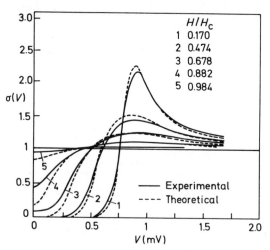

Fig. 6.11 Normalised differential conductance as a function of voltage for the indicated field strengths. After Levine [201].

6.6 Effect of transport current

A current flowing parallel to the junction will also reduce the energy gap. The development of the theory parallelled that concerned with the effect of magnetic field. Bardeen [204] derived a theory on the basis of the Ginzburg–Landau equations which was followed by Fulde's theory [205] in which it was shown that in the limit of short mean-free-path the density of states function will be the same as that for magnetic impurities in the Abrikosov–Gorkov theory [184]. Hence a sufficiently large transport current will induce gaplessness.

Experimental work was done by Levine [71] and Mitescu [206]. The geometry of the junction used by the former author is shown in Fig. 6.13.

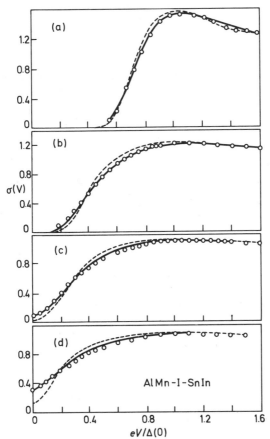

Fig. 6.12 Normalised differential conductance as a function of normalised voltage in parallel magnetic field at $T = 0.361°K$. Dashed curve is Maki's theory [195–198] $(l/\xi_0 = 0)$; solid curve is Strässler–Wyder theory [203] for $l/\xi_0 = \pi/10$. The values of the magnetic field are (a) $H/H_{c\parallel} = 0.43$, (b) $H/H_{c\parallel} = 0.7$, (c) $H/H_{c\parallel} = 0.82$, (d) $H/H_{c\parallel} = 0.89$. After Millstein and Tinkham [202].

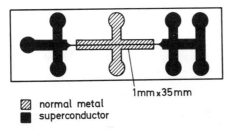

Fig. 6.13 Geometry of tunnelling junction. Tunnelling measurements are made between the normal metal and the superconductor while a transport current is passed between the terminals. After Levine [71].

The tunnel current flows between the normal metal (Al in the normal state) and the superconductor (Sn or In) while the transport current is passed between terminals 1 and 2. Both authors show that the reduction in the energy gap is proportional to the square of the transport current (as it follows from theory) but no attempt is made to prove gaplessness or to compare the measured density of states with the theoretical predictions.

6.7 Dependence of the energy gap on film thickness

The energy gap for ultrathin films was determined by Wilson [72] in an experiment where the $I-V$ characteristics of tunnel junctions were monitored during deposition of the top film. The results for an Al–I–Bi junction, normalised to take into account the varying thickness, are shown in Fig. 6.14. It may be seen that the gap slowly increases to its full value [207] at around 4 nm. Similar results were obtained for Al–I–Sn junctions as well.

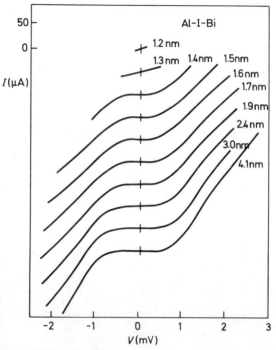

Fig. 6.14 $I-V$ characteristics of an Al–I–Bi junction for various values of Bi thickness. After Wilson [72].

6.8 Gap anisotropy; the effect of crystal orientation

It seems obvious that the assumption of isotropic conditions in the BCS model cannot apply to single crystal specimen so one may expect different energy gaps in different directions. In polycrystalline specimens several gaps are simultaneously present which is one of the early [69] explanations for the observed smearing of the energy gap.

Zavaritskii [73–76] made a number of tunnelling measurements on single crystal tin, obtaining values from 4·3 to 3·1 for $2\Delta/kT_c$. He explained the results with the anisotropy of the Fermi surface. Bennet's semiempirical approach [208] uses Zavaritskii's data for building up a theory according to which the principal contribution to gap anisotropy comes from the anisotropic phonon spectrum of the metal.

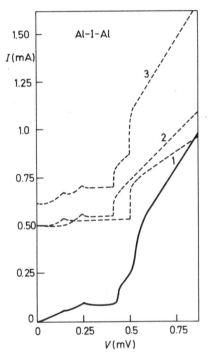

Fig. 6.15 I–V characteristic of an Al–I–Al junction at $T = 1·14°$K. Assuming that one of the Al films (the thicker one) displays two distinct gaps, the sum of the two theoretical characteristics (curves 1 and 2) reproduces well (curve 3) the experimental curve (solid line). After Campbell and Walmsley [209].

Tunnelling between two thin films, and between a thin and a thick (1·6 μm) film of Al was measured by Campbell and Walmsley [209]. They found no sign of anisotropy in the first case; the current increased rapidly at the gap. In the second case the *I–V* characteristic was of a more complicated shape, as shown in Fig. 6.15. It is interesting to see that the assumption of two distinct energy gaps in the thick film (due to crystallites of different orientations) is sufficient to reproduce the main features of the measured curve. Similar results were found for thick films of lead, indium and tin.

The effect of the mean free path was investigated by Campbell, Dynes and Walmsley [210] by adding bismuth (5, 10 and 15 atomic percent) to lead. The added impurity causes additional scattering, hence a smaller mean free path and a reduction in anisotropy. The measured anisotropy did indeed decrease with increasing bismuth concentration.

Nb single crystals grown by electron-beam zone melting were used by MacVicar and Rose [211–215]. They could measure tunnelling in different crystallographic directions by the simple method of evaporating indium stripes upon different parts of the single crystal cylinder as shown in Fig. 6.16. They find that the anisotropy depends on the resistivity ratio $\rho_r = [\rho(300°K) - \rho(4°K)]/\rho(300°K)$. Anisotropy was observed for crystals with $\rho_r > 178$ but no anisotropy was found (probably due to impurity scattering) for $\rho_r = 65$. For one of the crystals ($\rho_r = 267$) grown approximately in the $\langle 110 \rangle$ direction the energy gaps measured in the various directions are shown in Fig. 6.17. The other specimen with high resistivity ratios behaved similarly. The lowest gap observed was 2·84 meV in the $\langle 100 \rangle$ direction and the highest about 3·19 meV in the $\langle 311 \rangle$ and $\langle 111 \rangle$ directions.

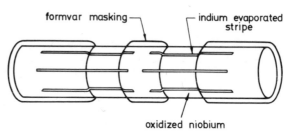

Fig. 6.16 Nb–I–Sn junctions on single crystal Nb placed at various crystallographic directions. After MacVicar and Rose [213].

Further measurements with high resistivity ratio (up to 3000) Nb samples were performed by Hafstrom et al. [216] who found evidence for a second energy gap in the $\langle 110 \rangle$ and $\langle 111 \rangle$ directions. The experimental data were evaluated by MacVicar [217] who concluded that the observed energy values are grouped non-randomly in k-space indicating some relationship with the topology of the Fermi surface.

The anisotropy of the energy gap has now been firmly established (not only by tunnelling; ultrasonic and infrared measurements were equally useful and some evidence was found in nuclear relaxation and surface resistance measurements) but apparently neither the experimental technique nor the theories are as yet sufficiently refined for expecting detailed agreement between them.

In conclusion we wish to note the predictions of Dowman et al. [52] that the selection rules (that is, which electrons are allowed to tunnel across the junction) depend on the nature of the barrier (different rules apply to single crystal and amorphous oxides). Thus for a correct interpretation of the anisotropy experiments more information is needed about the properties of the barrier.

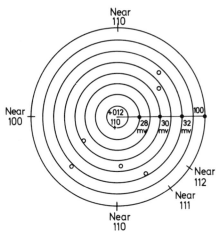

Fig. 6.17 Polar plot of energy gap versus angular position of junction, $\rho_r = 267$. After MacVivar and Rose [213].

6.9 Strong coupling superconductors

The most convincing proof of the validity of the BCS theory came from tunnelling measurements but it became clear from the same measurements that some superconductors behaved somewhat differently. Giaever and Megerle [69] observed that the temperature dependence of the energy gap of lead obeyed the BCS theory only if they used the experimentally obtained value of $\Delta(0)$. The ratio $2\Delta(0)/kT_c$ was found to be 4·3, well above the BCS value of 3·52. A year later Giaever, Hart and Megerle [66] found a small but significant deviation from the BCS density of states when plotting $\sigma(V)$ as shown in Fig. 6.18. The crossover point is at about $k\theta_D$ the Debye energy suggesting immediately (low Debye energy implies strong electron–phonon interaction) the cause and the direction in which the BCS theory should be modified. In fact, more general theories were already available. Eliashberg [68] had already derived his gap equation taking into account both the electron–phonon matrix element and the phonon spectrum, but before the strategic attack upon this equation a number of tactical advances were necessary to make it sure that the experimental results were in line with the theoretical predictions. The story of these efforts is reviewed by Rowell and McMillan [165]; we shall briefly outline here the major steps in the process.

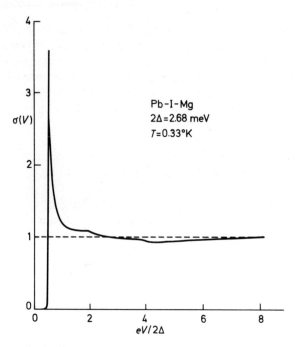

Fig. 6.18 Normalised differential conductance as a function of normalised voltage for a Pb–I–Mg junction. The structure shown in the curve signifies deviation from the BCS theory. The cross-over point corresponds to the Debye energy. After Giaever, Hart and Megerle [66].

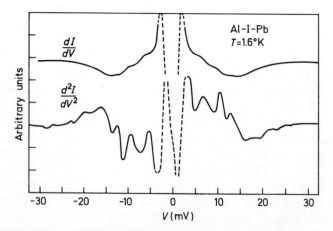

Fig. 6.19 First and second derivatives of the I–V characteristic as a function of voltage for an Al–I–Pb junction. After Rowell, Chynoweth and Phillips [219].

Morel and Anderson [218] approximated the effective phonon density by an Einstein peak at the longitudinal phonon frequency ω_L and predicted structure in the tunnelling density of states at energies $n\hbar\omega_L$. Subsequent measurements by Rowell *et al.* [219] (on Al–I–Pb junctions at $1\cdot6°K$) of the second derivative of the *I–V* characteristic did indeed produce the expected structure as shown in Fig. 6.19. The peak positions are given empirically by the formula

$$E_n = \Delta^1 + n\theta \qquad (6.2)$$

where Δ^1 agrees roughly with the value for half the energy gap and $\theta = 3\cdot7$ meV is in the range of appreciable transverse phonon energies measured by neutron diffraction.

A refinement of the theory by Schrieffer *et al.* [220] was matched by further refinement of the experimental techniques [221]. By assuming the phonon density in the form of two Lorentzian peaks (representing the transverse and longitudinal phonons respectively) shown in Fig. 6.20 (*b*) and taking reasonable

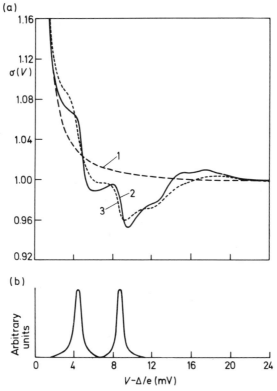

Fig. 6.20 (a) The normalised differential conductance as a function of voltage (zero shifted by Δ/e), 1: BCS theory, 2: theory by Schrieffer *et al.* [220], 3: experimental results by Rowell, Anderson and Thomas [221]. (*b*) The phonon spectrum assumed by Schrieffer *et al.* [220].

values for α the electron–phonon coupling and μ^* the Coulomb pseudo-potential (a parameter entering the Eliashberg gap equation) a theoretical curve was obtained for $\sigma(V)$ in excellent agreement with tunnelling data (Fig. 6.20 (a)). Second derivative measurements [221] (Fig. 6.21) did in fact show considerably more structure for which the theory of Schrieffer et al. [220] could not account. These came to be identified with Van Hove singularities* measured by neutron diffraction [222]. Although the relationship of this last structure to the density of states was explained by Scalapino and Anderson [223] the conviction grew that a frontal assault on the Eliashberg equation had become feasible.

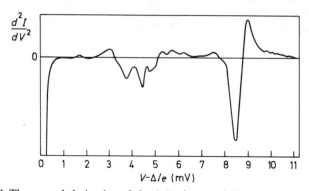

Fig. 6.21 The second derivative of the I–V characteristic as a function of voltage (zero shifted by Δ/e). Arrows indicate the voltage for Van Hove singularities on the basis of the neutron diffraction experiments of Brockhouse et al. [222]. After Rowell, Anderson and Thomas [221].

The method used by McMillan and Rowell [67] is based on the fact that in Eliashberg's theory the electron–phonon coupling constant weighted by the phonon density of states $\alpha^2(\omega)F(\omega)$ is uniquely related to the electronic density of states (as measured by tunnelling experiments). Hence besides working out the density of states from phonon data it is also possible to invert the Eliashberg equation and get the phonon data from the measured tunnelling characteristics. With the aid of a computer $\alpha^2(\omega)F(\omega)$ and μ^* are adjusted until the computed density of states accurately fits the measured density of states for $E < k\theta_D$. Examining the agreement for $E > k\theta_D$ the accuracy of the Eliashberg equation can be tested. The results of this rather difficult exercise may be seen in Fig. 6.22 where theory and experiments are compared. The experimental points below $E - \Delta = 11$ meV were used for determining the 'input' quantities, and the points above that are to be compared with the theory. It is remarkable that an experimental curve as complicated as that can be theoretically reproduced. McMillan

---

*It was predicted by Van Hove [224] that for three-dimensional crystals there are at least two (usually more than two) discontinuities in the phonon spectrum occurring at saddle points, maxima and minima in the dispersion curves.

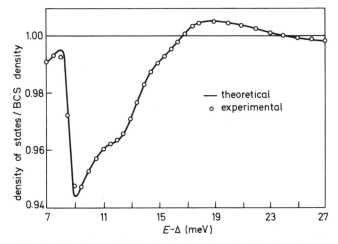

Fig. 6.22 The density of states for lead related to the BCS density of states as a function of energy (zero shifted by Δ). In the experiment the sharp drop near 9 meV is affected by thermal smearing. After McMillan and Rowell [165].

and Rowell [166] conclude that our present theories of superconductivity are accurate to a few percent.*

All the experiments mentioned so far were concerned with the properties of lead. The other strong-coupling superconductor, mercury (with $2\Delta/kT_c = 4\cdot6$), was less thoroughly investigated but the available data [225–226] suggest similar conclusions.

Next we shall briefly review the tunnelling experiments on lead-based alloys. The effect of a small amount of indium on the phonon spectrum was investigated by Rowell, McMillan and Anderson [227] using the inversion technique described above. The results for $\alpha^2(\omega)F(\omega)$ are shown in Fig. 6.23. The additional structure at $\hbar\omega = 9\cdot5$ meV is due to the presence of an 'impurity band'.** The impurity band is still present at higher indium concentrations as reported by Adler *et al.* [228]. The value of $2\Delta/kT_c$ slowly decreases with increasing indium concentration; it is $4\cdot34$ at 2 atomic percent indium reducing to $4\cdot20$ at 70 atomic percent indium.

Alloying lead with thallium is of considerable interest since the atomic mass of thallium is about the same as that of lead and it is possible to dissolve more than 85 atomic percent Tl into Pb and still maintain the f.c.c. Pb structure. Hence one may study the effect of electron density on the electron–phonon

---

*This does not mean that we can already predict all properties of all superconductors to that accuracy. It only implies that if we had all the relevant data available about the properties of the normal metal, we could predict from them the superconducting properties.

**A light substitutional impurity atom in a cubic crystal causes a triply degenerate bound vibrational state to appear above the high frequency cut-off of the phonon spectrum. By analogy with the electronic states we may refer to this as an 'impurity band'.

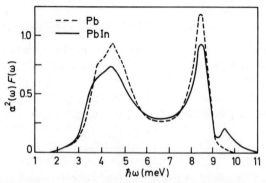

Fig. 6.23 The phonon spectrum $\alpha^2(\omega)F(\omega)$ obtained by the gap inversion technique. The localised phonon mode of the light indium impurity may be seen at about 9·5 meV. After Rowell, McMillan and Anderson [227].

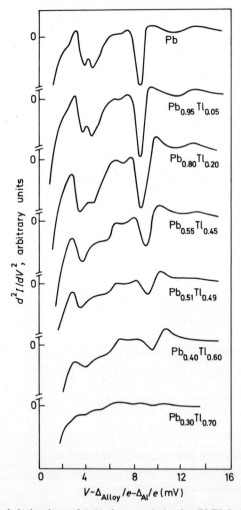

Fig. 6.24 Second derivatives of $I–V$ characteristics for PbTl–I–Al junctions. Structure is decreasing with increasing Tl concentration. After Claeson and Grimwall [231].

coupling strength and on the phonon spectrum. As the concentration of Tl increases the critical temperature decreases [229] which already indicates some weakening of the coupling responsible for superconductivity. $2\Delta/kT_c$ decreases as well, reaching the BCS value at around 60 atomic percent Tl [229–230]. The decrease in coupling strength is reflected in the second derivative structure [231]; it gets smoother with increasing Tl concentration as may be seen in Fig. 6.24.

Lead–bismuth alloys behave in a more complicated manner. Two energy gaps were observed by Giaever [232] who attributed the effect to the presence of two phases. In a more detailed study by Adler and Ng [233] the results are explained with the aid of the proximity effect, and tunnelling from three different regions (two superconducting, one normal) is claimed. The structure in $\sigma(V)$ was found to be decreasing, nevertheless the ratio $2\Delta/kT_c$ *increased* with increasing Bi concentration.

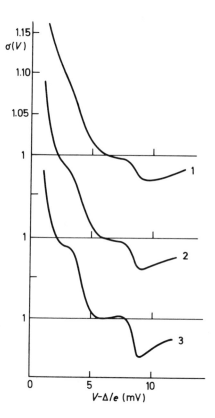

Fig. 6.25 Normalised differential conductance as a function of voltage (zero shifted by Δ/e) for 1: lead condensed at 1·6°K, 2: annealed to 80°K, 3: annealed to 300°K. After Zavaritskii [234].

6.10 Amorphous superconductors

Low temperature condensation (that is, evaporation upon substrates cooled to a few degrees Kelvin) of superconductors leads usually to higher critical temperatures and stronger coupling. Lead was investigated by Zavaritskii [234, 235] who measured the derivative curves of Al–I–Pb tunnel junctions with the lead deposited at $1\cdot6°K$ and subsequently annealed to $80°K$ and $300°K$. The measured $\sigma(V)$ curves for the three cases are shown in Fig. 6.25. It may be seen that the principal features remain unchanged. The curves of $\alpha^2(\omega)F(\omega)$ (calculated by the inversion method from the second derivative measurements) also retain their general appearance but there are some pertinent changes as may be seen in Fig. 6.26. There is a broadening of the maxima (due to the

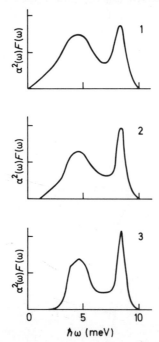

Fig. 6.26 The phonon spectrum, $\alpha^2(\omega)F(\omega)$ obtained by the gap inversion technique of 1: lead condensed at $1\cdot6°K$, 2: annealed to $80°K$, 3: identical to the dashed curve of Fig. 6.23 given here for comparison. After Zavaritskii [234].

smearing of the Van Hove singularities) and a low energy tail down to zero energies. No change in the critical temperature was found but the ratio $2\Delta/kT_c$ increases indicating the effect of the modified phonon spectrum.

The effect of low temperature condensation upon other superconductors, Ga [178, 235, 236], Al [237–239, 235], Bi [235, 240, 207], Sn [235, 241] and In [235] has been to increase both T_c and $2\Delta/kT_c$. The increase is definitely due to the amorphous character of the solid; when the film is heated to room temperature the proper crystal structure appears and the superconductor reverts to its usual (weak-coupling) properties. Exceptions are the films prepared by evaporation in an oxygen atmosphere [237] where the precipitation of oxygen at the grain boundaries prevents recrystallisation and grain growth.

6.11 Influence of the phonon spectra on the critical temperature

There are a number of theories concerned with the prediction of the critical temperature from the detailed knowledge of the normal state parameters. Most of these are rather complicated and offer no chance to separate the effect of various factors. We shall mention here a theory by McMillan [242] which gives the critical temperature in the simple form

$$T_c = \frac{\theta_D}{1 \cdot 45} \exp\left[-\frac{1 \cdot 04(1+\lambda)}{\lambda - \mu^*(1 + 0 \cdot 62\lambda)} \right] \tag{6.3}$$

where

$$\lambda = 2 \int_0^\infty \left[\alpha^2(\omega)F(\omega)/\omega \right] d\omega \tag{6.4}$$

is a dimensionless electron–phonon coupling constant corresponding to the $N(0)V$ of the BCS model.

Leger and Klein [238] calculated the phonon spectrum of three different types of Al films (ordinary, $T_c = 1 \cdot 3°K$; granular, $T_c = 2 \cdot 30°K$ and $T_c = 3 \cdot 66°K$) from second derivative measurements. They found a low frequency tail in the phonon spectrum of the films possessing the higher critical temperatures and explained it by the large surface to volume ratio in granular Al. Substituting the obtained phonon spectrum into McMillan's equations yielded good agreement for the critical temperature.

A more extensive study by Dynes [243] on In–Tl alloys gave less encouraging results. Comparison of the critical temperatures of 11 samples (covering the whole range from zero to 100% concentration) with the McMillan and the Garland–Allen [244] theory (a modification of the McMillan theory containing an extra parameter) leads to errors as much as 50%. It is some consolation that the shape of the T_c versus concentration plot is accurately predicted by Garland and Allen although the predicted value of T_c is consistently below the observed one.

6.12 Effect of pressure

The effect of pressure on the various properties of superconductors has been widely investigated [245] but it was left again to the technique of tunnelling to yield information about the pressure-dependence of the energy gap and of the phonon spectrum.

The results of Zavaritskii *et al.* [246, 247] are shown in Fig. 6.27 where T_c (as determined experimentally by Smith and Chu [248]) and Δ are plotted against pressure (similar results were reported by Galkin and Svistunov [249]).

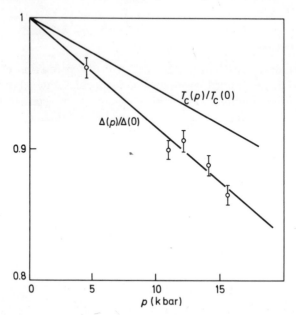

Fig. 6.27 The reduced critical temperature and reduced gap as a function of pressure. After Zavaritskii *et al.* [247].

It is notable that the energy gap reduces faster than the critical temperature; this is in accord with the predictions of Geilikman and Kresin [250] whose formula for strong coupling superconductors

$$\frac{2\Delta}{kT_c} = 3.52\left[1 + 5.3\left(\frac{T_c}{\theta_D}\right)^2 \ln\frac{\theta_D}{T_c}\right] \tag{6.5}$$

is satisfied by the experimentally determined values of T_c and Δ provided θ_D is replaced by $\hbar\omega_\perp/k$ where $\omega_\perp$ is the peak transverse phonon frequency.

Similar results were obtained by Franck and Keeler [77] who came to the conclusion that the reduction in $2\Delta/kT_c$ is due mainly to two effects, namely the reduction with pressure of $N(0)$ the single particle density of states at the Fermi level, and to the increase in phonon frequencies with pressure. The latter was determined by Franck and Keeler [251] and Zavaritskii *et al.* [246, 247] from derivative measurements whereas Franck, Keeler and Wu [252] obtained the complete phonon spectrum with the aid of the gap inversion technique. This is shown in Fig. 6.28 where the shift to higher frequencies can be clearly seen.

The phonon spectrum obtained may be used in McMillan's formula for the critical temperature. Though the direction of change is given correctly, there is considerable discrepancy (about 50%) as far as the numerical values are concerned.

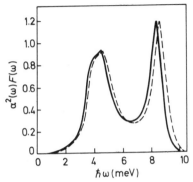

Fig. 6.28 The phonon spectrum, $\alpha^2(\omega)F(\omega)$, of lead at $p = 0$ (solid line) and $p = 3445$ bar (dashed line). After Franck, Keeler and Wu [252].

6.13 The proximity effect

The first experiments proving the proximity effect were performed by Meissner [253–254] well before the existence of any theory. Measuring the I–V characteristics of crossed copper-coated tin wires he found that for sufficiently low current and sufficiently thin copper coating the junction resistance is zero, that is, Cooper pairs can move unhindered across the normal metal.

We may look at Meissner's experiments either as a manifestation of a sort of tunnelling (it will be discussed in a little more detail in Section 21.2 as a variant of Josephson tunnelling) or we might simply say that the properties of the normal metal change owing to the proximity of the superconductor. A logical next step is to expect the normal metal to display an energy gap and of course tunnelling can again be relied on to provide an answer. The first experiments were performed by Smith, Shapiro, Miles and Nicol [79] who could show the presence of an energy gap by measuring the I–V characteristics of Pb–I–Ag–Pb junctions. Conversely, the energy gap of a superconductor was found to diminish [80, 186] when backed by a normal metal.

The present state of affairs is that the existing theories give only qualitative agreement with experiments. The main reason is that the theories of Fulde and Maki [255] and de Gennes and Mauro [256]* are valid only in the limit

---

*Note that Fulde and Maki [255] came to the conclusion that a nonmagnetic metal may induce gaplessness while de Gennes and Mauro [256] permit gaplessness only for an infinitely thick normal metal or for a magnetic metal.

$T \to T_c$ where the predicted variation of $\sigma(V)$ hardly differs from BCS. The gap in that region is small anyway so it is not so easy to measure its magnitude or show convincingly that it is zero. The only theory valid at any temperature is that of McMillan [257] which has a double connection with tunnelling. It not only predicts the density of states measured by tunnelling but also looks upon the flow of particles between the normal metal and the superconductor as if there was a thin barrier between them. Assuming further that both the N and the S layers are thin compared with the characteristic superconducting length, he can take the properties of each film space independent and solve the problem with the aid of the tunnelling Hamiltonian formalism. Since practical samples rarely satisfy the assumptions, lack of agreement with McMillan's theory is not conclusive.

There are numerous difficulties on the experimental front as well (for a brief review see Clarke [258]). It is difficult to get well-defined boundaries between the metals. Room temperature deposition may lead to some interdiffusion (and alloying if the metals are not carefully chosen) whereas low temperature deposition prevents an accurate determination of the thicknesses.

A qualitative comparison of the experimental results for $\sigma(V)$ with theory was given by Claeson and Gygax [259]. Fig 6.29 (a) shows the BCS dependence, Fig. 6.29 (b) gives the experimental results, while Fig. 6.29 (c) is calculated from

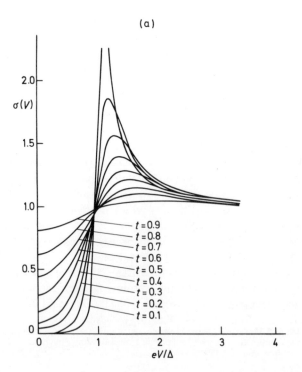

(a)

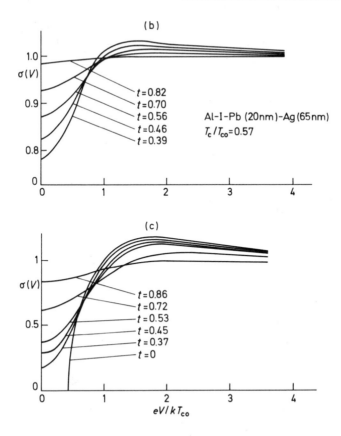

Fig. 6.29 Normalised differential conductance as a function of normalised voltage calculated from the BCS theory for several values of the reduced temperature $t = T/T_c$. (b) Experimental curves for $\sigma(V)$ as a function of eV/kT_{co} for an Al–I–Pb–Ag junction. The off-set zero is attributed to leakage currents. (c) Theoretical curves of $\sigma(V)$ versus eV/kT_{co} for a pair breaking parameter $T_c/T_{co} \cong 0.60$ for several values of the reduced temperature $t = T/T_c$. The calculation is based on an analogy with the paramagnetic impurity case. After Claeson, Gygax and Maki [267].

Maki's pair-breaking theory mentioned before. The rather crude assumption is here that the reduction in the critical temperature of the superconductor caused by the proximity of the normal metal is of the same type as that caused by the other pair-breaking mechanisms. It may be easily seen that the pair-breaking theory is much nearer to the experimental results.

One of the theoretical predictions of Fulde and Maki [255] for tunnelling into the superconducting side of proximity sandwiches is that $\sigma(0)$ tends linearly to unity as T tends to T_{cNS} (the critical temperature of the proximity sandwich). This was proven experimentally by Hauser [260, 261] on Al–I–Pb–M junctions where M is Ni, Fe and Pt. He found that $\sigma(0)$ approached unity in the same manner* both for magnetic and non-magnetic metals.

Comparisons with McMillan's theory were made by Adkins and Kington [262], Freake and Adkins [263], and Vrba and Woods [264, 265]. Adkins and Kington [262] find good agreement for the energy gap on the normal side if they make the not quite justified assumptions that (i) the energy gap may be calculated from the measured value of $\sigma(0)$ with the aid of the BCS density of states and (ii) the barrier penetration probability (a parameter in McMillan's theory) may be adjusted so as to give a fit to the experimental results. The agreement for $\sigma(V)$ is considerably worse although the general shapes of the curves are similar. Somewhat better agreement (especially for tunnelling into the superconducting side) is obtained by Freake and Adkins [263] whose results are shown in Fig. 6.30.

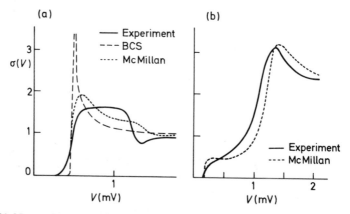

Fig. 6.30 Normalised differential conductance as a function of voltage for (a) N–I–Cu (44 nm)–Pb (500 nm) and (b) N–I–Pb (80 nm)–Cu (200 nm) junctions compared with theory. After Freake and Adkins [263].

The measurements of Vrba and Woods [264, 265] have a better chance to agree with McMillan's theory because their samples have a thin oxide layer between the normal metal and the superconductor comprising the proximity sandwich. The barrier penetration probability is taken by them as well as an adjustable parameter but its relative magnitude can now be varied by controlling the thickness of the barrier. The measured $\sigma(V)$ for an Al–I–AlSn junction (Al normal, Al and Sn weakly coupled) is compared with the theory in Fig. 6.31. As may be seen the BCS result (for the same energy gap) is considerably different but the agreement with McMillan's theory may be considered good.

We wish to mention the empirical rule enunciated by Guyon et al. [266]

---

*In other respects there are, of course, considerable differences between the effects of magnetic and non-magnetic metals. A magnetic metal has a larger effect on the density of states even if measured with sandwiches of the same critical temperature.

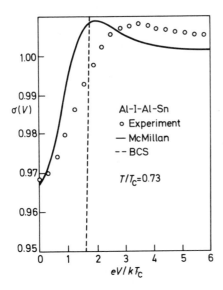

Fig. 6.31 The normalised differential conductance as a function of normalised voltage for an Al–I–Al–Sn proximity sandwich. The films of Al and Sn are lightly coupled. After Vrba and Woods [265].

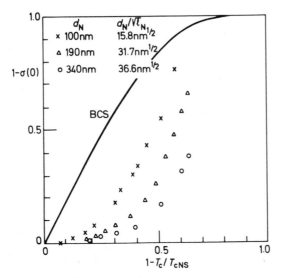

Fig. 6.32 $1-\sigma(0)$ as a function of $1-T/T_{cNS}$ for 'dirty' Al–I–Zn–InBi junctions. After Guyon *et al.* [266].

for tunnelling into the normal side. They find that in the vicinity of T_{cNS} the relationship

$$1 - \sigma(0) = (1 - T/T_{cNS})^n \qquad (6.6)$$

applies. Their experimental results (the BCS variation is shown for comparison) on 'dirty' Al–I–Zn–InBi junctions (both Al and Zn normal) are shown in Fig. 6.32. The value of n is apparently dependent on the parameter $d_N/\sqrt{l_N}$ (where d_N is the thickness of the normal layer) and it varies from 2·3 for $d_N/\sqrt{l_N} = 15\cdot3$ nm$^{1/2}$ to about 3 for $d_N/\sqrt{l_N} = 63$ nm$^{1/2}$. Claeson $et\ al.$ [267] find a value of about 3, while Vrba and Woods (who could control the barrier penetration probability) obtained values from $n = 1$ to 6.

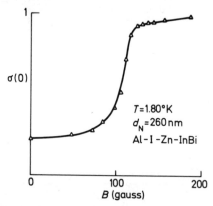

Fig. 6.33 The normalised differential conductance at zero bias as a function of magnetic field for an Al–I–Zn–InBi junction. After Burger $et\ al.$ [269].

The effect of a magnetic field has also been investigated [260, 268, 269]. As expected the energy gap decreases with increasing magnetic field but no general relationship has yet been established. Burger $et\ al.$ [269] found a sharp decrease in the induced gap (sharp increase in the value of $\sigma(0)$ as shown in Fig. 6.33) at a certain magnetic field in their measurements on Al–I–Zn–InBi junctions at low temperatures. The phenomenon was explained by de Gennes and Hurault [270] by introducing the concept of a wall dividing the normal metal into two parts, one (near the surface) having low concentration, and the other (near the S side) having high concentration of Cooper pairs.

6.14 Determination of normal electron lifetime

In a superconductor at any finite temperature a dynamic equilibrium exists, that is, normal electrons recombine to create pairs and are continuously created by the break up of pairs. The rate of these processes can be determined by tunnelling as it was first done by Ginsberg [287].

We shall describe here a later experiment by Miller and Dayem [88] based on similar principles but measuring simultaneously two different types of lifetime, namely (i) τ_T, relaxation time to the gap and (ii) τ_R, recombination time.

We shall consider two tunnel junctions in series as shown in Fig. 6.34 where the voltages are V_1 and V_2 and the first and second superconductors have the same energy gap. Normal electrons injected into superconductor 2 will be present with large concentration at the energy level $E = eV_1 - \Delta_1$ which gives a maximum in the tunnel current I_2 when

$$V_2 = \frac{\Delta_1 + \Delta_2}{e} - V_1. \tag{6.7}$$

Due to the high density of states at $E = \Delta_1$ a large proportion of these injected normal electrons will relax to Δ_1 by the emission of a phonon (a process more probable than the emission of a photon). Hence I_2 at a voltage V_2 satisfying Equation (6.7) is a measure of the electrons' stay in excited states above the gap.

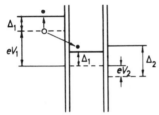

Fig. 6.34 Energy diagram of two superconductive tunnel junctions in series for the bias conditions given by Equation (6.7).

When $V_1 = 2\Delta/e$ the normal electrons injected into superconductor 2 are just above the gap. They may now recombine (by creating a Cooper-pair) or tunnel into superconductor 3. Hence the number of electrons reaching superconductor 3 is a measure of the recombination time.

The measured results for the relaxation time were expressed by Miller and Dayem as

$$\tau_T(E - \Delta) = 1.11 \; 10^{-7} \exp\left[-3.34(E - \Delta)/\Delta\right] \tag{6.8}$$

and for the recombination time as

$$\tau_R \sim \exp\left(0.3\,\Delta/kT\right). \tag{6.9}$$

Typical figures are $\tau_R = 10^{-7}$ sec and about an order of magnitude less for τ_T. This means that the injected normal electrons first relax to the top of the gap and then recombine.

Later measurements by Levine and Hsieh [271] yielded a different temperature variation for the recombination time, namely

$$\tau_R \sim \exp\left(\Delta/kT\right) \tag{6.10}$$

which is in agreement with the theoretical prediction of Schrieffer and Ginsberg [272].

We shall further discuss here the experiments of Gray *et al.* [273] performed on Al junctions by two different techniques:

(*i*) An a.c. method using the same three-electrode structure as shown in Fig. 6.34. A square wave is superimposed on a d.c. bias so that current is injected in the positive periods and the resulting excess current is measured.

(*ii*) A pulse method using an ordinary two electrode structure biased at $V < 2\Delta/e$. A current pulse is then superimposed driving the voltage above $2\Delta/e$ so that normal electrons are injected raising the population above the equilibrium level. After the pulse is removed the decay of the excess population is obtained by observing the decay of current on an oscilloscope. An advantage of this method is that the recombination time can be directly measured.

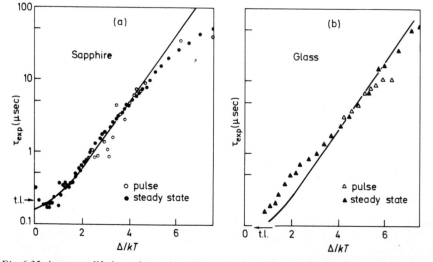

Fig. 6.35 Apparent lifetimes determined by experiment. The solid curves show the temperature dependence required by theory and t.l. marks the corresponding theoretical limit as $\Delta/kT \to 0$. (*a*) Sapphire substrate. Aluminium film thicknesses: 102, 76, 64 nm. $\Delta_0 = 195$ μeV. Electron mean free path = 30 nm. (*b*) Glass substrate. Aluminium film thicknesses: 30, 32, 30 nm. $\Delta_0 = 200$ μeV. Electron mean free path = 21 nm. After Gray *et al.* [273].

The results for two particular specimens are shown in Fig. 6.35. There is remarkable agreement between the steady state and the pulse method but some discrepancy may be noticed when the results on the two different samples are compared. Gray *et al.* [273] managed to resolve this discrepancy by considering the reabsorption of phonons (first discussed by Rothwarf and Taylor [274]) that is the break-up of Cooper-pairs by some of the phonons, themselves created by recombination. The size of the effect depends on the phonon lifetime against Cooper-pair breaking and on the rate of phonon escape to the substrate, the latter one dependent on the substrate material (or rather on the acoustic mismatch between the superconducting film and the substrate). The total correction due to these factors is typically about 1·1 on glass but as large as 2 on sapphire.

The comparison between experiments and theory (a more accurate theory by Gray [275] which takes into account the details of the Fermi surface of Al) is made by starting with the value predicted by the theory and then calculating the corrections due to the experimental set-up. This leads to a value τ_{calc} which is to be compared with the experimentally measured value τ_{exp}. The comparison at $\Delta/kT = 4$ is shown in Fig. 6.36. The agreement between theory and experiment

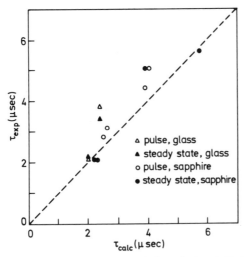

Fig. 6.36 Comparison of the experimental and the calculated lifetimes at $\Delta/kT = 4$. The overall inaccuracies in the measurements lead to estimated errors of $\pm 20\%$ on both axes. After Gray et al. [273].

is good and may be regarded as generally valid because of (i) the agreement between the pulse and steady state methods; (ii) the agreement among the large number of specimens with widely differing parameters; and (iii) the agreement of the results for specimens in which the phonon escape correction is large with those in which it is small. The lifetime in clean, bulk Al with $\Delta = 176 \ \mu eV$ is estimated to be about 2·4 μsec at $\Delta/kT = 4$.

6.15 Determination of normal electron diffusion constant

The experimental arrangement used by Levine and Hsieh [89, 276] for studying normal electron diffusion consisted of six closely spaced tunnel junctions as shown in Fig. 6.37. Electrons are generated at a steady rate at each of junctions 1–5 and junction 6 serves as a detector. Because of the finite lifetime of the electrons the number reaching the detector decreases exponentially with junction separation, with a characteristic length δ called the diffusion length, related to the diffusion constant, D, by the formula

$$\delta = \sqrt{(D\tau_R)}. \tag{6.11}$$

By measuring the decay of the response received from subsequent junctions, δ can be determined and hence (assuming that τ_R is available from a separate experiment) D. Levine and Hsieh [89] obtain the numerical value $D = 22 \cdot 5$ 10^{-4} m^2 sec^{-1} in good agreement with that calculated from the mean-free-path and the group velocity.

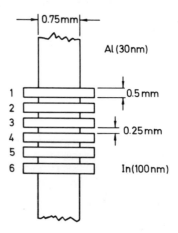

Fig. 6.37 Tunnel junction array (not to scale). The Al strip is about 20 mm long. After Levine and Hsieh [89].

6.16 Investigations of vortex structure

In studying the effect of a parallel magnetic field on the tunnelling characteristics of Type II superconductors Tomasch [81–83] noticed some structure which he tentatively identified with the entry of vortices.

Similar experiments in parallel fields were performed by Sutton [84] whose measurements on Al–I–PbIn junctions may be briefly summarised as follows:

(*i*) for sufficiently thin films $\Delta I(H) = I_{NN} - I_{SN}(H)$ (measured at a voltage about half the gap value) decreases monotonically to zero;

(*ii*) for thicker films there is a dip in the $\Delta I(H)$ curve which occurs at progressively lower fields as the film thickness increases;

(*iii*) with increasing film thickness more dips appear;

(*iv*) there is a hysteresis region in decreasing magnetic field.

Sutton [84] interprets these results by a model in which vortices nucleate at the surface and move into the interior of the films. The corresponding conclusions are:

(i) vortices do not enter a thin film;

(ii) at a field $H_{e0} > H_{c1}$ a vortex enters a thicker film, H_{e0} decreases with increasing thickness;

(iii) if the film is sufficiently thick more vortices enter;

(iv) the exit of the vortices is delayed by surface trapping.

A detailed study of vortices in a perpendicular magnetic field near the upper critical field was conducted by Nedellec et al. [85] on Al–I–InBi junctions. They find that the agreement with the existing theories [277, 278] (using the measured value of $\sigma(0)$ for comparison) can be largely improved by including into the model (i) macroscopic screening currents due to the large demagnetising effect of the thin film geometry and (ii) the inhomogeneous magnetic field distribution outside the sample caused by the vortex structure.

Donaldson and Brassington [86] studied the opposite limit when the perpendicular magnetic field (up to 25 gauss) is small in comparison with the upper critical field but well above the lower critical field which for their film was 0·2 gauss. Their measured results in the 0·1°K range could be approximately expressed with the formulae

$$\sigma_1(0) \sim H \qquad \text{for N–I–S junctions}$$

$$\sigma_2(0) \sim H^2 \qquad \text{for } S_1\text{–I–}S_2 \text{ junctions}$$

The linear variation found for the N–I–S junction is in accordance with the model that up to a certain field the vortices are independent of each other. The quadratic variation for the S_1–I–S_2 junction may be explained by the increasing alignment of vortices on the two sides. Using a simple model that the core (of area S_c) can be located with equal probability anywhere within the area of the vortex, S_v and making use of the relationship $S_v \sim H^{-1}$, we get

$$\sigma_2(0) \sim \frac{S_c}{S_v}\sigma_1(0) \sim S_c H^2$$

where it is assumed that the only contribution to the zero voltage conductance occurs when normal materials are opposite to each other.

Finally we wish to mention the point contact measurements of Sharvin [279, 280] which were done in the intermediate state and not in the mixed state but present an interesting way of observing flux motion. It is based on the idea that the tunnel current increases whenever a normal region moves across the point contact, and the expected oscillations were indeed found. Other workers [281, 282] repeating the experiment claim, however, that the oscillations are due to thermal instability.

6.17 High resolution spectroscopy

We have so far assumed that the electrons tunnelling across the barrier do not lose energy in the barrier itself. It was shown by Jaklevic and Lambe [283] for tunnel junctions between normal metals that in the range 0·1 to 0·5 meV the energy of the tunnelling electrons may be lost by exciting vibrations of molecules trapped in the oxide.* This loss of energy appears as higher conductivity because it provides an additional path for the electrons. The onset of higher conductivity is at discrete energies at which the electrons can transfer their energies to molecular vibrations. Hence a break appears in the I–V characteristic at the corresponding voltage, showing up as a peak in the second derivative curves. By this technique (often referred to as tunnelling spectroscopy) both

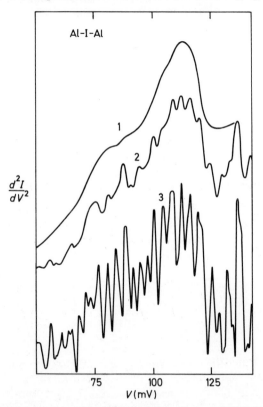

Fig. 6.38 Second derivative d^2I/dV^2 against V for an Al–I–Al junction. (1) Broad maximum attributed to the Al–O–H bending mode [285]. (2) Improved resolution by Klein and Leger [92]. (3) Further improvement in resolution when Al becomes superconducting.

* In fact some electrons (about 1%) do lose energies below this range by exciting vibrations of the oxide (50–100 meV) and of the surfaces of the metal films (0–30 meV). This effect is usually negligible but can be corrected for when high accuracy is required [286, 287].

infrared and Raman spectra have been measured to a high resolution (for a review see Lambe and Jaklevic [284]).

First indications of absorption peaks with superconducting electrodes were obtained by Marcus [90] and Clark [91]. Absorption due to optical phonons in the barrier was investigated by Giaever and Zeller [119] using a number of barrier materials. In one of their experiments performed on Zn–ZnO–Pb junctions at 4·2°K they found that the resolution deteriorated when Pb was made normal by the application of a high magnetic field. Better resolution by superconducting electrodes was also reported by Klein and Leger [92] for Al–I–Al junctions as shown in Fig. 6.38. The broad maximum in curve (1), found by Lambe and Jaklevic [285], is attributed to an Al–O–H bending mode (Fig. 6.39) of a hydrate. Curves (2) and (3) illustrate the tunnelling measurements

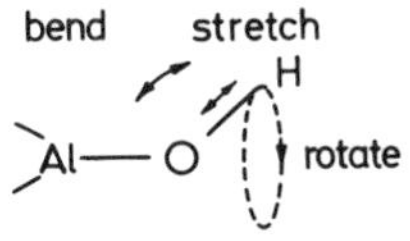

bend   stretch

$\nearrow$ $\nearrow\nwarrow$ H

$\searrow$ Al —— O $\left|\right.$ $\uparrow$ rotate

Fig. 6.39 Schematic representation of the bending and rotation modes of Al(OH)$_3$.

of Klein and Leger [92] in the normal and superconducting states respectively. It may be seen that the resolution is very much improved when aluminium becomes superconducting; the fine structure of OH rotation is clearly resolved.

6.18 Superconductivity of small particles; zero bias anomalies

It was shown by Giaever and Zeller [288, 289] that the technique of tunnelling can be used for the study of small superconducting particles embedded in the oxide barrier. The stages in the fabrication of this composite junction may be seen in Fig. 6.40. First the Al electrode is evaporated on a glass slide and lightly

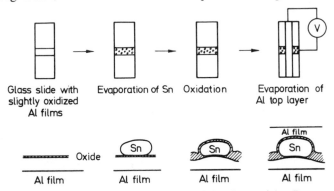

Glass slide with   Evaporation of Sn   Oxidation   Evaporation of
slightly oxidized   Al top layer
Al films

$-----$ Oxide

Al film   Al film   Al film   Al film

Fig. 6.40 Fabrication of the tunnel junction containing Sn particles. Because of the faster oxidation of Al with respect to Sn, the oxide in the space between the particles is much thicker than the oxide at the surface of the particles. After Zeller and Giaever [289].

oxidised (Fig. 6.40(a)). Next Sn is evaporated which agglomerates into small particles (Fig. 6.40(b)) depending mostly upon the amount of Sn evaporated. Then follows another longer oxidation process at the end of which (Fig. 6.40(c)) the space between the Sn particles is filled with a thick Al oxide while the Sn particles (because of slower growth of the oxide) are covered with only a thin layer of Sn oxide. Finally the Al counterelectrode is evaporated (Fig. 6.40(d)). This configuration is suitable for the study of the Sn particles because tunnelling is now more likely to occur via the Sn particles than directly through the oxide layer between the Al electrodes.

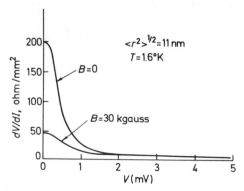

Fig. 6.41 Differential resistance as a function of voltage for particles in the normal and superconducting state at $T = 1.6°K$. Practically all particles are normal at $B = 30$ kgauss. After Giaever and Zeller [288].

A typical experimental result is shown in Fig. 6.41. It may be seen that the resistance is maximum at zero bias and that the resistance becomes considerably lower when the superconductivity of the Sn particles is suppressed by a sufficiently large magnetic field. Further experiments showed (Fig. 6.42) that the zero bias resistance is dependent both on particle size and on magnetic field. The decrease of zero bias resistance with magnetic field is attributed to the quenching of superconductivity. Since smaller particles have higher critical fields the zero bias resistance reaches constant value at a higher magnetic field for smaller particles. Hence one may argue that as long as a decrease in zero bias resistance is found, one may take it as a proof that at that value of the magnetic field some of the Sn particles are still making transitions to the normal state. Since some decrease in zero bias resistance was seen even for particle distributions (as measured by electron micrographs) with a maximum radius of about 2.5 nm, Giaever and Zeller [288, 289] conclude that superconductivity still exists in such small particles. This is in agreement with the theoretical predictions of Anderson [290] and is not in serious disagreement with Markowitz's calculations [291] which yield $r = 10$ nm for the lower limit of superconductivity for a free (not embedded in a lattice) particle.

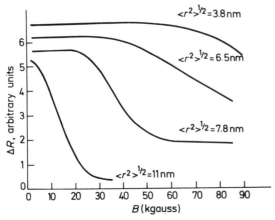

Fig. 6.42 Magnetic field dependence of the junction resistance at zero bias and $T = 1°K$. After Zeller and Giaever [289].

The maxima (or minima) in resistance occurring at zero bias have been extensively studied (see Duke [292, 293]) under the name of zero bias anomalies. Giaever and Zeller [288, 289] maintain that a large class of those experimental results can be explained with a simple capacitor model. The essential point in that model is that because of the small size of the particles the energy needed to accommodate one extra electron is fairly large, of the order of meV's. Hence for a current to flow via the Sn particles a certain minimum energy needs to be provided, and that is responsible for the large zero bias resistance. The effect of higher temperatures is to reduce this resistance because a new mechanism for current flow opens up due to the presence of excited electrons. The increase in resistance which occurs when the particles become superconducting is explained by the energy gap structure. The minimum energy needed to put an electron on the Sn particle increases then by Δ.

7. Device applications

7.1 Negative resistance devices

From the device point of view a superconducting tunnel diode may be classified as a two-terminal negative resistance device. Thus, quite naturally, soon after the discovery of the negative resistance region the question of practical applications arose. The process was helped by the remarkable similarity between the superconductor and the semiconductor tunnel diodes. In both cases electron tunnelling is involved, the negative resistance region originates in the specific distribution of electron states, and both possess a significant junction capacitance. It was, therefore, logical to expect that the superconductor tunnel diode will operate in the same type of circuits and will have the same type of limitations (e.g. maximum frequency limited by junction capacitance) as its semiconductor counterpart.

The first attempt to build switches and oscillators was made by Giaever and Megerle [294]. They found that superconductor tunnel diodes could be used both as mono-stable and bi-stable switches. They also got oscillations in two different circuits. The highest frequency observed was 4 MHz.

Miles et al. [93] employed an Al–Al$_2$O$_3$–Sn junction operation at $T = 1 \cdot 2°$K yielding a negative resistance between $1\,\Omega$ and $10\,\Omega$. The circuit (designed again in analogy with circuits known to work for the semiconductor tunnel diode) oscillated at a frequency of 72·5 MHz and in a modified version (as an amplifier) gave 23 db gain at a frequency of 50 MHz. Though no higher frequency operation was reported it is believed that the device could work in the lower microwave region.

The basic handicap of all these devices is that they operate at low temperatures. They must be superior to room temperature devices in several respects before their practical application will be seriously considered. There is just one application for which there is a glimmer of hope, namely low noise amplification at microwaves. The amplifier would have low thermal noise because the operating temperature is low and there is practically no series resistance in the diode, and the shot noise will be small as well because the current is small [93, 295].

There is a somewhat better chance for applications as a distributed device because of its possible use in computers. We shall therefore investigate this case in a little more detail.

A distributed tunnel junction is in fact a strip transmission line as shown in Fig. 7.1. In the passive case (no d.c. voltage applied) the propagation problem

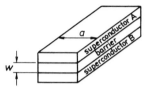

Fig. 7.1 A distributed tunnel junction (not to scale).

was treated by Swihart [136] solving Maxwell's equations in conjunction with the phenomenological equations of London for the superconductors. We shall follow here Scott's treatment [296] where the distributed junction is represented by the equivalent circuit shown in Fig. 7.2 in the best traditions of transmission line theory. Z and Y in the figure are the series impedance per unit length and shunt admittance per unit length respectively.

Fig. 7.2 The equivalent circuit of a transmission line applicable to a distributed tunnel junction.

It is fairly straightforward to express the shunt admittance

$$Y = (g_s + j\omega c_s)a \tag{7.1}$$

where g_s and c_s are the conductance per unit surface and capacitance per unit surface respectively, and a is the width of the strip. For the 'active' transmission line, that is when the appropriate d.c. voltage is applied across the junction, g_s is negative.

It is somewhat more difficult to express the series impedance because both the resistance and the inductance are different for a superconductor than for a normal metal. The resistance is different (smaller of course) because only the normal electrons contribute to resistivity, and the inductance is different because in a superconductor there is a new energy storing mechanism, the kinetic energy of the Cooper pairs. It turns out [296] that for frequencies for which the superconducting penetration depth is smaller than the electromagnetic skin depth, the series impedance per unit length of the line may be obtained by the *parallel* combination of a frequency independent resistance per unit length

$$R = \frac{\mu_0(\lambda_A + \lambda_B + w)^2}{a(\lambda_A/\omega_{sA} + \lambda_B/\omega_{sB})} \tag{7.2}$$

and a frequency independent inductance per unit length

$$L = \frac{\mu_0(\lambda_A + \lambda_B + w)}{a} \tag{7.3}$$

where μ_0 is free space permeability, λ_A, λ_B penetration depth at the operating temperature in superconductors A and B respectively, w the width of the insulating layer. The parameters ω_{sA} and ω_{sB} have dimensions of frequency and may be expressed as

$$\omega_s = \frac{\omega^2 \mu_0 \lambda}{R_s} \tag{7.4}$$

where R_s is the high frequency surface resistance of the superconductor approximately proportional to ω^2 (so that ω_s is independent of frequency).

The propagation constant for a sinusoidal signal $\exp j(\omega t - \gamma_z)$ is given by the general expression

$$\gamma = \alpha + j\beta = (ZY)^{1/2}. \tag{7.5}$$

Noting that Z is given by the parallel combination of R and L, and Y by Equation (7.1) we get after a few approximations the expressions

$$\alpha \cong -\tfrac{1}{2}[\mu_0 c_s(\lambda_A + \lambda_B + w)^{1/2}(\omega_d - \omega^2/\omega'_s) \tag{7.6}$$

and

$$\beta \cong \omega[\mu_0 c_s(\lambda_A + \lambda_B + w)]^{1/2} \tag{7.7}$$

where

$$\omega_d = -g_s/c_s \quad \text{and} \quad \omega'_s = (\lambda_A + \lambda_B + w)/(\lambda_A/\omega_{sA} + \lambda_B/\omega_{sB}) \tag{7.8}$$

and the assumption

$$\omega_d \ll \omega \ll \omega'_s \tag{7.9}$$

was used. The upper frequency limit for amplification is given by the highest frequency for which α is still negative, i.e.

$$\omega_{max} = (\omega_d \omega'_s)^{1/2}. \tag{7.10}$$

The magnitude of this frequency may be estimated by considering some available data on tin–oxide–lead junctions with an oxide width of about 2 nm. The capacitance per unit area is about [294] 5×10^{-2} F/m², and the negative conductance per unit area is about [102] 25×10^4 Ω/m², giving

$$f_d = \frac{\omega_d}{2\pi} \cong 0.8 \text{ MHz.} \tag{7.11}$$

Further data at $T = 2 \cdot 1°\text{K}$ from Simon [297] are

	$\dfrac{\omega^2}{R_s}, \Omega^{-1}\text{s}^{-1}$	λ, m
tin	$1 \cdot 4 \times 10^{-4}$	$4 \cdot 65 \times 10^{-8}$
lead	$7 \cdot 35 \times 10^{-5}$	$5 \cdot 4 \times 10^{-8}$

yielding

$$f_s' \cong 330\,\text{GHz} \quad \text{and} \quad f_{max} = 510\,\text{MHz}. \tag{7.12}$$

This frequency limit can be pushed further up into the microwave region by reducing the temperature and thereby reducing the surface resistance R_s. There is, however, another limitation which definitely sets an upper limit to the frequency and that is photon absorption across the gap. For tin at $0°\text{K}$ this frequency comes to about 140 GHz.

It follows from the previous analysis that a length of line terminated by a load may oscillate. The condition (besides having the right phase relationship) is that the propagating wave should have enough gain to compensate for the loss of amplitude due to reflection from the load.

Both the amplification and the onset of oscillation are linear problems for which the foregoing analysis is sufficient. There is, however, another mode of interest when a signal propagates without attenuation and without change of shape. This type of operation was suggested by Crane [298] who called the lines capable to maintain such propagation *neuristors*. The term is derived from the neuron which carries information in the same manner in the nervous system. Crane proved that all types of logical operation can be performed with the aid of such lines, and that they can be realised by distributed devices possessing negative resistance. In the aforementioned paper [296] Scott gives an approximate analysis of the neuristor mode in distributed superconductor tunnel junctions. With out previous notations the velocity of propagation comes to $u = (Lc_s a)^{-1/2}$ which in a typical line is about 10^7 m/s.

Both oscillation and neuristor propagation were experimentally observed by Yuan and Scott [94] on a Sn–SnO–Pb line 95 mm long. The observed frequency was 77 MHz and the neuristor pulse velocity, $1 \cdot 5 \times 10^7$ m/s.

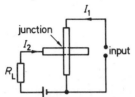

Fig. 7.3 A three-terminal amplifier using a tunnel junction. Varying I_1 the energy gap varies and hence I_2 varies as well. The output is taken from the load resistance R_L. After Ogushi *et al.* [300].

Finally, we would like to mention Giaever's suggestion [299] for a three-terminal device. The essence of the proposal was to use an external magnetic field to change the energy gap and thereby change the tunnelling current. A modified form of this device (Fig. 7.3) in which the magnetic field is due to the current flowing through the junction was analysed by Ogushi *et al.* [300] Amplification is possible in principle but it is difficult to envisage any practical applications.

7.2 Detection of electromagnetic waves

As far as the detection of electromagnetic waves is concerned there is a striking similarity between a p–n junction and a superconductor tunnel junction. In the former case the incident electromagnetic wave creates electron-hole pairs increasing thereby the current flowing across the junction. In the latter case Cooper-pairs are broken up increasing the number of normal electrons available for tunnelling. The essential difference between the two cases is that the energy gaps are different. For semiconductors* the energy gap is in the visible or in the near infrared region. For superconductors the energy gap is considerably smaller and happens to cover the region** where good detectors are not available. So the competition is against other low temperature devices.

Following Burstein *et al.* [95, 301] we shall consider a tunnel junction made of identical superconductors and shall assume that the bias voltage V is a little below the value $2\Delta/e$, where the sudden rise in current begins.

The tunnelling current is given, in general, by Equation (4.14). Since in the present case the lower limit of integration is

$$E = eV + \Delta \tag{7.13}$$

and eV is nearly equal to Δ, we may use the approximations

$$f(E - eV) \gg f(E) \tag{7.14}$$

and

$$\frac{E}{(E^2 - \Delta^2)^{1/2}} \cong 1 \tag{7.15}$$

Hence the current may be written as

$$I_{SS} = A N_{2N}(0) \int_{eV + \Delta}^{\infty} N_1(E - eV) f(E - eV)\, dE. \tag{7.16}$$

It may be recognised that the integral gives the number of particles residing in states above the gap. Hence we may argue that the current will always be

---

*At least for those which can be prepared with sufficient purity.
** From about 0·1 mm to 1 mm wavelengths, called the submillimeter region by microwave people relationship was formulated by Fulton [376].

proportional to the total number of excited particles independently of the way of excitation.

For the density of excited particles $\mathcal{N}_N$ we may write the following differential equation

$$\frac{d\mathcal{N}_N}{dt} = G_T + G_B - \frac{\mathcal{N}_N}{\tau} \qquad (7.17)$$

where G_T is the rate at which particles are generated per unit volume by thermal processes, G_B is the rate at which they are generated by background radiation and τ is the lifetime of the excited particles. Hence the steady-state density of excited particles is

$$\mathcal{N}_{N0} = (G_T + G_B)\tau. \qquad (7.18)$$

The change in the density of particles produced by the signal (the electromagnetic wave to be detected) is similarly given by

$$\Delta\mathcal{N}_N = G_S\tau \qquad (7.19)$$

where G_S is the rate at which particles are generated per unit volume by the signal radiation. The fractional change in the density of excited particles $\Delta\mathcal{N}_N/\mathcal{N}_{N0}$ which determines the increase in the tunnelling current is then

$$\frac{\Delta\mathcal{N}_N}{\mathcal{N}_{N0}} = \frac{\Delta I_{SS}}{I_{SS}} = \frac{G_S}{G_B + G_T}. \qquad (7.20)$$

G_T depends on the ambient temperature so it can be decreased by lowering the temperature. The background radiation, however, remains. The rate of generation of excited particles by this background radiation is given by [302]

$$G_B = \frac{F_B}{d} \int_{\lambda_1}^{\lambda_2} A(\lambda) Q_B(\lambda)\, d\lambda \qquad (7.21)$$

where $Q_B(\lambda)$ is the rate at which background radiation photons of wavelength λ are incident on a unit area of the superconductor for a full 2π steradian solid angle from blackbody radiation at ambient temperature, F_B is a geometrical factor for the background radiation which depends, among other things, on the solid angle through which the background radiation photons are incident, A is the absorptivity of the superconductor, d is the thickness of the superconductor, and $\lambda_2 - \lambda_1$ is the wavelength range of radiation.

The rate of generation of carriers by the radiation signal is similarly given by

$$G_S = \frac{F_S}{d} \int_{\lambda_1}^{\lambda_2} A(\lambda) Q_S(\lambda)\, d\lambda \qquad (7.22)$$

where the subscript s refers to the signal.

Making the superconductor tunnel junction in thin film form with an overall thickness much less than the skin depth, the reflections from the front and back surfaces largely cancel. The absorptivity (the fraction of incident radiation absorbed by the superconductor) can be calculated by introducing the complex conductivity $\sigma_s = \sigma_1 + i\sigma_2$. For a normal metal $\sigma_1 = \sigma_N$ and $\sigma_2 = 0$. In the present case when we are considering radiation above the frequencies corresponding to the gap energy, $\sigma_1 < \sigma_N$ and σ_2 is finite due to the breaking up of Cooper pairs. The values of σ_1 and σ_2 calculated by Mattis and Bardeen [303] (which are in reasonable agreement with experimental values) were used by Burstein et al. [95] in calculating A for the thin film structure. The results are plotted in Fig. 7.4. The optimum thickness is about 3 nm. For thicker films the absorptivity may be shown to be inversely proportional to film thickness.

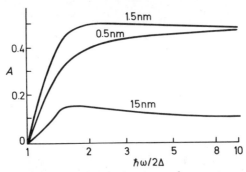

Fig. 7.4 Theoretical absorptivity for superconducting lead films of 0·5, 1·5 and 15 nm thickness. After Burstein et al. [95, 301].

The ratio G_s/G_B will be further influenced by the geometrical factors and the spectrum of the background radiation. The former may be reduced by decreasing the angular aperture to the minimum necessary to admit the radiation signal, and the latter may be narrowed to the wavelength range of the signal by appropriate filters.

The junction, as mentioned before, is biased to a voltage just below the gap energy and is fed by a constant current source. The voltage signal is then

$$\Delta V_s = \Delta I_{ss} R_{ac}$$

$$= V_s \frac{R_{ac}}{R_{dc}} \frac{\Delta \mathcal{N}_N}{\mathcal{N}_{NO}} \qquad (7.23)$$

where $R_{ac} = dV/dI$ is the a.c. resistance of the junction for small signals, $R_{dc} = V_s/I_{ss}$ is the d.c. resistance of the junction. Since $R_{ac} \gg R_{dc}$ we have a large compensating factor for the smallness of V_s (about 1 mV).

The signal to noise ratio was also discussed by Burstein et al. [95]. They conclude that at sufficiently low temperatures both the thermal noise and the shot noise are negligible and the dominant noise is due to the random generation and breaking up of Cooper pairs. In this case the signal to noise ratio is given by

$$\frac{\Delta V_s}{V_{noise}} = \left(\frac{G_s^2 v}{4\Delta f\, G_B}\right)^{1/2} \tag{7.24}$$

where v is the volume of the superconductor. Thus an additional good feature of this detector is its small volume.

7.3 Generation of electromagnetic waves

The emission of electromagnetic waves from an N–I–S point contáct configuration was first observed by Gregory et al. [97, 304]. The normal metal was an aluminium (at $T = 4.26°K$) plate with an oxide layer on the surface upon which a specially formed (Fig. 7.5) tantalum probe was pressed. The radiation

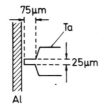

Fig. 7.5 The junction between a thin film of oxidised Al and a Ta probe of special geometry. After Gregory et al. [303].

was identified by Leopold et al. [305] in a companion paper as stimulated emission due to the recombination of normal electrons into Cooper-pairs. As mentioned before in the previous section the probability of phonon emission is much higher (lifetime shorter) than photon emission so one might rather expect stimulated emission of phonons than photons. However, it is difficult to make a resonant structure for short wavelength phonons whereas the junction would readily serve as an *electromagnetic* resonator in exactly the same way as in a p–n junction injection laser. Thus the semiconductor analogy may again be expected to come useful. Leopold et al. [305] take indeed over Lasher's equation [306] for the threshold current density of a semiconductor injection laser and modify it by including the small probability (10^{-7}) of radiative recombination in a superconductor. They come to the conclusion that there is a range of current densities which are high enough so that stimulated emission is possible and low enough so that the superconducting state will not be destroyed. This is the main consideration in choosing L the lateral dimension of the junction. For large values of L the total current may be too large but L must not be smaller than the half-wavelength of the electromagnetic wave in the medium. The final choice was $L = 25\ \mu m$. The corresponding current density for a radiation of 75 GHz was calculated to be 10^9 A/m² which was very near to the experimental value.

The experimental and theoretical results showed good agreement for the output power as well which was about 10^{-7} W.

Further experimental results by the same group were reported later [307, 308] in which some further proofs were given in favour of the proposed radiation mechanism:

(i) The frequency of the emitted radiation followed the change in energy gap as the temperature was varied (Fig. 7.6).

(ii) The radiation disappeared at the transition temperature of the tantalum probe.

(iii) The radiation disappeared when the tip was distorted, implying the significance of the chosen geometry.

For shorter wavelengths the required current density is higher so realisation is more difficult.

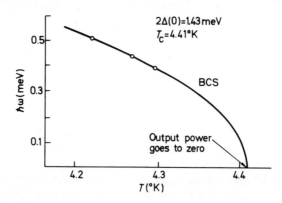

Fig. 7.6 The measured frequency of output radiation as a function of temperature. Solid curve is the BCS theory. Output power disappears at the transition temperature of Ta. After Leopold et al. [307].

7.4 Detection and generation of high frequency phonons

As discussed in Section 5.3 an incident acoustic wave will give rise to an extra current in the I–V characteristic. Hence phonon-assisted tunnelling may be used for the detection of phonons in the GHz range. A schematic representation of the detector developed by Goldstein, Abeles and Keller [309] is shown in Fig. 7.7. The tunnel junction (Al–I–Pb) is deposited on one end of a 1 inch long rod of single crystal germanium or quartz. The extra tunnelling current flows through the load resistor R_L and the voltage V_a is measured. The extra current $I_a = V_a/R_L$ through the load resistor is given in first approximation by

$$I_a = -\frac{dI}{dV}V_a + \Delta I_s \tag{7.25}$$

where ΔI_s is the current change due to the acoustic wave for short circuit conditions ($R_L = 0$). Hence

$$V_a = \Delta I_s \frac{R_L(dV/dI)}{R_L + (dV/dI)}. \tag{7.26}$$

Note that ΔI_s is strongly dependent on bias voltage, the maximum occurring at $V \cong \Delta_1 + \Delta_2$ as it was shown in Figs. 5.16 and 5.17. The optimum load resistance is $R_L = dV/dI$.

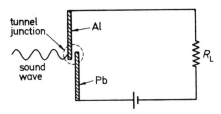

Fig. 7.7 Detection of acoustic waves by a superconducting tunnel junction. After Goldstein *et al.* [309].

The voltage created by unit acoustic input power V_a/P is a measure of the quality of the detector. The value of V_a/P under the above outlined optimum conditions was experimentally measured [309] at a frequency 9·3 GHz to be $2\cdot5 \times 10^{-7}$ Vm² W⁻¹ for longitudinal waves and 2×10^{-7} Vm² W⁻¹ for transverse waves.

Another interesting experiment on phonon detection was performed by Schulz and Weis [310] on Sn–I–Sn junctions deposited on the end of a sapphire single crystal. The phonon emitter at the other face of the sapphire crystal consists of a deposited constantan layer heated by 100 nsec current pulses. The spectrum of the emitted phonons depends on the radiation temperature. The authors estimated the mid-frequency of the Planck spectrum to be in the THz range. The pulse length of the received voltage signal was equal to the length of the exciting electric current pulse. Longitudinal and transverse phonons were both detected and could be resolved owing to their different group velocities.

As these two examples show, a tunnel junction may be used as a phonon detector in a very wide frequency range. Its further advantages are that the acoustic wave is directly converted into a d.c. electric signal (there is no need for an electromechanical transducer) and since it is an energy detector it is insensitive to phase variations across the wavefront.

Tunnel junctions may also be used for the generation of high frequency phonons. The basic mechanism is phonon emission by decaying normal electrons. This was already discussed in Section 6.14 in connection with normal electron lifetime and is shown again in Fig. 7.8. A voltage $V_1 > 2\Delta$ is applied to the tunnel junction; a Cooper-pair breaks up and an electron tunnels across.

The next step is that the electron drops into a state just above the gap by emitting a phonon of frequency ω_T. Finally, two electrons residing just above the gap recombine and emit a phonon of frequency $2\Delta/\hbar$. We cannot produce coherent phonons this way because failing to make an acoustic resonator* the phase relationships cannot be controlled. It is, however, possible by this method to emit phonons in a narrow band of frequencies. Since the upper limit of producing coherent phonons by conversion of microwaves is not far above 50 GHz, any method which can produce phonons in a narrow band is of great value.

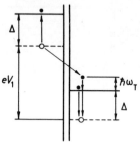

Fig. 7.8 Energy diagram showing the break-up of a Cooper-pair and subsequent tunnelling, relaxation and recombination of electrons.

An elegant measurement in which both phonon generation and detection are demonstrated was performed by Eisenmenger and Dayem [96]. In their experimental arrangement two Sn–I–Sn junctions were deposited at the opposite ends of a 1·04-cm-long sapphire rod. Applying a voltage in excess of the energy gap at one end, the generated phonons travelled down the sapphire rod and were detected at the opposite end. The receiver junction was biased $V_2 < 2\Delta$ in analogy to the photon detector in Section 7.2. Note that the detection mechanism is *not* phonon-assisted tunnelling (though there is a small contribution from it) but the straightforward breaking up of Cooper pairs, leading to excess electrons and excess tunnelling current. The evidence for the described processes is provided by

(i) the experimentally observed [96] receiver output against time which shows three peaks corresponding to longitudinal, fast transverse and slow transverse phonons;

(ii) the experimentally observed current–voltage characteristic shown in Fig. 7.9. It may be seen that in the region $2\Delta < eV_1 < 4\Delta$ the received voltage V_2^1 increases linearly with I_1 the current in junction 1. At $eV_1 = 4\Delta$ there is a sharp break and for $4\Delta < eV_1 < 6\Delta$ the current varies again linearly. The interpretation is that up to $eV_1 = 4\Delta$ the phonon of frequency ω_T (obtained by the relaxation of the electron

*The wavelength of the phonons is so small that reflection at the boundaries is diffuse, consequently standing waves cannot be built up.

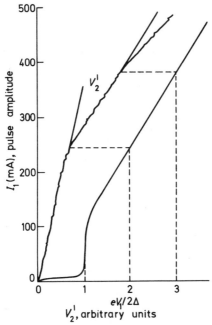

Fig. 7.9 I–V characteristic of junction 1 and the received voltage V_2' obtained across junction 2. The slope of the V_2' curve changes at currents corresponding to $eV_1/2\Delta = 2$ and 3 because at those values a new group of phonons can produce an output. After Eisenmenger and Dayem [96].

to the top of the gap) produces no output in the receiver. But when $eV_1 > 4\Delta$ the phonon due to the relaxation of the electron also contributes to the current.

Analysing the results Eisenmenger and Dayem further conclude that,

(i) about 80% of the recombination occurs via emission of phonons of energy 2Δ;

(ii) the overall quantum efficiency, defined by $2\pi r^2 I_2^1/AI_1$ is typically of the order of 1% where I_2^1 is the current increment in the receiver due to phonon interaction, A is the area of the tunnel junction and r is the distance between the diodes.

Similar results were obtained in a further study by Kinder *et al.* [311]. Comparison with numerical solutions of Tewordt's theory [312, 313] yielded qualitative agreement.

7.5 Nuclear detection

The energy resolution of a semiconductor nuclear particle spectrometer is limited by the mean energy loss suffered by an excited electron hole pair, equal to a few times the energy gap. If a superconductor is used for the same purpose one may expect a better resolution due to the smaller energy gap. The first experiments were performed by Wood and White [98] who exposed a Sn–I–Sn tunnel junction to bombardment by 5·1 MeV energy α particles. The mechanism of detection is the creation of normal electrons and phonons by the incident α particle. The resulting rise in temperature also leads to the excitation of normal electrons and an increase in the tunnelling current. The extra current is maximum immediately after the passage of the alpha particle and decays with a characteristic time constant τ as the temperature distribution relaxes back to equilibrium. The measured pulse is shown in Fig. 7.10.

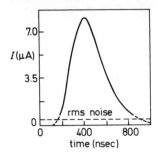

Fig. 7.10 Tracing from photograph of output pulse induced by an alpha particle traversing the junction. After Wood and White [98].

The maximum value of the energy loss per normal electron which actually tunnelled in the experiment was 0·141 eV. Comparing this with 3·6 eV valid for Si detectors the improvement in the statistically limited energy resolution is $(0·141/3·6)^{1/2} = 0·198$. The limit in the improvement which one might achieve is $(\Delta_{sem}/\Delta_{sup})^{1/2} \approx 0·03$. It should be noted that for spectrometric purposes the *total* energy of the particle should be absorbed. Wood and White suggest the use of a three-dimensional array (produced by microelectronic techniques) of tunnel junctions in a practical spectrometer.

7.6 Thermometry

As we have discussed many times before, the tunnelling current depends strongly (at least in certain voltage ranges) on temperature. Hence it is possible to invert the relationship and deduce the temperature from the measured tunnelling current. Giaever *et al.* [66] used a numerical solution of Equation (4.14) to obtain the temperature from the current measured on an Al–I–Al junction. The calculated and measured (with the aid of a paramagnetic salt thermometer) temperatures were in good agreement as shown in Fig. 7.11.

Temperatures in the range 0·1°K to 1°K were measured by Bakker *et al.*

[315] with the aid of Al–I–Ag junctions. They relied on a numerical solution of Equation (4.8) and used both $I(V)$ and $\sigma(V)$ to derive the temperature.

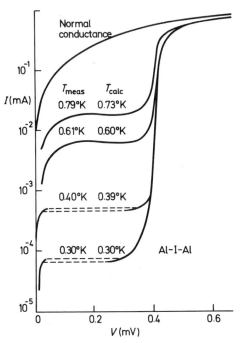

Fig. 7.11 I–V characteristic of an Al–I–Al junction. The current at an applied voltage Δ/e is used to calculate the temperature. Measured and calculated temperatures are in good agreement. After Giaever et al. [66].

Another region of interest is where

$$T < 0\cdot5\, T_c \quad \text{and} \quad 2kT < eV < \Delta - 1\cdot5\, kT \qquad (7.27)$$

in which case Equation (4.9) reduces to

$$I_{NS} = G_{NN} A(\Delta, T) \exp{(eV/kT)} \qquad (7.28)$$

where $A(\Delta, T)$ is a slowly varying function of temperature. Measurements in this region were first made by Giaever and Megerle [69] who got excellent agreement between theory and experiment. Donaldson and Band [316] have suggested that tunnelling may be used in this region for the measurement of temperature. If $\ln I_{NS}$ is plotted against V a straight line should be obtained with a slope e/kT from which T can be determined. The suggested method has several advantages:

(i) The exponential dependence is not sensitive on the smearing of the gap (due to some non-ideal effects).

(ii) If excess currents flow the departure from ideal behaviour can be detected and the linear contribution separated from the exponential one.

(iii) The measurement depends only on the temperature of the normal metal hence thermodynamic equilibrium between the normal metal and the superconductor is not a requirement.

Besides the d.c. method of plotting points in the I–V characteristics Donaldson and Band have made use of an a.c. method as well in which the junction is fed by a current

$$I = I_0(1 + \alpha \cos \omega t). \tag{7.29}$$

The voltage response of the junction to first order is

$$V(I_0) + \left(\frac{dV}{\partial I}\right)_{I_0} \alpha I_0 \cos \omega t. \tag{7.30}$$

For the range where the current is an exponential function of voltage the amplitude of the a.c. component of the output voltage is

$$V_1 = \frac{\alpha kT}{e}. \tag{7.31}$$

Thus the value of V_1 appropriately scaled may directly give the absolute temperature. The results of measurements on Pb–I–Mg and Pb–I–Al junctions are shown in Fig. 7.12. The accuracy is a few percent of the absolute temperature.

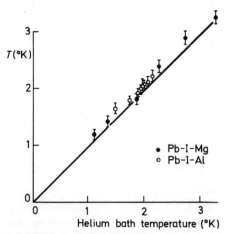

Fig. 7.12 Temperature, obtained with the aid of Equation (7.31), as a function of measured bath temperature. After Donaldson and Band [316].

III. Josephson tunnelling

8. Introduction

8.1 The discovery of Josephson tunnelling

Josephson [47] predicted in 1962 a new type of tunnelling which became later known as Josephson tunnelling. It is instructive to ask again the question we discussed in connection with normal electron tunnelling; could Josephson tunnelling have been discovered earlier? The answer to it is an unambiguous no; it could have easily been discovered five to ten years later but hardly earlier. Could it have been discovered experimentally? Yes, there is one aspect of Josephson tunnelling which could have been discovered experimentally and, as we mentioned before, it *was* discovered by Holm and Meissner [54] in 1932. They came to the conclusion that the contact resistance between two metals vanishes as soon as both become superconductors; and this is indeed the simplest manifestation of Josephson tunnelling.

These first experiments did not excite much interest because the existing theoretical apparatus was far too inadequate for explaining the effect. It is interesting to contrast this with another experiment of Meissner concerning the expulsion of the magnetic field by a superconductor. That was a significant experiment in the sense that it completely upset all the ideas reigning at the time; it showed superconductivity in an entirely new light as something incompatible with Classical Physics. The contact resistance experiment was performed too early, the theoreticians were not ready yet.

Interestingly, a student of Meissner, I. Dietrich [317], repeated the experiment twenty years later in 1952. She measured the current between a tantalum probe and a flat piece of tantalum coated by a thin layer of TiO_2 or CeO_2. She found that below a certain temperature the resistance disappeared and that for a higher current the resistance disappeared at a lower temperature. But 1952 was still too early. The mainstream of research in superconductivity was oriented towards finding the basic mechanism. The isotope effect had already been found, Fröhlich had already suggested the interaction responsible for the effect, so theoreticians were not interested to explore little side-streets. Dietrich's paper shared the fate of Holm and Meissner's; it sank into oblivion.

In 1961 Giaever and Megerle [69] found a supercurrent in one of their samples but explained it away as due to metallic shorts. They had no incentive to investigate the phenomenon in more detail.

The climate was, however, more favourable in the beginning of the 1960s. The theory of superconductivity had recently been put into an elegant mathematical form by Bogoliubov and Gorkov, tunnelling was made part of many-body theory by Bardeen, and Cohen *et al.* so a more general attack, for the first time, became possible.

Josephson was unaware of previous systematic studies but it would have essentially made no difference had he known about them, firstly, because all experimental results are suspect which are not designed specifically to test the validity of a particular theory and secondly because the d.c. supercurrent was only one of a number of predicted phenomena. It is probably true to say that there have never been more predictions crowded in the space of two pages [47].

The existence of a d.c. supercurrent follows more or less from common sense; a sufficiently small disturbance can always be neglected. So we may argue that the Cooper pairs will simply ignore an insulator of no more than a few atomic layers thick. But taken together with the other predicted effects they were not only unexpected but far-reaching and significant. It is, of course, not easy to define which advances are significant and which are not. A predicted effect (or even a set of predicted effects) may be found without bearing any influence on further progress. It seems worthwhile to quote G. H. Hardy on this topic who, working in a related science (or shall we say art?), expressed the opinion that 'a mathematical idea is "significant" if it can be connected in a natural and illuminating way, with a large complex of other mathematical ideas' [318]. Assuming the same to be true in Physics, Josephson's discovery does satisfy the criterion. It has many facets, brings together many different concepts and may be approached from many different directions. It both connects and separates electromagnetic and superconductive phenomena, gives new insight into long-range order, displays quantum effects on a macroscopic scale and relates easily-measurable physical quantities by some of the basic constants of Physics. To make the story more romantic, Josephson made his predictions while still a graduate student at Cambridge, and ironically (ironical because the original formulation was highly abstract) there turned out to be a large number of potential applications, some of them already used in practice.

Owing to the significance of its predictions, Josephson's theory could have easily become an object of controversy. As it happened there was only one objection. On the basis of a similar formalism Bardeen [319] concluded that 'pairing does not extend into the barrier, so there can be no such superfluid flow'. There was not much time left for further objections because some important aspects of the theory were very soon (within nine months) confirmed by the carefully designed experiments of Anderson and Rowell [320].

8.2 A brief history of Josephson tunnelling

The first break made, and significance recognised, further work was rapidly started in a number of laboratories around the world. The magnetic field dependence was proved by Rowell [321], steps in the I–V characteristics by Shapiro [322]. Self-excited steps were found by Fiske [323], quantum interference by Mercereau and co-workers [324], microwave emission by Yanson et al. [325], frequency multiplication by Shapiro [326], mixing by Grimes and Shapiro [327]. Determination of e/h was first reported by Langenberg et al. [328].

Theoretical work also flourished; some of the major advances were: new microscopic theories by Anderson [329], Ferrel and Prange [330] and de Gennes [331], temperature dependence of the Josephson current by Ambegaokar and Baratoff [332], rigorous inclusion of a.c. phenomena by Riedel [127] and Werthamer [128], vortex solutions by Owen and Scalapino [333] and Lebwohl and Stephen [334], I–V characteristics by McCumber [335], Stewart [336] and Scott [337].

Some of the devices developed on the basis of the Josephson effects are as follows: voltmeters by Clarke [338] and Zimmerman and Silver [339], magnetometers by Clarke [340], Zimmerman and Silver [341], Mercereau [342] and Nisenoff [343], memory elements by Matisoo [344], infrared detectors by Grimes et al. [345].

Considerable part of the work was carried out by larger groups which conglomerated at Bell Telephone Laboratories, Ford Scientific Laboratory, University of Pennsylvania, Physico–technical institute of the Ukrainian Academy of Sciences, Kammerlingh Onnes Laboratorium.

In the eight years since the original discovery the study of Josephson tunnelling has spread into the four corners of the world. It is of interest to follow the course of its development because it gives considerable insight into the present-day laws governing the growth of Science. We shall return to this question in Chapter 22 where a few statistical data are discussed.

8.3 A qualitative discussion of the Josephson effects

In its simplest form the Josephson effect claims that up to a certain current the voltage across a sufficiently thin tunnel junction is zero. One should not, in general, expect an insulator to change its complexion and turn into a superconductor but after all if the insulator is no more than a few atomic layers thick this is not unreasonable; it can be easily incorporated into one's picture of the physical world. The corresponding energy diagram looks also exceedingly simple. All what happens is that a Cooper-pair tunnels across the junction as shown in Fig. 8.1. This is true up to a certain current, I_J. Above that current there will be a finite voltage; the junction will revert to 'normal' behaviour. This is again reasonable. All superconductors change into the normal state above a certain current so the same could well apply to our insulator. In the 'normal' state of the insulator we have normal tunnelling of electrons as discussed in Part II. We know the I–V characteristics associated with normal tunnelling and we know also the *I–V* characteristics for Josephson tunnelling (just a vertical line at the origin) so we may combine the two. Assuming that the junction is fed by an ideal current generator, there is no voltage across the junction (line a in Fig. 8.2) up to the value I_J, the maximum supercurrent. As

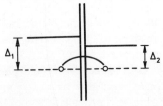

Fig. 8.1 A Cooper-pair tunnels across the junction in the absence of a potential difference.

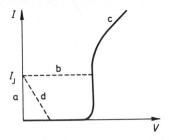

Fig. 8.2 *I–V* characteristic of a Josephson junction.

we increase the current above I_J the characteristic switches over (line b) to the single electron tunnelling curve and follows it (curve c). If we do not prescribe the current then the switch to the single electron tunnelling curve occurs along the circuit load line d.

It is not very difficult even at this stage to discern some chances of practical applications. If the junction can switch from a zero voltage state to a finite voltage state then these may represent the two states of a binary code. We have a potential memory element as it was first realised by Rowell [346].

The dependence of I_J on magnetic field is yet another property of the d.c. Josephson effect (there is an a.c. Josephson effect as well which we shall presently discuss). There is nothing unexpected in this. Magnetic fields do quench super-conductivity so they could be expected to have a deteriorating effect on the maximum supercurrent. What is surprising, however, is that I_J is *not* a mono-tonic function of the applied magnetic field. The function looks more like an interference pattern as shown in Fig. 8.3. It is fairly lengthy to derive this curve from first principles but we can compensate for the lack of rigour by referring to an analogue. Note that Fig. 8.3 not only *looks* like an interference pattern, it *is* an interference pattern. So even if the reason for the appearance of macroscopic quantum interference may not be immediately obvious, the result at least

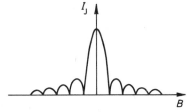

Fig. 8.3 The maximum supercurrent I_j as a function of magnetic field.

looks familiar. The physical concepts are known, the mathematical formulation is ready, all we need to do is some scaling.

The next question of interest is what happens when two thin tunnel junctions (we shall refer to them as Josephson junctions if they display a supercurrent) are connected in parallel in the presence of a magnetic field. This problem was investigated in detail by Mercereau and co-workers [324, 347, 348]. They discovered another quantum interference effect which (following Anderson's suggestion [349]) we shall call the Mercereau effect. The role of the magnetic field is somewhat different here; the relevant quantity is the magnetic flux enclosed by the *parallel circuit*. We can make this effect more comprehensible (or at least more familiar) by making our interference analogy a little more specific and considering the radiation pattern of antennas.

A single junction is analogous to an antenna with a continuous field distribution in its aperture, as for example a parabolic antenna. We get the radiation pattern of the antenna by adding the radiation from each surface element taking account of the proper phase. Similarly, for the junction we must add the current crossing each surface element in the proper phase (this time though the phase has nothing to do with time; it depends on magnetic field as will be discussed in Chapter 10). Two junctions in parallel are analogous to two of these antennas. There are now two kinds of interference patterns; firstly the interference pattern of the individual antennas and superimposed upon them the interference pattern of two point radiators. A very elegant measurement illustrating the double interference effect is discussed in Chapter 13 and is shown in Fig. 13.6. Here we shall dwell a little longer on the analogy with discrete radiators because that gives us the right functional relationship too.

If two discrete radiators are at a distance a from each other (Fig. 8.4) and are

Fig. 8.4 An antenna analogue. Rays emanating from two discrete radiators have a phase difference.

fed in phase then the relative field strength at a distant point in the θ direction is

$$E(\theta) = \left| 1 + \exp j\left(\frac{2\pi}{\lambda} a \cos \theta\right) \right| = 2\left| \cos\left(\frac{2\pi}{\lambda} a \cos \theta\right) \right| \qquad (8.1)$$

where λ is the wavelength of electromagnetic radiation.

When two identical Josephson junctions are connected in parallel (Fig. 8.5)

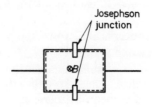

Josephson
junction

===== Superconducting path

Fig. 8.5 Two Josephson junctions in parallel in the presence of a magnetic field. The junctions are connected by a superconducting path.

then the maximum supercurrent flowing through the two junctions is given by the formula

$$I_{max} = 2I_{J}\left| \cos\frac{\pi\Phi}{\Phi_0} \right| \qquad (8.2)$$

where Φ is the magnetic flux enclosed by the circuit and Φ_0 is the flux quantum equal to $2\pi\hbar/q \cong 2 \times 10^{-7}$ gauss cm^2 = 2×10^{-15} Weber, where $q = 2e$ (from now on we shall work in terms of twice the electronic charge).

When the magnetic flux changes by a small amount, the maximum supercurrent will change as well. A change in Φ equal to Φ_0 will move the interference pattern by a whole period. If we can see where we are in the pattern to (say) one tenth of the period, then the change in the magnetic flux can be measured with an accuracy of $\Phi_0/10$. Assuming that we use a circuit with an enclosed area of 1 cm^2, the accuracy of the magnetic field measurement can be 10^{-8} gauss = 10^{-12} Tesla. There has been a fair amount of work on these lines to make devices capable to detect small magnetic fields. Zimmerman and Silver [341] detected field changes of less than 10^{-9} gauss, Mercereau [350] talked of a digital magnetometer counting increments of 10^{-7} gauss at a rate of 10^4/sec, Beasley and Webb [351] measured field changes less than 10^{-7} gauss with a 1 sec time constant, in an applied field of 2500 gauss.

Next we shall discuss the a.c. Josephson effect. In its simplest form it says that if there is a d.c. voltage, V_0, across the junction, then the junction will radiate at the angular frequency

$$\omega_0 = \frac{q}{\hbar}V_0. \qquad (8.3)$$

Can this effect be expressed in terms as simple as its d.c. counterpart? Yes, the production of an a.c. signal from a d.c. voltage may be understood in terms of elementary quantum mechanics. We know that the energy of a photon of angular frequency ω, is $\hbar\omega$, and that the energy of a particle of charge q in a potential V_0 is qV_0. All we need to do now is to postulate that the energy of our electron-pair may be converted into photons, and Equation (8.3) follows.

The a.c. Josephson effect and its relationship to the d.c. Josephson effect may also be understood in terms of a concept, called long-range order. Roughly speaking, this long-range order means that the phase of the superconducting wavefunction at one point is uniquely related to the phase at another point a long distance away. It is like in a coherent electromagnetic wave where the phase difference between any two points is fixed.

If we have a rather thick insulator between two superconductors then the phase coherence is broken. The phases of the two superconductors bear no relation to each other. If the insulator is infinitely thin (that is, absent), the phases must be locked. If the insulator is sufficiently thin then there is still a strong tendency to fix the phase in spite of the presence of the insulator. If phase locking wins, there is a zero-voltage current. If phase locking loses, the energies of electron-pairs will be different on the opposite sides of the insulator; the phase difference across the junction will not be a constant but they will be related to each other.

Another consequence of the a.c. effect is that if we impress an a.c. signal upon the junction, we will observe a step structure in the I–V characteristics. The quantitative explanation is a little lengthy but again (yet again) there is a close analogy. The incident a.c. signal will frequency modulate the supercurrent and we end up with the familiar formulae of frequency modulation. It is easy to show then that if the carrier frequency is an integral multiple of the modulation frequency then there are d.c. terms in the current.

For our junction the condition is that ω_0, the frequency at which the junction radiates, should be equal to an integral multiple of ω, the frequency of the impressed a.c. signal

$$\omega_0 = n\omega \qquad (n = 1, 2, 3, \ldots). \qquad (8.4)$$

Hence the voltages at which the steps appear are

$$V_{0n} = \frac{n\hbar}{q}\omega \qquad (n = 1, 2, 3, \ldots). \qquad (8.5)$$

Remarkably, the dot-dot-dot after $n = 1, 2, 3$ must be taken seriously; n may reach quite high values. Irradiating the junction by microwaves of 70 GHz McDonald et al. [352] observed structure as far as $n = 103$, whereas at an input frequency of about 10 GHz Finnegan et al. [353] found steps even above five hundred.

The emission of radiation and the steps induced by incident radiation are two phenomena which offer obvious applications as oscillators and electromagnetic detectors. It may be seen from Equation (8.3) that the frequency of radiation is proportional to voltage, that is (in engineering terms) we have a voltage tunable oscillator. If we put in the values of the constants we get

$$f = 483 \cdot 6 \, \text{MHz}/\mu V. \tag{8.6}$$

A few microvolts would lead us into the microwave region and a few millivolts into the far infrared (e.g. $V_0 = 2$ mV gives a wavelength of 310 μm).

An oscillator which needs to be cooled to liquid helium temperatures to give a small amount of power is not much of a practical proposition at frequencies where there are cheap and efficient oscillators available – and working at room temperature. However, at the far infrafred, where one needs to rely on mono-chromators to filter out the required band from a hot body spectrum, a tunable oscillator would indeed come useful. The same applies to detectors as well. There are cheap and efficient detectors available down to millimeter wave-lengths but not below that. In the infrared region the Josephson junction already provides the best detector available. For incoherent input the effect upon I_J the maximum Josephson current is measured [345, 354], for coherent input the induced steps are utilised [355].

There are a few more applications we could discuss but we shall mention here in the Introduction only one more which is in a different class. We use the Josephson effect for gaining information about something else. It may be seen from Equation (8.5) that if both the voltage at the nth step, and ω the frequency of the impressed a.c. signal are measured then e/h can be determined. The actual experiment was performed [356] by measuring the voltage change of a high order step when the junction current was reversed. Since frequency can be measured with great accuracy (say 1 part in 10^8), the accuracy of determining e/h depends on the voltage measurement. The most recent result [353] is an accuracy of a few nV in measuring about 10 mV, leading to a determination of e/h better than 1 part per million.

Conversely, if a sufficiently accurate independent measurement of e/h is available then we may define the volt with the aid of the Josephson effect in terms of frequency. Unfortunately, no independent determination of e/h is sufficiently accurate at present so we must be satisfied with maintaining the volt as the emf of a certain group of batteries. However, the Josephson effect may already be used for relative measurements. The drift in the emf of the batteries (a few ppm per year) may be detected by measuring the Josephson frequency voltage ratio as a function of time. Any change of this ratio must be attributed to a change in time of the reference standard (provided the measuring apparatus is unchanged). Another possible application is to infer the relationship between the standards of emf maintained by various laboratories around the world.

All the applications mentioned so far and indeed all the actual and potential applications will be discussed in later chapters. We have to realise, however, that many of the experiments (in fact the majority of them) were performed on geometries which cannot be strictly classified as tunnel junctions. It was slowly realised (it cannot be attributed to any single author) that these other geometries display effects very similar to those in thin tunnel junctions. Significantly, in Josephson's second paper [357], the transition is already made from the 'pure' Josephson effect to 'coupled superconductors', and the term 'weak super-conductivity' coined by Anderson [329] also indicated that we are dealing with a wider set of phenomena than straightforward tunnelling. On the experimental front the first experiment which may be commented upon in terms of weak superconductivity was that of Little and Parks [358] who measured the dependence of critical temperature on magnetic field for a thin film cylinder. Further experiments by Parks et al. [359] and Parks and Mochel [360] were explained in terms of quantised vortices but as it was discussed by Anderson and Dayem [361] the effects resemble strongly the Josephson effects.

In addition it was found [253, 254] (the first experiments were actually done prior to Josephson's analysis) that a normal metal placed between two super-conductors may also carry a supercurrent, and later experiments [362] showed clearly that the behaviour is very similar to Josephson tunnelling.

Summarising, the present state of affairs is that besides thin tunnel junctions and point contacts, discussed in connection with normal electron tunnelling, we have some other geometries as well displaying the characteristic effects (like induced steps in the $I-V$ characteristics, quantum interference, radiation, etc.).

The various geometries are shown in Fig. 8.6. The superconducting thin film bridge (Fig. 8.6 (a)) was developed by Dayem and it is often referred to as the Dayem bridge. It consists essentially of a narrow constriction of the order of $(1 \ \mu m)^2$ between two superconducting films.

A variation of the thin film bridge is the Notarys bridge [363] shown in Fig. 8.6 (b). This relies on the proximity effect to cause weak superconductivity. There is first a layer of normal metal evaporated, and then the superconductor forming the bridge. The 'weakness' is controlled by the relative thicknesses of the normal and superconducting layers.

The solder junction (Fig. 8.6 (c)) was developed by Clarke [338]. It is very simple to make; one simply has to dip a piece of oxidised wire into molten solder. When the solder freezes tight mechanical contact is established and the junction is ready. Its disadvantage is that the junction is not clearly defined.

The point contact junction is also widely used but not necessarily in the same form as for single particle tunnelling. Then the presence of the insulating layer is essential for displaying tunnelling characteristics. The Josephson effects may, however, be observed whether there is an insulating layer or not. If the two superconductors are in contact then we have a bridge. In actual experiments very often we do not even know whether we have a weak link or a tunnel junction. Take for example the point contact junctions developed by Zimmerman and Silver [341]. They used superconducting screws with lock nuts pressed against a superconducting surface through a Mylar spacer (Fig. 8.6 (d)). There is no way of knowing whether there is a direct contact between the superconductors or not. All the measurements can tell us is that there is a difference in the *I–V* characteristics (the increase in the maximum supercurrent can be clearly observed) as more pressure is applied, leading eventually to a 'strong link', i.e. a proper short-circuit.

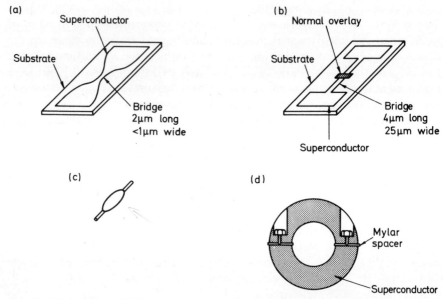

Fig. 8.6 Weak links exhibiting properties similar to those of Josephson junctions. (*a*) A superconducting thin film bridge (Dayem bridge), (*b*) a superconducting thin film bridge with a normal metal overlay (Notarys bridge), (*c*) Clarke's solder drop junction (*d*) double point-contact junction developed by Zimmerman and Silver [341].

Let us discuss now briefly that in what sense does a weak link resemble to a tunnel junction. There is no difficulty as far as the d.c. Josephson effect is concerned. At low currents there is no voltage across the weak link but as the current is increased the weak link will first turn resistive. This is because it has a smaller cross-section and thus a higher current density or (as in the Notarys bridge) it has a lower critical temperature. So we may say that it is true both for tunnel junctions and for weak links that there is no voltage up to a certain current and finite voltage above a certain current.

It is somewhat more difficult to envisage the a.c. effects. We can no longer appeal to the simple physical picture that the tunnelling Cooper pairs radiate out their energy. There is though a simple heuristic argument (due to Mercereau [364]) which can justify the presence of an a.c. supercurrent. The argument (in an abbreviated form) runs as follows.

A voltage across the weak link will effect both the normal and the super-conducting electrons. The superconducting electrons will be accelerated according to Newton's Law

$$\frac{dv}{dt} = \frac{q}{m}\mathscr{E}$$ (8.7)

where the electric field $\mathscr{E}$ is sustained by the normal electrons. It is known, however, that when the superconducting electrons reach a certain velocity (say v_c) their density vanishes. But then there is only an electric field across the weak link which on its own cannot quench superconductivity; hence the superconducting electrons reappear and the whole process repeats itself.

Let us now make a rough calculation of the time needed to accelerate the superconducting electrons to a velocity v_c. For a constant $\mathscr{E}$ we get

$$\tau = \frac{mv_c}{q\mathscr{E}}.$$ (8.8)

Assuming that the electric field is constant over one coherence length (the shortest distance over which the density of superconducting electrons can change), the voltage across this region is given by

$$V = \mathscr{E}\xi$$ (8.9)

Expressing further the final momentum in terms of the coherence length by the relation

$$mv_c = \frac{h}{\xi}$$ (8.10)

where mv_c is the uncertainty in momentum, we get

$$\tau = \frac{h}{qV}$$

for the period of this 'relaxation oscillation'. Hence the angular frequency is

$$\omega = \frac{qV}{h}$$ (8.11)

which agrees with Equation (8.3). So we managed to show (by hook or by crook) that weak links have the same radiation properties as tunnel junctions. In fact, the difference between tunnel junctions and weak links is not so much in the basic processes but rather in the circuits they represent to the outside world.

For most purposes a junction may be adequately represented by the parallel combination of an 'ideal' junction (one which carries a supercurrent only) a resistance and a capacitance. The main differences between tunnel junctions, point contacts and bridges are in the respective values of these circuit elements.

For a Josephson current of $I_J = 10$ μA, the range for practical junctions is shown in Table 8.1 [365]. It is then obvious that, for example, for high frequency operation (microwaves or above) the point contact is superior to the tunnel junction. The relative values of the circuit elements play an important role in determining the I–V characeristics as well. This will be discussed in Chapter 11.

Table 8.1

	I_J	$1/G$ (*for $V < 2\Delta$*)	C
Thin film tunnel junction	10 μA	> 100 ohm	100 to 1000 pF
Point contact	10 μA	10 to 100 ohm	1 to 1000 pF
Thin film bridge	10 μA	< 1 ohm	1 pF

The theory will be derived for tunnel junctions in Chapters 9 and 10 but subsequently the type of junction used will always be mentioned, and modifications introduced if necessary.

Before going further to start the more rigorous treatment let us summarise the properties of Josephson junctions. For an electromagnetic input it can give a d.c. output; for a d.c. input it can give an electromagnetic output; it has two stable states; the maximum supercurrent across it is sensitive to magnetic fields; it can produce harmonics and subharmonics; it has a variable reactance and it can be produced in a number of forms (including the thin film form suitable for large scale manufacturing). Looking at this impressive list we can have no doubts about the merits of Josephson junctions. It may be said without exaggeration that it is the most versatile single device that has come our way since the invention of the p–n junction.

Its major handicaps are: the need for low temperature operation (unless someone will oblige humanity by inventing a high temperature superconductor), and not sufficiently developed technology. The latter can only be changed by more effort, and more effort will only be forthcoming if more people will be convinced that Josephson junctions will turn out to be useful devices – made perhaps in the billions one day.

9. Derivation of the basic equations

9.1 Microscopic theories

Josephson's basic equations may be obtained in a number of different ways. We shall be concerned in this book with the simplest derivation only which involves no more than the phenomenological theory of superconductivity combined with electrodynamics and a little imagination. It seems, however, worthwhile to give a brief review of the various microscopic approaches before setting up our model.

Josephson's original derivation [47] was based on the assumption that the Hamiltonian of the system can be written as

$$H = H_0 + H_T \tag{9.1}$$

where H_0 is the Hamiltonian of the unperturbed system consisting of two isolated superconductors, and H_T is the tunnelling Hamiltonian of Cohen et al. [50] shown in Equation (2.26).

First order perturbation leads to the formulae for normal electron tunnelling. In addition there is a zero voltage δ-function singularity for the transfer of Cooper pairs. The singularity arises because the state with one Cooper pair transferred is degenerate with the initial state and it means that the calculation was not started with the correct linear combination of the degenerate states. The correct choice leads directly to Josephson tunnelling which turns out to be proportional to $|T|^2$ in contrast to two-particle tunnelling which is proportional to $|T|^4$.

An approach which does not rely on a perturbation solution was also used by Josephson [366]. In this the propagation of electrons is described in terms of the electron Green's function which takes into account the finite time needed by an electron to traverse the junction. The calculation was, however, suitable only for the equilibrium case, that is for the supercurrent across the junction.

In the method used by de Gennes [331, 367] the link between the two superconductors was regarded as modifying the boundary conditions to be applied to the Ginzburg–Landau equations.

Ferrel and Prange [330] restricted the generality by assuming that only Cooper-pairs tunnel across the junction, one at a time. Hence the perturbation connects only states for which the number of Cooper-pairs differs by unity. The resulting formulation is analogous to that obtained for a one-dimensional chain of atoms in the tight-binding approximation leading to the formula

$$E_\phi = E - 2M \cos \phi \qquad (9.2)$$

where E_ϕ and E are the energies of the perturbed and unperturbed states respectively, $-M$ is the matrix element for transfer of Cooper-pairs, ϕ is the phase difference across the junction and $-2M \cos \phi$ is the coupling energy. In the absence of a current source, ϕ adjusts itself so as to minimise the coupling energy. In the presence of a current source and of a d.c. voltage across the junction the mathematical relationships are obtained from Hamilton's equations written for the conjugate variables, ϕ and $\mathcal{N}_s$, the latter being the number of Cooper-pairs.

Those interested in further details may first consult the reviews of Anderson [349], Josephson [368] and Scalapino [369].

9.2 Macroscopic theories

We shall assume here the validity of the Ginzburg–Landau theory which describes the spatial variation of the order parameter, ψ, and of the magnetic vector potential, A. An alternative way of looking at the order parameter is to regard it as a wavefunction (very much like the one-electron wavefunction obtained by solving Schrödinger's equation) where $|\psi|^2$ is equal to the density of superconducting electrons. ψ is in general complex but may be expressed in terms of two real functions as (Equation (1.34))

$$\psi(x, y, z) = [\rho(x, y, z)]^{1/2} \exp iv(x, y, z) \qquad (9.3)$$

where $[\rho(x, y, z)]^{1/2}$ is the amplitude function and $v(x, y, z)$ is the phase function. Long range order in a superconductor means that once the phase is fixed at a certain point, it is determined everywhere.

If two superconductors are separated from each other by a sufficiently thick barrier (insulator, semiconductor, normal metal or may even be a different superconductor) then the wavefunctions on the two sides are not related in any way. The phases may vary independently. If there is no barrier at all, that is, we have one single block of superconductor, then the relative phases are fixed at each point. We may then ask the question, what happens when the barrier is thin and the wavefunctions on the two sides of the barrier are in a certain sense coupled to each other? How will the relative phases vary and what will be the values of the currents, of the electric and magnetic fields in the barrier?

In order to answer these questions we shall follow Josephson [366] in investigating a simple geometrical case where the barrier is flat and infinitely thin. The two superconductors extend from $z = -\infty$ to 0 and from 0 to $+\infty$ respectively. The barrier lies in the $z = 0$ plane and the two sides will be distinguished by referring to points at $z = +0$ and $z = -0$.

The quantities required to specify the problem are as follows:

(i) $\phi(x, y, z, t) = v(x, y, +0, t) - v(x, y, -0, t)$, the change in the phase of the wavefunction as one crosses the barrier from negative z to positive z,

(ii) $B_x(x, y, t)$ and $B_y(x, y, t)$ the components of the tangential magnetic field in the barrier,

(iii) $V(x, y, t)$, the potential difference across the barrier,

(iv) $j_z(x, y, t)$, the current density through the barrier.

First we need a relationship between the current flowing in the superconductor and the wavefunction in the presence of a magnetic field. This is provided by the Ginzburg–Landau theory or the macroscopic Schrödinger equation in the form

$$\mathbf{j} = \frac{q}{m}\left\{\tfrac{1}{2}i\hbar(\psi\nabla\psi^* - \psi^*\nabla\psi) - q\mathbf{A}|\psi|^2\right\}. \tag{9.4}$$

Substituting Equation (9.3) into Equation (9.4) and rearranging, we get

$$\nabla v = \frac{q}{\hbar}\left(\mathbf{A} + \frac{m}{q^2\rho}\mathbf{j}\right) \tag{9.5}$$

which is valid in the superconductor but not in the barrier. Next we derive an expression relating the values of ϕ at two points of the barrier, P_1 and P_2. Denoting the coordinates of the points by $(x_1, y_1, 0)$ and $(x_2, y_2, 0)$ respectively, we may write

$$\phi(P_1) - \phi(P_2) = v(x_1, y_1, +0) - v(x_1, y_1, -0) - v(x_2, y_2, +0) + v(x_2, y_2, -0). \tag{9.6}$$

Integrating ∇v along the curves γ_- and γ_+ (Fig. 9.1) we get

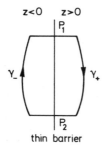

Fig. 9.1 Integration contour used to evaluate the space dependence of ϕ.

$$\int_{\gamma_-} \nabla v\, ds = v(x_2, y_2, -0) - v(x_1, y_1, -0) \tag{9.7}$$

and

$$\int_{\gamma_+} \nabla v\, ds = v(x_1, y_1, +0) - v(x_2, y_2, +0). \tag{9.8}$$

Substituting Equations (9.7) and (9.8) into (9.6) and taking account of (9.5) we obtain

$$\phi(P_1) - \phi(P_2) = \frac{q}{h} \int_{\gamma_- + \gamma_+} \left(A + \frac{m}{q^2 \rho} j \right) ds. \tag{9.9}$$

Noting that the flux quantum is $\Phi_0 = 2\pi h/q$ and the line integral of the vector potential is just the magnetic flux enclosed, the first term of the integral comes to $2\pi\Phi_\gamma/\Phi_0$. The second term may be made small by choosing a suitable integration contour. This is possible because, owing to the Meissner effect, the current density declines rapidly with distance from the barrier; furthermore its direction is essentially parallel to the plane of the barrier in the penetration region. Hence the correct choice for γ is a curve which lies outside the penetration region everywhere except near P_1 and P_2, and in the neighbourhood of these points crosses the penetration region in a direction normal to the barrier as shown in Fig. 9.1. For such a contour Equation (9.9) may be integrated to give

$$\phi(P_1) - \phi(P_2) = 2\pi \frac{\Phi_\gamma}{\Phi_0}. \tag{9.10}$$

If P_1 and P_2 are close to each other we may write

$$\Delta\phi = \frac{\partial\phi}{\partial x}\Delta x + \frac{\partial\phi}{\partial y}\Delta y. \tag{9.11}$$

Next we derive an expression for the magnetic flux enclosed by the curve in Fig. 9.1. Note that the magnetic field penetrates only to a distance d which may be conveniently defined as*

$$d = \frac{1}{B(0)} \int_{-\infty}^{\infty} B(z) \, dz. \tag{9.12}$$

The magnetic flux enclosed may then be obtained with the aid of Fig. 9.2 in the form

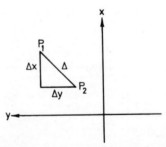

Fig. 9.2 Points P_1 and P_2 of the barrier in the x–y plane; used to evaluate magnetic flux.

* If the thickness of the barrier is taken a finite value w then $d = w + 2\lambda$ where λ is the penetration depth of the superconductors.

$$\Phi_y = \mathbf{B}(\mathbf{i}_z \times \mathbf{\Delta})d$$

$$= \mathbf{i}_z(\mathbf{\Delta} \times \mathbf{B})d \tag{9.13}$$

where $\mathbf{\Delta} = \Delta \times \mathbf{i}_x + \Delta_y \mathbf{i}_y$ and $\mathbf{i}_x$, $\mathbf{i}_y$, $\mathbf{i}_z$ are unit vectors.

Equation (6.14) may be reduced to the form

$$\Phi_y = (B_y \Delta x - B_x \Delta y)d \tag{9.14}$$

which, compared with Equations (9.10) and (9.11) leads to the differential equations*

$$\frac{\partial \phi}{\partial x} = \frac{qd}{h}B_y \tag{9.15}$$

$$\frac{\partial \phi}{\partial y} = -\frac{qd}{h}B_x. \tag{9.16}$$

A further relationship between the relative phase ϕ and the potential difference between the superconductors V may be obtained by the following simple consideration. The relative energy difference of Cooper-pairs between the opposite sides of the junction is qV. Since the rate of change of the phase of a wavefunction varies as its energy divided by $\hbar$, the relative phase difference varies as $qV/\hbar$. The time dependence of ϕ is therefore given by the equation

$$\frac{\partial \phi}{\partial t} = \frac{q}{\hbar}V. \tag{9.17}$$

We shall need one more equation expressing the relationship between the current and the relative phase. Feynman [21] obtained this equation by extending his method of coupled modes (successful in problems as diverse as the bond in a hydrogen molecule, the operation of an ammonia maser or the turning of the K° particle into its own anti-particle) to Josephson tunnelling. According to this picture the rate of change of the wavefunction on one side depends on the actual values of the wavefunctions on *both* sides. The coupled differential equations look like

$$i\hbar\frac{\partial \psi_1}{\partial t} = U_1\psi_1 + K\psi_2 \tag{9.18}$$

$$i\hbar\frac{\partial \psi_2}{\partial t} = U_2\psi_2 + K\psi_1 \tag{9.19}$$

where U_1 and U_2 are the self energies and K is a constant characteristic of the junction. When K is zero Equations (9.18) and (9.19) may be solved separately;

*A heuristic derivation of Equations (9.15) and (9.16) may be obtained from the London equation discussed in Section 1.4

$$\nabla \times \mathbf{v} = -\frac{q}{m}\mathbf{B}.$$

Identifying mv/h with λ, the de Broglie wavelength of a Cooper-pair, and taking $(2\pi/\lambda)d$ as the phase difference ϕ, Equations (9.15) and (9.16) follow.

the two superconductors are independent of each other. When K is finite, the two superconductors are coupled.

In the general case when there is a voltage across the junction, the difference in self energies must be $U_2 - U_1 = qV$. We may shift then the chosen zero of energy and rewrite Equations (9.18) and (9.19) in the form*

$$i\hbar\frac{\partial\psi_1}{\partial t} = \frac{qV}{2}\psi_1 + K\psi_2 \tag{9.20}$$

$$i\hbar\frac{\partial\psi_2}{\partial t} = K\psi_1 - \frac{qV}{2}\psi_2. \tag{9.21}$$

Writing now both wavefunctions in the form of Equation (9.3) and substituting them into Equations (9.20) and (9.21) we get the following four differential equations

$$\frac{\partial\rho_1}{\partial t} = \frac{2}{\hbar}K\sqrt{(\rho_1\rho_2)}\sin\phi$$

$$\frac{\partial\rho_2}{\partial t} = -\frac{2}{\hbar}K\sqrt{(\rho_1\rho_2)}\sin\phi$$

$$\frac{\partial v_1}{\partial t} = -\frac{K}{\hbar}\sqrt{\frac{\rho_2}{\rho_1}}\cos\phi - \frac{qV}{2\hbar} \tag{9.22}$$

$$\frac{\partial v_2}{\partial t} = -\frac{K}{\hbar}\sqrt{\frac{\rho_1}{\rho_2}}\cos\phi + \frac{qV}{2\hbar}$$

where we identified $v_2 - v_1$ with ϕ. Note that

$$\frac{\partial\rho_1}{\partial t} = -\frac{\partial\rho_2}{\partial t} \tag{9.23}$$

that is one side starts to lose charge at the same rate as the other side accumulates it. But whatever charge is lost it will be immediately replenished by the active element (voltage or current source) in the circuit. Taking further $\rho_1 = \rho_2 = \rho_0$ we may write for the current

$$j_z = j_J \sin\phi \tag{9.24}$$

where

$$j_J = \frac{2K\rho_0}{\hbar}. \tag{9.25}$$

---

* This is for the case of no magnetic field in the junction. The generalisation to include magnetic fields was done by de Waele and de Bruyn Ouboter [369]; K needs then to be multiplied by an extra factor involving the integral of the vector potential across the junction. A very similar derivation was also given by Zawadowski [371].

From Equation (9.22) we may also get

$$\frac{\partial v_2}{\partial t} - \frac{\partial v_1}{\partial t} = \frac{qV}{\hbar} \qquad (9.26)$$

which is an alternative derivation of Equation (9.17).

The only unknown is the value of K in Equation (9.25). It may be expected to depend on the properties of the superconductors on both sides of the barrier and above all on the thickness of the insulator. For thick barriers the coupling is small so K is small and hence the supercurrent flowing across the barrier must also be small. In fact when the barrier is thick we find no supercurrent at all. The junction exhibits only normal electron tunnelling with the associated high resistance. The reason for this behaviour is the presence of noise we have so far neglected. When the coupling energy becomes comparable with the noise energy the supercurrent is no longer observable (we shall say a little more about this problem in Chapter 20 concerned with fluctuations).

There have been several attempts to determine K (or rather j_J) from macroscopic theories. Jacobson [372] used the Ginzburg–Landau theory for the superconductors and Schrödinger's equation with a repulsive potential for the insulator and matched the respective wavefunctions at the boundaries. He obtained an expression of the form of Equation (9.24) but not the correct value for j_J. For superconducting barriers (like superconducting bridges or point contacts) there is more justification for using the Ginzburg–Landau equations. Unfortunately, one has to make another allowance there, namely to restrict the problem to one dimension because otherwise the mathematics becomes untractable. Solutions obtained for that case by Baratoff [373], Baratoff et al. [374] and Yamafuji et al. [375] displayed the right functional dependence on phase in the limit of weak coupling. Experimental proofs for superconducting bridges for the correctness of the $j_z = j_J \sin \phi$ relationship have been provided by Fulton [376] and Fulton and Dynes [377].

The magnitude of j_J for an insulating barrier was derived from microscopic theory by Ambegaokar and Baratoff [332] in the form*

$$j_J = \frac{\pi \Delta(T)}{q R_{NN}} \tanh \frac{\Delta(T)}{2kT} \qquad (9.27)$$

(where R_{NN} is the resistance per unit area of the junction in the normal state) for identical superconductors. If the superconductors on the opposite sides of the junction are not identical j_J must be determined by numerical methods. The curves of $j_J(T)/j_J(0)$ for tin–tin oxide–tin and lead–lead oxide–tin are shown in Fig. 9.3. The experimental points were obtained by Fiske [323].

We have now obtained all the necessary relationships for the junction

---

*Strong-coupling effects reduce j_J to 78·8% and to 91·1% of this value for Pb and Sn respectively [378].

originating from the properties of the superconductors. However, we should not forget that Maxwell's equations are still valid in the insulator, and in the general time-dependent case a displacement current density

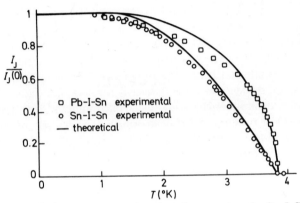

Fig. 9.3 The maximum supercurrent as a function of temperature for Sn–I–Sn and Pb–I–Sn junctions. The solid lines represent the theoretical curves of Ambegaokar and Baratoff [332], the experimental points were measured by Fiske [323].

$$c_s \frac{\partial V}{\partial t} \tag{9.28}$$

is present as well (c_s, capacitance per unit area of the junction) leading to the equation

$$\frac{\partial B_y}{\partial x} - \frac{\partial B_x}{\partial y} = \mu_0 j_z + \mu_0 c_s \frac{\partial V}{\partial t}. \tag{9.29}$$

Equations (9.15), (9.16), (9.17), (9.24) and (9.29) give now a complete description of the behaviour of the junction. Expressing B_y and B_x from Equations (9.15) and (9.16), V from Equation (9.17), j_z from Equation (9.24) and substituting them into Equation (9.29) we obtain a single differential equation in ϕ

$$\frac{\partial^2 \phi}{\partial x^2} + \frac{\partial^2 \phi}{\partial y^2} - \frac{1}{v^2} \frac{\partial^2 \phi}{\partial t^2} = \lambda_J^{-2} \sin \phi \tag{9.30}$$

where

$$\lambda_J = (\hbar / \mu_0 q j_J d)^{1/2} \tag{9.31}$$

and*

$$v = (\mu_0 c_s d)^{-1/2} = c(w/\varepsilon_r d). \tag{9.32}$$

* Equation (9.32) was first derived by Swihart [136] for the passive propagation of electromagnetic waves in an insulator sandwiched between two superconductors. This reduction in velocity is responsible for example for the standing waves in small junctions as discussed in Section 5.2.

Note that so far we have been concerned with the supercurrent only. If we want to be more general we may add to Equation (9.24) the current due to the tunnelling of normal electrons as well. The complete formula derived by Josephson is of the form

$$j_z = j_J(V) \sin \phi + \{g_0(V) + g_1(V) \cos \phi\} V. \tag{9.33}$$

For most purposes $g_1(V)$ may be neglected and $g_0(V)$ taken as a constant, that is the form assumed for the current density is

$$j_z = j_J \sin \phi + g_0 V. \tag{9.34}$$

We may now obtain a more general differential equation for the phase by substituting Equation (9.34) into (9.29) yielding

$$\left[\frac{\partial^2}{\partial x^2} + \frac{\partial^2}{\partial y^2} - \frac{1}{v^2}\frac{\partial^2}{\partial t^2} - \frac{\beta}{v^2}\frac{\partial}{\partial t}\right]\phi = \lambda_J^{-2} \sin \phi \tag{9.35}$$

where $\beta = g_0/c_s$. The above differential equation has no analytical solution. It is, however, possible to obtain solutions in some special cases which will be discussed in subsequent chapters.

10.   The Josephson effects

10.1 The d.c. Josephson effect

The equations derived in the previous chapter become considerably simpler for the time-independent case, giving rise to the so-called d.c. Josephson effect. It follows from Equation (9.17) that if ϕ is independent of time then $V = 0$. Similarly, we can immediately see from Equations (9.15) and (9.16) that in the absence of a magnetic field* $\phi = $ constant. So we are left with the sole equation

$$j_z = j_J \sin \phi. \tag{10.1}$$

This equation tells us that if $\phi \neq 0$ (and there is no reason why ϕ should be equal to zero) then a finite current flows across the insulator and this can happen without causing any voltage drop. So, in fact, the insulator behaves as a super-conductor. It follows also from Equation (10.1) that the maximum current density which can flow across the insulator is equal to j_J; multiplying with the area of the junction we get the maximum permissible current I_J. For any value of current below this, the actual current flowing is determined by the external circuit. The relative phase ϕ will adjust itself in such a way as to give the right amount of current. When $\phi = \pi/2$ the junction carries the maximum current which can flow across it without causing a voltage.

Let us investigate now what happens when the external circuit consists of an ideal current generator and we force a current in excess of I_J across the junction. In the simplest case the voltage will jump from the Josephson tunnelling characteristic to the normal tunnelling characteristic as shown in Fig. 8.2. The supercurrent, its maximum value, and the voltage jump to the normal tunnelling curve were first observed by Anderson and Rowell [320] in 1963, a mere nine months after the publication of Josephson's original prediction.

Next we shall look at the case when there is a constant (both temporally and spatially) magnetic field, $\mathbf{B} = B_x \mathbf{i}_x + B_y \mathbf{i}_y$ in the plane of the junction. Then Equations (9.15) and (9.16) may be integrated to give

$$\phi = \frac{qd}{\hbar}(B_y x - B_x y) + \alpha \tag{10.2}$$

where α is an integration constant. It may be seen from the above equation that

---

*Of course, there is always a magnetic field when a current flows across the junction but it may be neglected in certain cases.

depending on the applied magnetic field ϕ may change its sign. This means that not only will the magnitude of the current density vary across the junction but its direction may change as well. So it is not particularly surprising that for certain magnetic fields the total current vanishes. Mathematically, we may express the total current as

$$I = \int j_J \sin \phi \, d\phi = \text{Im} \int j_J \exp \left\{ i \left[\frac{qd}{h} (B_y x - B_x y) + \alpha \right] \right\} dS$$

$$= \text{Im} \left[\exp (i\alpha) \int j_J \exp \left\{ i \frac{qd}{h} (B_y x - B_x y) \right\} dS \right] \qquad (10.3)$$

where Im denotes the imaginary part and the integration is taken over the area of the junction. The maximum supercurrent through the insulator is given by the maximum value of this expression with respect to changes in α. Obviously, the expression is a maximum when the complex vector is in the direction of the imaginary axis. Hence

$$I_J = \left| \int j_J \exp \left\{ i \frac{qd}{h} (B_y x - B_x y) \right\} dS \right|. \qquad (10.4)$$

This integral, the Fourier transform of j_J, may be calculated if the spatial variation of j_J is known. Usually, one intends to make the junction uniform so that for most cases of interest j_J may be taken as a constant. For a rectangular junction of dimensions a_x and a_y, Equation (10.4) reduces to

$$I_J = I_{J0} \left| \frac{\sin u a_x}{u a_x} \right| \left| \frac{\sin v a_y}{v a_y} \right| \qquad (10.5)$$

where

$$u = \frac{qd}{2h} B_y, \qquad v = \frac{qd}{2h} B_x \qquad (10.6)$$

and $I_{J0} = j_J a_x a_y$, is the maximum supercurrent in the absence of a magnetic field.

If the applied magnetic field is parallel with one of the edges of the junction (say $B_x = 0$) then we get the particularly simple form

$$I = I_{J0} \left| \frac{\sin \pi \Phi/\Phi_0}{\pi \Phi/\Phi_0} \right| \qquad (10.7)$$

where Φ denotes again the magnetic flux enclosed. Thus whenever the magnetic flux is an integral multiple of Φ_0 the flux quantum, no supercurrent will flow across the junction. This effect was first observed by Rowell [321] in 1963. Here we are going to show (Fig. 10.1) a more detailed experimental curve taken by Langenberg et al [379]. It may be immediately seen that the zeros in the experimental curve are the same distance ($B = 1.25$ gauss) from each other.

For a quantitative comparison with theory (Equation (10.7)) we need to know the magnetic flux Φ. Since that was not measured we could choose instead the geometry of the junction so as to make Φ equal to Φ_0 at $B = 1\cdot25$ gauss. If we do that and identify the measured current maximum with I_{J0}, the resultant theoretical curve is hardly distinguishable from the experimental one.

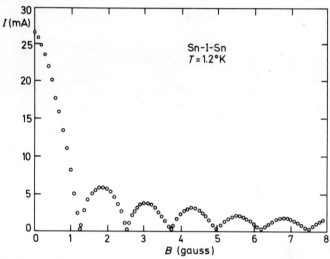

Fig. 10.1 The maximum supercurrent as a function of magnetic field for an Sn–I–Sn junction. After Langenberg et al. [379].

We may now check whether the above assumption is reasonable by working out the penetration depth d, which makes

$$Ba_xd = \Phi_0 \tag{10.8}$$

for $B = 1\cdot25$ gauss and $a_x = 0\cdot25$ mm. The result is $d = 64$ nm which is certainly near to the expected value for tin at $1\cdot2°$K.

Let us return to the case of no applied magnetic fields. Equation (10.2) gives then $\phi =$ constant but we have to remember that Equation (10.2) is a result of an approximation where the self-magnetic fields were neglected. Mathematically this appeared as ignoring Equation (9.29) which relates currents to magnetic fields. If we include the latter relationship we finish up with Equation (9.30). Confining our attention to the time-independent case and assuming a one-dimensional geometry, the differential equation to be solved is as follows

$$\frac{\partial^2\phi}{\partial y^2} = \lambda_J^{-2}\sin\phi. \tag{10.9}$$

An exact solution of Equation (10.9) is possible but first we shall investigate the case when ϕ is small and replace $\sin\phi$ in Equation (10.9) by ϕ. Then Equation

(10.9) comes to

$$\frac{\partial^2 \phi}{\partial y^2} - \lambda_J^{-2} \phi = 0 \tag{10.10}$$

which is the same as London's equation for a bulk superconductor. It has the solution

$$\phi \sim \exp\left(-y/\lambda_J\right) \tag{10.11}$$

that is, λ_J may be regarded as the penetration depth (usually referred to as the Josephson penetration depth) in analogy with the Meissner effect. Thus as long as λ_J is large in comparison with the dimensions of the junction ϕ can be regarded a constant and the total current can be calculated by multiplying the current density by the area of the junction. If λ_J is small the current is confined to the edge of the junction and the maximum supercurrent is drastically reduced.

Typical values for j_J and d are 0.5×10^6 A/m^2 and 80 nm respectively, giving for λ_J (from Equation (9.33))

$$\lambda_J = \left(\frac{\hbar}{\mu_0 q j_J d}\right)^{1/2} = \left(\frac{1.05 \times 10^{-34}}{4\pi \times 10^{-7}\, 3.2 \times 10^{-19}\, 0.5 \times 10^6\, 80 \times 10^{-9}}\right)^{1/2}$$

$$= 0.084 \text{ mm} \tag{10.12}$$

We may say, in general, that λ_J is of the order of 0.1 mm but, of course, it can be made much larger by restricting the value of j_J.

It must be noted that the exclusion of the magnetic field from the junction (in analogy with Type II superconductors) will be destroyed by sufficiently large fields and quantised flux lines will enter the barrier. This shall be discussed in Chapter 12 by solving Equation (10.9) without assuming that ϕ is small.

10.2 The a.c. Josephson effect

The a.c. Josephson effect – as the name implies – is concerned with the temporal variation of the phase across the barrier. All the relevant properties may be deduced from the solution of Equation (9.35) but it is much easier to look first at some approximate solutions which would immediately give a physical picture and point at the possibility of applications.

The simplest case is when a d.c. voltage V_0 appears across the junction but not magnetic fields are applied and the self-magnetic fields are negligible. It is possible then (as a first approximation) to ignore all the equations with the exception of Equations (9.17) and (9.24). Integrating Equation (9.17) we get

$$\phi = \alpha + \frac{q}{\hbar} V_0 t \tag{10.13}$$

which substituted into Equation (9.24) leads to

$$j_z = j_J \sin\left(\alpha + \frac{q}{\hbar} V_0 t\right) \tag{10.14}$$

where α is a constant. The current is oscillating at the frequency

$$\omega_0 = \frac{q}{\hbar} V_0 \tag{10.15}$$

that is, we have an oscillator whose frequency is proportional to d.c. voltage (a voltage tunable oscillator in engineering terms).

The radiation from the junction was first observed by Giaever [380] using for detection the phenomenon of photon-assisted tunnelling. As discussed in Section 5.2 there will be sudden rises in the normal electron tunnelling characteristics between two identical superconductors whenever the condition

$$V_n = \frac{1}{e}(2\Delta \pm n\hbar\omega)$$

is satisfied. Hence a 'normal' tunnel junction may detect the radiation generated by a Josephson tunnel junction.

The physical arrangement is shown in Fig. 10.2 where 1, 2 and 3 denote layers of tin; the junction between 1 and 2 is normal whereas the one between 2 and 3 displays the Josephson effects. Applying a voltage between 2 and 3 the junction radiates at the frequency $\omega_0 = (q/\hbar) V_{23}$. This radiation falls upon junction 1,2 causing current rises separated by $2V_{23}$ as shown in Fig. 10.3.

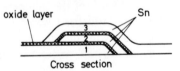

oxide layer   Sn

Cross section

Fig. 10.2 Schematic drawing of two Sn–I–Sn tunnel junctions coupled to each other. Junction 2,3 displays Josephson tunnelling whereas junction 1,2 can only exhibit normal electron tunnelling. The radiation created by junction 2,3 is detected by junction 1,2. After Giaever [380].

The first direct observation of the radiation was made by Yanson *et al.* [325] who placed the Josephson junction in a waveguide and detected the radiation in the 10 GHz region by an external detector.

Next (based again on Equations (9.17) and (9.24)), we are going to investigate the case when besides the d.c. voltage there is an a.c. voltage as well across the junction. So we take for the voltage

$$V = V_0 + V_1 \cos \omega t \tag{10.16}$$

leading to

$$\phi = \frac{q}{\hbar} V_0 t + \frac{qV_1}{\hbar\omega} \sin \omega t + \alpha \tag{10.17}$$

and

$$j_z = j_\text{J} \sin\left\{\frac{q}{\hbar}V_0 t + \frac{qV_1}{\hbar\omega}\sin\omega t + \alpha\right\}. \tag{10.18}$$

The above equation may be expanded into a Fourier–Bessel series [381] giving*

$$j_z = j_\text{J} \sum_{m=-\infty}^{\infty} J_m(qV_1/\hbar\omega)\sin\left\{(m\omega + qV_0/\hbar)t + \alpha\right\} \tag{10.19}$$

where J_m is the mth order Bessel function. It may be seen that if $qV_0/\hbar\omega = n$ where n is an integer then j_z has a d.c. component

$$(j_z)_\text{dc} = (-1)^n j_\text{J} J_n(qV_1/\hbar\omega)\sin\alpha = (-1)^n j_\text{J} J_n(nV_1/V_0)\sin\alpha. \tag{10.20}$$

The above equation for the d.c. component bears strong similarity to the Josephson supercurrent given by Equation (10.1) in the sense that in both cases the currents are determined by quantum mechanical phase differences. As may

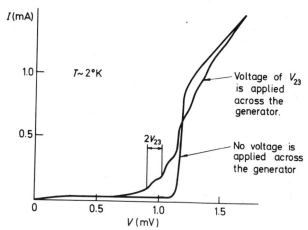

Fig. 10.3 The measured I–V characteristics of junction 1,2 in the absence and presence of voltage V_{23} applied across the junction 2,3. The structure due to the incident photons is clearly discernible. After Giaever [380].

be seen in Equation (10.19) α is the phase difference between two oscillators which have identical frequencies. One is the nth harmonic of the applied microwave signal and the other is the Josephson oscillation. If the phases of the two oscillators are locked a d.c. current appears whose magnitude varies between $\pm I_\text{J}|J_n(qV_1/\hbar\omega)|$. Thus the current is in the form of a series of spikes as shown in Fig. 10.4 for $qV_1/\hbar\omega = 3\cdot5$. In practice this structure has never been obtained.

*The same as obtained for frequency modulation.

The experimentally observed *I–V* characteristics look considerably different. The reason is that the d.c. current depends on the external circuit and there are some other contributions* too which have not all been properly identified.

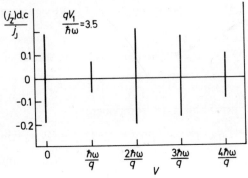

Fig. 10.4 Spikes in the d.c. *I–V* characteristic caused by an a.c. voltage, according to Equation (10.20).

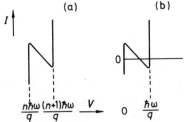

Fig. 10.5 Sections of measured *I–V* characteristics [322] in the presence of microwaves. (*a*) Two vertical steps are connected by a negative resistance region. (*b*) The initial section of an *I–V* characteristic. The current may flow in a direction opposite to the applied voltage.

The first observation of structure in the *I–V* characteristics at $V_0 = n\hbar\omega/q$ was reported by Shapiro [322]. The general appearance of the discontinuity in current was as shown in Fig. 10.5 (*a*). Two abrupt current steps were connected by a negative resistance region. The negative resistance following the zero-voltage current step was sometimes (depending on the microwave excitation) so low that the current at $V_0 = \hbar\omega/q$ was negative as shown in Fig. 10.5 (*b*). This is in accordance with the prediction of Equation (10.20) and it is energetically possible because power is absorbed from the microwave field.

The height of the current steps does follow a Bessel function dependence as it was first confirmed by Shapiro *et al.* [382]. We shall show here some later measurements by Grimes and Shapiro [327] which were carried out with the

---

*We did see in Section 5.2 that there can be considerable contribution to the d.c. current by photon-assisted tunnelling.

aid of a high impedance source (current generator) so the current was a monotonically increasing function of voltage as shown in Fig. 10.6. The height of the
current steps at a given value of V_0 is plotted in Fig. 10.7. Since the amplitude
of the microwave field is not known, only relative values can be compared. It is
done here by fitting the theoretical curve to the experimental data at the point
denoted by a double circle.

It is interesting to note that the microwaves needed to bring about the steps
may be produced by the junction itself. This will be discussed in more detail in
Section 11.8.

Before proceeding further it is worth pausing here for a moment to consider
the physical mechanism responsible for the electromagnetic output of the
junction. Since the energy of a photon is $\hbar\omega$ and the energy of a Cooper-pair at a
potential V_0 is qV_0, we need to evoke no more than a conversion of energy from

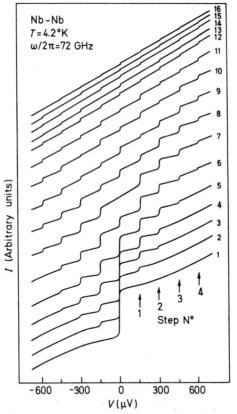

Fig. 10.6 I–V characteristics of a Nb–Nb point-contact junction taken by a high impedance
source. 1: no microwave power, 2–16: microwave power increasing gradually by 26 db.
Frequency: 72 GHz, $\hbar\omega/q = 149$ μV. First four steps at multiples of 149 μV are clearly
discernible. After Grimes and Shapiro [327].

one form to another in order to arrive at Equation (10.15). If, in addition, we assume that the energy of the Cooper-pair may be converted into m photons of equal energy, we get

$$\omega = \frac{qV_0}{\hbar m} \qquad (10.21)$$

that is, we may predict the presence of subharmonics which were observed experimentally [383–385] in weak links.

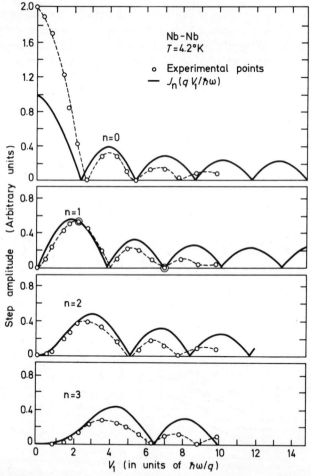

Fig. 10.7 Data from Fig. 10.6 plotted to display how the current in several constant-voltage steps varies as the applied rf voltage ~ (microwave power)$^{\frac{1}{2}}$ is varied. The data points from the nth step are compared with the amplitude of the nth order Bessel function. The data were fitted to the theoretical curves at the two points denoted by double circles. The rf voltage across the junction is expressed in units of $\hbar\omega/q$. After Grimes and Shapiro [327].

Usually, the radiation from the junction and the presence of current steps in the *I–V* characteristic are regarded as manifestations of the a.c. Josephson effect but we shall treat here under this heading one more time-varying case when an electromagnetic wave propagates along the junction in the absence of applied fields. Assuming that the electromagnetic wave represents a small disturbance of the reigning conditions we may put

$$\phi(x, y, t) = \phi_0(x, y) + \phi_1(x, y, t) \tag{10.22}$$

into Equation (9.30) yielding

$$\frac{\partial^2 \phi_1}{\partial x^2} + \frac{\partial^2 \phi_1}{\partial y^2} - \frac{1}{v^2} \frac{\partial^2 \phi_1}{\partial t^2} = (\cos \phi_0) \lambda_J^{-2} \phi_1 \tag{10.23}$$

Assuming further that $\phi_0(x, y)$ is a slowly varying function, the solution is

$$\phi_1 = \exp i(\omega t - \mathbf{k}\mathbf{r}) \tag{10.24}$$

leading to the dispersion equation

$$\omega^2 = k^2 v^2 + \omega_p^2 \tag{10.25}$$

where $k = |\mathbf{k}|$ and

$$\omega_p^2 = v^2 \lambda_J^{-2} \cos \phi_0. \tag{10.26}$$

Taking $\cos \phi_0 \approx 1$, $\lambda_J = 10^{-4}$ m and $v = c/25$ we get about 2 GHz for f_p.

Plotting Equation (10.25) in Fig. 10.8 it may be seen that there is no real k solution for $\omega < \omega_p$. This is analogous to the behaviour of a warm plasma or of a dielectric filled waveguide with a cut-off frequency ω_p. For sufficiently high frequencies the dispersion relationship reduces to

$$\omega = \pm kv. \tag{10.27}$$

In the static limit, $\omega = 0$, $k = i\lambda_J$ which is equivalent to the condition, stated before, that the fields are confined to the edge of the junction.

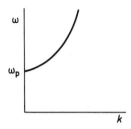

Fig. 10.8 The dispersion characteristic of a Josephson junction. Propagation is cut off at $\omega = \omega_p$.

At $k = 0$ there is no spatial variation, the magnetic field is zero, the electric field is perpendicular to the plane of the junction and there is a periodic exchange of energy between the electric field and the junction coupling energy. The mode thus strongly resembles to the oscillations of a cold plasma.* This effect (known as the Josephson plasma resonance) was experimentally observed by Dahm et al. [386, 387, 388]. They irradiated the junction by a signal and obtained significantly increased second harmonic output when the resonance condition $\omega = \omega_p$ was satisfied.

---

* This is consistent with the concept of weak superconductivity. The plasma frequency is reduced and, as shown by Equation (10.12), the penetration depth is increased.

11. Current–voltage characteristics

11.1 Introduction

The most easily measurable property of a junction is its I–V characteristic so it is hardly surprising that so many have been measured and so many types have been found. We have already discussed some aspects of I–V characteristics: we know that the junction is capable to carry a d.c. supercurrent up to a certain magnitude and that an impressed a.c. field can cause d.c. steps in the current. But the measured I–V characteristics displayed a much wider variety. Many measurements showed for example hysteresis; voltage and current became multivalued. Other measurements on weak links and on superconductor–normal metal–superconductor junctions showed steps at subharmonics of the impressed a.c. signal, and finally, to add insult to injury, steps appeared in the d.c. current without any impressed signal whatsoever.

It cannot be claimed that all the measured I–V characteristics have been accounted for by the theories developed, but it looks like (and this chapter purports to show it) that we can adequately approximate all the measured characteristics by resorting to relatively simple models.

In Sections 11.2 and 11.3 we shall investigate the circuits shown in Fig. 11.1. The essential difference from our previous attempts is that here we take into account the feeding arrangement. The first circuit corresponds to a tunnel junction fed by a constant current source and the second to a weak link (e.g. point contact geometry having a certain inductance) fed by a constant voltage source. The I–V characteristics (Figs. 11.3 and 11.4) obtained in the two cases are different but both of them display multivalued regions.

Section 11.4 briefly reviews a more complicated model where the proper current–voltage relations of normal electron tunnelling are also taken into account. The numerical results show very good agreement with measured characteristics.

Section 11.5 looks at the problem of finite source conductance which may be important for small tunnel junctions. Section 11.6 is concerned with the situation when both d.c. and a.c. currents are impressed upon the junction. The results show not only the steps at harmonics but at subharmonics as well. Section 11.7 treats the a.c. current source on its own and derives the a.c. impedance of the junction. It leads to the realisation that a Josephson junction is equivalent to a lumped element with a variable inductance, so may be used for parametric amplification.

In Section 11.8 we shall investigate a rather complicated situation when the a.c. current flowing across the junction interacts with the fields propagating in the junction. It will be shown that this interaction is responsible for the occurrence of steps in the absence of an a.c. source.

11.2 Capacitive loading; constant conductance

The circuit we are considering here (Fig. 11.1(a)) consists of an ideal current generator feeding current into the parallel combination of a Josephson junction,

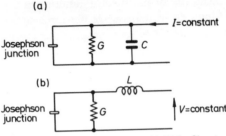

Fig. 11.1 Circuit representations of real junctions. (a) Thin film tunnel junction made up by an 'ideal' Josephson junction (satisfying the $I = I_J \sin \phi$ relationship), an ohmic conductance and a capacitance; analysed for a constant current input. (b) An approximation to a point-contact junction taking account of parallel conductance and series inductance; analysed for a constant voltage input.

an ordinary ohmic-conductance and a capacitance. We have then a supercurrent

$$I_J \sin \phi \qquad (11.1)$$

through the Josephson junction, an ohmic current

$$GV \qquad (11.2)$$

through the conductance and a capacitive current

$$C\frac{dV}{dt} \qquad (11.3)$$

through the capacitance. The sum of these three components should be equal to I, the current supplied by the generator. This is not really new. We have met all these components of current before; taking the current (more correctly the current density) in the form of Equation (9.34) and adding the displacement current, we get an equation expressing the same relationship

$$I = I_J \sin \phi + GV + C\frac{dV}{dt}. \qquad (11.4)$$

Expressing the voltage with the aid of the phase (Equation (9.17)) the above equation reduces to

$$I = \frac{\hbar C}{q} \frac{d^2\phi}{dt^2} + \frac{\hbar G}{q} \frac{d\phi}{dt} + I_J \sin \phi. \tag{11.5}$$

Assuming again that the junction is small enough so that ϕ is independent of the spatial coordinates, Equation (11.5) depends on time only and may be solved by numerical methods [335, 336]. Following McCumber [335] we introduce the dimensionless variables

$$\theta = \omega_m t = \frac{qI_J}{\hbar G} t \tag{11.6}$$

$$\beta_c = \frac{\omega_m C}{G} \tag{11.7}$$

and

$$\kappa = \frac{I}{I_J}, \qquad \eta(\theta) = \frac{GV}{I_J} = \frac{d\phi}{d\theta} \tag{11.8}$$

yielding the simpler form

$$\kappa = \beta_c \frac{d^2\phi}{d\theta^2} + \frac{d\phi}{d\theta} + \sin \phi. \tag{11.9}$$

We are interested in the d.c. current–voltage characteristic so we need to determine the time averaged voltage

$$\langle \eta(\theta) \rangle = \lim_{\theta \to \infty} \left[\frac{1}{\theta} \int_0^\theta \frac{d\phi}{d\theta} d\theta \right] = \lim_{\theta \to \infty} \frac{\phi(\theta) - \phi(0)}{\theta} \tag{11.10}$$

consistent with a given constant current.

For the special case $\beta_c = 0$, Equation (11.9) may be solved analytically [389]. For $|\kappa| \le 1$ the solution is $\phi = $ constant and consequently $\langle \eta \rangle = 0$. For $|\kappa| \ge 1$ Equation (11.9) may be brought to the form

$$\frac{d\phi}{\kappa - \sin \phi} = d\theta \tag{11.11}$$

from which $\phi(\theta)$ and $\eta(\theta)$ may be obtained (Appendix 6). The latter function takes the form

$$\eta(\theta) = \frac{\kappa^2 - 1}{\kappa + \sin[(\kappa^2 - 1)^{1/2}\theta + \gamma]} \tag{11.12}$$

where

$$\tan \gamma = (\kappa^2 - 1)^{-1/2}. \tag{11.13}$$

It may be seen that $\eta(\theta)$ is a periodic function with a period

$$2\pi(\kappa^2 - 1)^{-1/2}. \tag{11.14}$$

The average value of $\eta(\theta)$ (the normalised d.c. voltage) is also calculated in Appendix 6, coming to

$$\langle \eta(\theta) \rangle = (\kappa^2 - 1)^{1/2}. \tag{11.15}$$

When κ is just above unity most of the current still flows across the Josephson junction but ϕ is no longer a constant and η is no longer zero. For $\kappa = 1{\cdot}1$ the normalised voltage as a function of θ is plotted in Fig. 11.2. As may be seen the voltage waveform is rich in harmonics, the Josephson junction behaves as a harmonic generator.

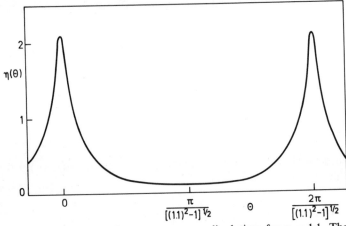

Fig. 11.2 The normalised voltage versus normalised time for $\kappa = 1{\cdot}1$. The function is periodic with a period $2\pi(\kappa^2 - 1)^{-\frac{1}{2}}$.

When $\kappa \gg 1$ the function $\eta(\theta)$ hardly varies, its average value may be taken as

$$\langle \eta(\theta) \rangle = \kappa \tag{11.16}$$

signifying that at high currents and high voltages we have effectively an ordinary conductance across a current generator satisfying Ohm's Law.

For $\beta_c \neq 0$ the solution

$$\phi = \text{constant}, \qquad \eta = 0 \tag{11.17}$$

is still valid provided $\kappa \leq 1$. Hence the region $0 < \kappa \leq 1$ is part of all the characteristics. For $\kappa > 1$ only numerical solutions are available [335] which are shown in Fig. 11.3. Note that in a certain range $\kappa_c \leq \kappa \leq 1$ (where κ_c is dependent on β_c) there are two solutions for a given current. This means that a different characteristic will be traced depending on whether the current is increasing or decreasing; there is a hysteresis region. The model investigated here is apparently the simplest one which will display a hysteresis region.

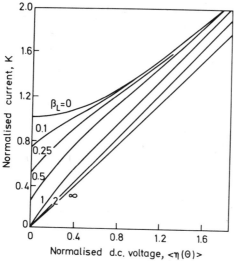

Fig. 11.3 Normalised *I–V* characteristics of a junction represented by the circuit of Fig. 11.1 (*a*). After McCumber [335].

Let us now return to the actual (not normalised) values of voltage and current for the $\beta_c = 0$ case, as it will throw some further light upon the inter-relations between the a.c. and d.c. quantities. Equation (11.12) modifies then to

$$V(t) = \frac{1}{G}\frac{I^2 - I_J^2}{I + I_J \sin(\omega_0 t + \gamma)}$$

(11.18)

where $\omega_0 = qV_0/\hbar$ has its usual meaning with V_0 being the d.c. voltage across the Josephson junction. The conclusion is that the voltage waveform $V(t)$ is periodic with a frequency ω_0 and has an average value, V_0, which is just $\hbar/q$ times the fundamental frequency. A Fourier analysis of Equation (11.18) would yield all the higher harmonics.

11.3 Inductive loading; constant conductance

Consider now the circuit of Fig. 11.1 (*b*) fed from a constant voltage source. The relevant equations for the voltage and current are

$$V = L\frac{dI}{dt} + \frac{\hbar}{q}\frac{d\phi}{dt}$$

(11.19)

$$I = \frac{\hbar G}{q}\frac{d\phi}{dt} + I_J \sin\phi.$$

(11.20)

Introducing again the dimensionless quantities defined by Equations (11.6) and (11.8) we obtain

$$\eta = \frac{1}{\beta_L}\frac{d\kappa}{d\theta}+\frac{d\phi}{d\theta} \tag{11.21}$$

and

$$\kappa = \frac{d\phi}{d\theta}+\sin\phi \tag{11.22}$$

where

$$\beta_L = \frac{1}{\omega_m LG} = \frac{\hbar}{qLI_J} \tag{11.23}$$

Now the voltage across the terminals is constant by definition and we need to determine the time averaged current

$$\langle\kappa\rangle = \lim_{\theta\to\infty}\left[\frac{1}{\theta}\int_0^\theta \kappa(\theta_1)\,d\theta_1\right]. \tag{11.24}$$

Introducing

$$\kappa_a = \kappa - \langle\kappa\rangle, \tag{11.25}$$

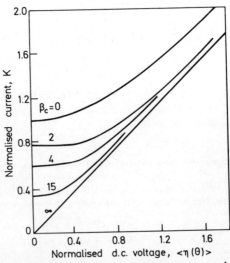

Fig. 11.4 Normalised I–V characteristics of a junction represented by the circuit of Fig. 11.1(*b*). After McCumber [335].

substituting it into Equation (11.21) and integrating we get

$$\kappa_a = \beta_L(\phi_c + \eta\theta - \phi) \tag{11.26}$$

where ϕ_c is an integration constant. Substituting now Equations (11.25) and (11.26) into Equation (11.22) we obtain

$$\langle\kappa\rangle = \frac{d\phi}{d\theta} + \sin\phi + \beta_L(\phi - \phi_c - \eta\theta). \tag{11.27}$$

For the special case $\beta_L = 0$ the above equation is identical with Equation (11.9) for $\beta_c = 0$, hence the analytical solution given by Equation (11.15) applies. The numerical solution for $\beta_L \neq 0$ is far from being straightforward [335] but eventually it leads to the curves of Fig. 11.4. There are again regions where the current is multivalued; the main difference from the previous case is in the slope $d\eta/d\langle\kappa\rangle$ at $\eta = 0$.

There are as yet no quantitative measurements available for comparison but the qualitative behaviour was confirmed [390]. It may be noted that Pankove's voltage–current characteristics for pressure contacts [391] look also similar to the curves of Fig. 11.4.

11.4 Capacitive loading; variable conductance

Let us consider again the circuit of Fig. 11.1 (a) but instead of taking a constant conductance let us acknowledge the fact that normal tunnelling may take place as well and take the conductance voltage dependent so as to satisfy the I–V relationship of normal electron tunnelling. The sum of the three currents may then be written as

$$I = I_J \sin\phi + I_q + C\frac{dV}{dt} \tag{11.28}$$

where

$$I_q = G(V)V \tag{11.29}$$

is the current produced by the tunnelling of normal electrons, and the differential relationship between phase and voltage is of course still valid.

The numerical solution of the above differential equation was obtained by Scott [337]. He measured first the normal electron tunnelling curve and compared it with theory. The agreement was excellent as shown in Figs. 11.5 (a) and (c) for two different temperatures. Having determined the $I_q(V)$ curve and got an experimental value for I_J he could then obtain numerical solutions of Equation (11.28) which are plotted in Figs. 11.5 (b) and (d). The experimentally found hysteresis curve is very well reproduced by the theory though the experimental switch-back is at a somewhat higher current.

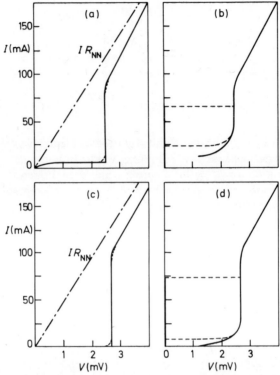

Fig. 11.5 Experimental (---) and theoretical (—) *I–V* characteristics of a Pb–I–Pb tunnel junction. (*a*) $T = 4\cdot2°K$; normal electron tunnelling (supercurrent suppressed), (*b*) $T = 4\cdot2°K$; Josephson tunnelling, (*c*) $T = 1\cdot39°K$; normal electron tunnelling (supercurrent suppressed), (*d*) $T = 1\cdot39°K$; Josephson tunnelling. After Scott [337].

11.5 Effect of the source conductance; small tunnel junctions

In this section we shall briefly consider the effect of a finite source conductance. The circuit to be analysed is shown in Fig. 11.6, that is we add a source conductance G_s to the circuit of Fig. 11.1 (*a*). The analysis of this circuit has been done in Section 11.2. Denoting the *I–V* characteristic obtained there by

$$\frac{I}{I_J} = F(\langle V(t)\rangle, G, C, I_J) \tag{11.30}$$

we get the I_s–V characteristic of the circuit of Fig. 11.6 by replacing I by I_s and G by $G+G_s$, that is

$$\frac{I_s}{I_J} = F(\langle V(t)\rangle, G+G_s, C, I_J). \tag{11.31}$$

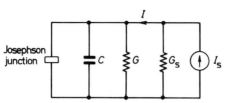

Fig. 11.6 Circuit representation of a real junction fed by a real current generator (G_s is finite).

Our aim is, however, to relate the voltage across the junction to the current I which flows into the junction in Fig. 11.6. Thus

$$\frac{I}{I_J} = \frac{I_s}{I_J} - \frac{G_s\langle V(t)\rangle}{I_J} = F(\langle V(t)\rangle, G+G_s, C, I_J) - \frac{G_s\langle V(t)\rangle}{I_J}. \tag{11.32}$$

The above equation is quite general. For any value of the source conductance G_s the d.c. I–V curve may be constructed from the curves of Fig. 11.3. We shall investigate here only the case corresponding to $\beta_c = 0$ when the solution of Equation (11.15) applies and

$$\frac{I_s}{I_J} = F(\langle V(t)\rangle, G+G_s, 0, I_J) = \left\{\left[\frac{(G+G_s)\langle V(t)\rangle}{I_J}\right]^2 + 1\right\}^{1/2}. \tag{11.33}$$

When $G_s \ll G$ we get back the previous solution but when the junction conductance becomes smaller than the source conductance the characteristic takes a different shape. This is shown in Fig. 11.7 for $G_s = G$ and $G_s = 10G$ where we retained the normalised variables of Equation (11.8).

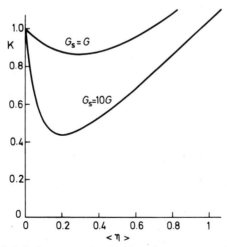

Fig. 11.7 Normalised I–V characteristics of a junction represented by the circuit of Fig. 11.6 for $\beta = 0$; $G_s = G$ and $G_s = 10$ G.

The conductance G may indeed become small for small tunnel junctions. Buckner et al. [392, 393] made junctions of less than 10^{-4} mm^2 area with normal resistances up to 100 ohm. The experimentally obtained $I-V$ characteristics did show the expected negative resistance region.

The problem of small tunnel junctions was first attacked by Ivanchenko and Zilberman [394, 395] who besides deriving the basic $I-V$ characteristics did also investigate the effect of noise; how thermal fluctuations can destroy the d.c. supercurrent. We shall briefly return to this problem in Section 20.5.

11.6 Impressed d.c. and a.c. current

Here we shall investigate the case treated by Clarke et al. [396] and Waldram et al. [397] when the circuit of Fig. 11.8 is fed by a constant current generator

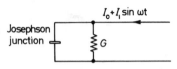

Fig. 11.8 Circuit of a junction consisting of an 'ideal' Josephson junction in parallel with a conductance G. The input currents (both d.c. and a.c.) are assumed constant produced by constant current generators.

producing both d.c. and a.c. current

$$I \doteq I_0 + I_1 \sin \omega t. \tag{11.34}$$

The differential equation may then be obtained, by substituting Equation (11.34) into (11.5) and neglecting C, in the form

$$\frac{\hbar G}{q} \frac{d\phi}{dt} = I_0 + I_1 \sin \omega t - I_J \sin \phi. \tag{11.35}$$

Generally the d.c. voltage, $(\hbar/q)\langle d\phi/dt \rangle$, is approximately equal to I_0/G for any trajectory ($\phi - \omega t$ relationship) because the second and third terms on the right-hand side of Equation (11.35) average to zero. However, when the slope $d\phi/d(\omega t)$ is close to unity the trajectory has a tendency to 'lock on' to the $\phi - \omega t$ 'lattice'. This lattice is shown in Fig. 11.9 where the

$$I_1 \sin \omega t - I_J \sin \phi = \text{constant} \tag{11.36}$$

curves are plotted for several values of the constant for $I_1 = I_J$. Trajectory AA (corresponding to $\phi = \omega t$) is a solution of the differential equation (11.35). The constant of Equation (11.36) is zero and we get for the d.c. voltage

$$V_0 = \frac{\hbar}{q} \left\langle \frac{d\phi}{dt} \right\rangle = \frac{\hbar \omega}{q} = \frac{I_0}{G}. \tag{11.37}$$

Let us look now at another solution of the differential equation given by the trajectory CC which passes close to the regions of maximum (denoted by $+$) and minimum (denoted by $-$) slope. As may be seen from Fig. 11.9 the trajectory

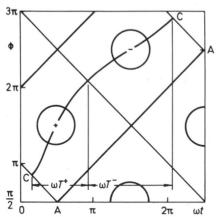

Fig. 11.9 A plot of Equation (11.36) for $I_1 = I_J$ and several values of the constant on the $\phi - \omega t$ plane. The $+$ and $-$ signs indicate the maximum and minimum values of the constant. Both trajectories AA and CC represent solutions of the differential equation (11.35). After Waldram *et al.* [397].

spends longer time in the negative than in the positive slope region ($\omega T^- > \omega T^+$), and therefore the time average of $I_1 \sin \omega t - I_J \sin \phi$ is negative. It follows that I_0 must be greater than $G\omega\hbar/q$ for this trajectory, to compensate. So both trajectories have the same long term slope (i.e. d.c. voltage) but different d.c. currents – and this means a vertical current step. Note that this explanation of the discontinuity in current predicts a step as observed in practice and *not* the symmetric spikes of Fig. 10.4. So we are getting nearer to the experimental results. In fact, the present model is capable to explain the experimentally observed *subharmonic* steps as well which occur at a voltage $V_0 = (n/m)(\hbar/q)\omega$ (where n and m are integers). Exactly the same considerations apply as discussed above; the trajectory locks on to the lattice when the long term slope is close to a rational number n/m though of course for large n and m the limiting trajectories run very close to each other and the corresponding steps in the characteristics are small. It is also proven from the trajectories that the d.c. voltage is a monotonically increasing function of d.c. current so there are no negative resistance regions.

11.7 Impressed a.c. current; parametric inductance

In contrast to the previous sections we shall not be concerned here with d.c. characteristics at all. We shall assume that the junction is fed by an ideal a.c.

generator and shall investigate the simplest possible case when both the capacitance and the resistance of the junction may be disregarded.

The capacitance of a tunnel junction plays an important role at high frequencies so by neglecting it we restrict the analysis to frequencies well below the Josephson plasma frequency. In fact, the Josephson plasma resonance may be looked upon as an LC resonance which would immediately give us for the effective inductance of the junction

$$L = \frac{1}{\omega_{pJ}^2 C} = \frac{\hbar}{qI_J}.$$

(11.38)

This is true for low excitation. For higher excitation the differential equation to be solved may be obtained from Equation (11.35) assuming $I_0 = 0$ and $G = 0$ leading to

$$I_J \sin\left(\frac{q}{\hbar}\int V\,dt\right) = I_1 \sin \omega t,$$

(11.39)

or solving for the voltage

$$V = \frac{\hbar}{q}\left[1 - \left(\frac{I_1}{I_J}\sin \omega t\right)^2\right]^{-1/2}\frac{1}{I_J}\frac{d}{dt}(I_1 \sin \omega t).$$

(11.40)

Hence we may define an inductance [398]

$$L = \frac{\hbar}{qI_J}\left[1 - \left(\frac{I_1}{I_J}\sin \omega t\right)^2\right]^{-1/2}$$

(11.41)

which for $I_1 = 0$ reduces to Equation (11.38) but otherwise represents a time varying inductance suitable for parametric applications.

Let us now further restrict generality by assuming that the junction is part of a circuit resonant at frequency ω so that all voltage components with the exception of the fundamental are suppressed. We may get the fundamental component of voltage from Equation (11.40) by the usual formula for the coefficients of a Fourier series

$$
\begin{aligned}
V_1 &= \frac{1}{\pi}\int_{-\pi}^{\pi} V \cos \omega t\, d(\omega t) \\
&= \frac{1}{\pi}\frac{\omega\hbar I_1}{qI_J}\int_{-\pi}^{\pi}\left[1 - \left(\frac{I_1}{I_J}\sin \omega t\right)^2\right]^{-1/2}\cos^2 \omega t\, d(\omega t) \\
&= \frac{4\hbar\omega k}{q\pi}\int_{0}^{\pi/2}\cos^2\phi\,(1 - k^2\sin^2\phi)^{-1/2}\,d\phi \\
&= \frac{4\hbar\omega k}{q\pi}B(k)
\end{aligned}
$$

(11.42)

where $k = I_1/I_J$ and $B(k)$ is one of the standard complete elliptic integrals [101]. Since $B(k)$ is a slowly varying function, the I_1–V_1 characteristic is not far from linear as shown in Fig. 11.10.

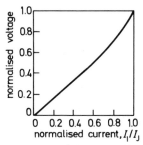

Fig. 11.10 The normalised voltage (proportional to the fundamental component) as a function of the normalised fundamental current I_1/I_J. After Silver et al. [398].

Considering now only the fundamental components of current and voltage we may define an effective inductance by the relationship

$$\omega L_1 = \frac{V_1}{I_1} \tag{11.43}$$

leading to

$$L_1 = \frac{4\hbar}{\pi q I_J} B(k). \tag{11.44}$$

It may be seen that the inductance depends little on I_1 and more on I_J. The variation of this inductance with I_1 and I_J was investigated experimentally by Silver et al. [398] by measuring the resonant frequency of the circuit shown in Fig. 11.11. I_1 was varied by varying the output of a signal generator (coupled lightly to the circuit) and I_J by applying a magnetic field to the junction (in view of Equation (10.7) the maximum supercurrent is strongly dependent on the flux threading the junction). The results are shown in Fig. 11.12. The abscissa

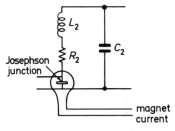

Fig. 11.11 A resonant circuit containing a Josephson junction. Resonant frequency is varied by changing I_J (and hence the inductance) with the aid of an applied magnetic field.

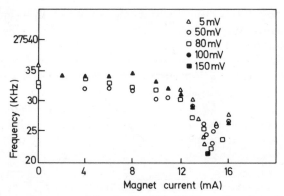

Fig. 11.12 Resonant frequency of the circuit of Fig. 11.11 as a function of magnetic current. The output voltage across the circuit is a measure of the impressed rf current I_1. After Silver *et al.* [398].

is the magnet current producing the magnetic field at the junction. In the measurements I_J is always larger than I_1 otherwise the junction becomes resistive spoiling the Q of the circuit. It may be seen that the resonant frequency is much more sensitive to changes in I_J than in I_1 in agreement with the prediction of Equation (11.44). For a quantitative comparison we may work out the maximum and minimum values of the inductance from Equation (11.44) for one series of experiments, reported by Silver *et al.* [398]. I_1 is kept constant at 10 μA and I_J is varied between 89 μA (at zero magnetic field) and 10 μA (at about 14 mA magnet current). The maximum of L_1 occurs at $I_1 = I_J$, i.e. $k = 1$, $B(k) = 1$ giving $\omega L_{1max} = 6\cdot 8 \times 10^{-3}$ Ω at a frequency 26·7 MHz. The minimum of ωL_1 is about nine times less, giving an inductance variation of about 6×10^{-3} Ω.

The measured fractional change in the resonant frequency is about 5×10^{-4} from Fig. 11.12 corresponding to a variation of 10^{-3} in the total circuit inductance $[\omega(L_1 + L_2)]$. Since ωL_2 was chosen 10 Ω in the experiment the measured variation in inductance is 10^{-2} Ω in contrast to the theoretical value 6×10^{-3} Ω; a satisfactory agreement.

It is doubtful that the Josephson junction will be used as a straightforward parametric amplifier but it is a property which may come useful in other types of application as for example in the magnetometer circuit of Zimmerman and Silver [339].

See Section 21.7 for a more detailed discussion of parametric amplification.

11.8 Coupling of fields and current in the junction

In deriving the various $I–V$ characteristics we have so far concentrated on the time variation of our variables and completely ignored the spatial variation. We do know, however, from Chapter 10 that in the presence of a magnetic field the phase difference across the junction is space dependent (Equation

(10.2)) and we have also derived solutions where a wave (Equation (10.24)) is travelling in the junction region.

In this section we shall derive the d.c. $I-V$ characteristic of a tunnel junction when the spatially varying a.c. current couples to the spatially varying a.c. fields. This coupling may be significant in two cases: (i) when the current travels with the same velocity as the fields and (ii) when there is a build-up of the fields in the junction due to reflection at the edges. In the first case we have a travelling wave interaction a little similar to that in a microwave travelling wave tube. In the second case a finite voltage may appear because that voltage produces a wave of just that frequency and wave number which can resonate in the junction and produce thereby a finite d.c. current.

The complete mathematical solution is very complicated so one must resort to approximations but the above physical picture will prove useful in choosing the type of approximations. Thus in the first step we shall derive an approximate formula for a travelling wave current. We may arrive there by first combining the solutions obtained in Equations (10.2) and (10.13). Retaining only the x component of the magnetic field we get

$$\phi_0 = \frac{q}{\hbar} V_0 t - \frac{qd}{\hbar} B_x y + \alpha. \tag{11.45}$$

The above equation does satisfy Equations (9.15) to (9.17) but it is only an approximate solution as already noted in the footnote on p. 154. We get then for the current

$$j_z = j_J \sin(\omega_0 t - k_0 y + \alpha) \tag{11.46}$$

where

$$\omega_0 = qV_0/\hbar \quad \text{and} \quad k_0 = B_x qd/\hbar \tag{11.47}$$

It may be seen from Equation (11.46) that the current is given by a wave travelling in the y direction with a phase velocity

$$v_\phi = \frac{\omega_0}{k_0} = \frac{V_0}{B_x d}. \tag{11.48}$$

We shall now regard this value of the current as the driving term in Equation (9.35). Hence the equation to solve is

$$\left[\frac{\partial^2}{\partial y^2} - \frac{1}{v^2} \frac{\partial^2}{\partial t^2} - \frac{\beta}{v^2} \frac{\partial}{\partial t} \right] \phi_1 = \lambda_J^{-2} \sin(\omega_0 t - k_0 y + \alpha). \tag{11.49}$$

It seems worthwhile to go through the steps taken so far. The expression given for ϕ_0 in Equation (11.45) represents a zero order approximation. Substituting that into the right-hand side of Equation (9.35) we obtain a first order solution, that is why we put ϕ_1 into the left-hand side of Equation (11.49).

However, before proceeding with the solution it is preferable to change the variable from the first order phase difference, ϕ_1, to the first order voltage, V_1, since that is a more familiar variable in propagation problems. We therefore differentiate Equation (11.49) with respect to time, and with the aid of Equation (9.17) we get

$$\left[\frac{\partial^2}{\partial y^2}-\frac{1}{v^2}\frac{\partial^2}{\partial t^2}-\frac{\beta}{v^2}\frac{\partial}{\partial t}\right]V_1 = \lambda_J^{-2}V_0\cos(\omega_0 t-k_0 y+\alpha). \qquad (11.50)$$

The simplest way to solve the above partial differential equation is to replace the cosine term by an exponential one, and to assume for the solution

$$V_1 = g\exp i(\omega_0 t-k_0 y+\alpha), \qquad (11.51)$$

the real part of which must be taken at the end. Substituting Equation (11.51) into (11.50) we get

$$g\left(-k_0^2+\frac{\omega_0^2}{v^2}-\frac{\beta i\omega_0}{v^2}\right) = \frac{V_0}{\lambda_J^2} \qquad (11.52)$$

whence

$$|g| = \frac{V_0}{\lambda_J^2}\frac{v^2}{\omega_0^2}\left\{\left[1-\frac{k_0^2 v^2}{\omega_0^2}\right]^2+\frac{\beta^2}{\omega_0^2}\right\}^{-1/2} \qquad (11.53)$$

and

$$\arg g = \theta = \tan^{-1}\frac{\beta/\omega_0}{1-(k_0 v/\omega_0)^2}. \qquad (11.54)$$

Thus, the solution* is

$$V_1 = |g|\cos(\omega_0 t-k_0 y+\alpha+\theta) \qquad (11.55)$$

or

$$\phi_1 = -\frac{|g|}{V_0}\sin(\omega_0 t-k_0 y+\alpha+\theta). \qquad (11.56)$$

Hence up to the first order the current may be written as

$$j_z = j_J\sin(\phi_0+\phi_1)$$

$$= j_J\sin\left[\omega_0 t-k_0 y-\frac{|g|}{V_0}\sin(\omega_0 t-k_0 y+\theta+\alpha)+\alpha\right]. \qquad (11.57)$$

* These equations were first derived by Eck, Scalapino and Taylor [399]; to get their expressions replace β/w_0 by $1/Q$.

If $|g|/V_0$ is small we do not need to expand Equation (11.57) into a Fourier–Bessel series; by simple trigonometrical operations we get for the d.c. component of current

$$j_{z_0} = j_J \frac{|g|}{2V_0} \sin \theta$$

$$= j_J \frac{v^2}{2\lambda_J^2 \omega_0^2} \frac{\beta/\omega_0}{[1-(k_0 v/\omega_0)^2]^2 + (\beta/\omega_0)^2}$$

$$= \frac{1}{2} \frac{wh}{q\varepsilon} \frac{j_J^2}{V_0^2} \frac{\beta/\omega_0}{[1-(k_0 v/\omega_0)^2]^2 + (\beta/\omega_0)^2} \qquad (11.58)$$

Thus there is always a d.c. component present due to the coupling of the current to the fields. However, this component may be expected to be very small unless the velocities are synchronous, that is

$$v = \frac{\omega_0}{k_0} = v_\phi \qquad (11.59)$$

in which the d.c. current is a maximum. Hence for a constant magnetic field the d.c. current will peak at the value of V_0, satisfying Equation (11.59).

Experimental results indeed show peaks in the $I-V$ characteristics as may be seen in Fig. 11.13. At higher magnetic fields the peak occurs at a higher value of the voltage to keep v_ϕ the same. The required linear relationship between the voltage and the magnetic field is confirmed by the experimental results shown in Fig. 11.14.

It is more difficult to prove the equality of the velocities because neither d, nor w, nor ε are known accurately under the experimental conditions. Taking $\varepsilon = 10$, $w = 1.5$ nm and $d = 60$ nm we get $v = 1.5 \times 10^7$ m/s and $v_\phi = 2.5 \times 10^7$ m/s. Doubling d to 120 nm would give somewhat closer results, namely $v = 1.06 \times 10^7$ m/s and $v_\phi = 1.25 \times 10^7$ m/s. So we can safely conclude that the velocities are of the right order of magnitude.

As the voltage increases beyond the value $\Delta(T)/e$ the observed current peak broadens, indicating the appearance of a new type of loss. Since in this case the photon energy is larger than the gap energy, it is very probable that the breaking up of electron pairs is the loss mechanism responsible.

It is interesting to note that in this interaction the geometrical dimensions of the junction play no role. This is only possible if there are heavy losses in the junction so that the reflected waves are negligible.

When the waves reflected from the open boundaries of the junction are not negligible, one may expect resonances to occur at specific wavelengths. Since the voltage is maximum at an open circuit, the allowed modes vary as

$$\cos \frac{n\pi}{a_y} y \qquad (11.60)$$

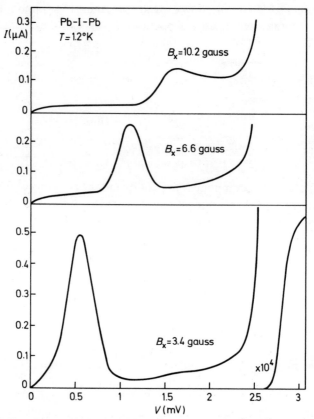

Fig. 11.13 *I–V* characteristics of a Pb–I–Pb junction showing the resonant peak and its dependence on magnetic field. After Langenberg *et al.* [379].

where n is an integer and a_y is the length of the junction in the y direction.

We need to solve Equation (11.50) again but now the solution will be attempted in a series form where all the allowed modes are represented. Hence the trial function is

$$V_1 = \sum g_n \exp i[\omega_0 t - (n\pi/a_y)y + \alpha]. \tag{11.61}$$

Substituting Equation (11.61) into (11.50) and following the same (but a lot more laborious) procedure* as before we get for the spatial average of the d.c. current density

---

*There is actually one more operation to perform than before. In contrast to Equation (11.58) the resulting d.c. current density is dependent on y in this case, so the relevant quantity is the spatial average of the d.c. current density, necessitating a further integration.

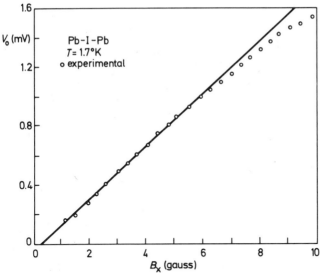

Fig. 11.14 Plot of the position of the resonant peak as a function of applied magnetic field taken from I–V characteristics of the junction shown in Fig. 11.13. After Langenberg et al. [379].

$$j_{z_0} = \frac{hw}{q\varepsilon} \frac{j_1^2}{V_0^2} \sum_{n=0}^{\infty} \frac{\beta/\omega_0}{[1-(n\pi v/\omega_0 a_y)^2]^2 + (\beta/\omega_0)^2} \times$$

$$\left[\frac{\sin(k_0 a_y - n\pi)/2}{(k_0 a_y - n\pi)/2} \right]^2 \frac{1}{(1+n\pi/k_0 a_y)^2} \times \begin{array}{ll} \frac{1}{2} & n = 0 \\ 1 & n \neq 0 \end{array} \quad (11.62)$$

It may be seen from the above equation that there is a peak in d.c. current whenever the resonance condition

$$\frac{\omega_0}{v} = \frac{n\pi}{a_y} \quad (11.63)$$

is satisfied. Hence the peaks will appear at voltages

$$V_0 = \frac{n\pi\hbar}{q a_y} \quad (11.64)$$

The distance between peaks does not depend on the magnetic field, only on junction geometry. The magnitude of the d.c. current, however, does strongly depend on the magnetic field. The contribution of the nth term may be maximised by choosing the magnetic field so that

$$k_0 a_y = n\pi. \quad (11.65)$$

The effect of junction geometry on d.c. current was first observed by Fiske [323] and explained quantitatively by Eck et al. [399] and Kulik [401]. Very neat experimental results were obtained by Dmitrenko *et al.* [402] and Langenberg *et al.* [379] as shown in Figs. 11.15 and 11.16. Since in both cases the junctions were fed by constant current generators there are current jumps instead of current peaks in the *I–V* characteristics. The jumps are of the same

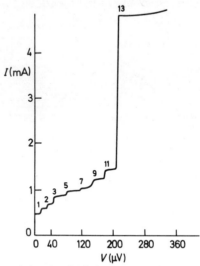

Fig. 11.13 I–V characteristic of a Sn–I–Sn junction in the presence of a magnetic field showing self-induced steps. For $n = 13$ the equality of phase velocities is satisfied. After Dmitrenko *et al.* [402].

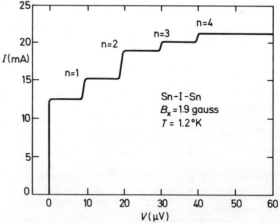

Fig. 11.16 Typical *I–V* characteristic of a Sn–I–Sn junction in the presence of a magnetic field showing self-induced steps. The voltage separation of the modes corresponds to a frequency separation of approximately 4·6 GHz. By adjusting the magnitude of B_x a greater portion of each mode could be observed as well as several other higher modes. After Langenberg *et al.* [379].

distance of each other (some may be missing) and they are of unequal height.

Working out the voltages at which the steps appear from Equation (11.64) does give good agreement with the experimental results. As far as the height of the steps is concerned it is interesting to see in Fig. 11.15 that the height of the 13th step is much greater than any of the others. According to our theory this should occur for the step number, n, for which

$$n = \frac{k_0 a_y}{\pi} = \frac{B_x q d a_y}{\hbar \pi}. \tag{11.66}$$

The magnetic field in the experiment reported was $B_x = 1 \cdot 12$ gauss and the length of the junction $a_y = 1 \cdot 25$ mm. If we take $d = 100$ nm, Equation (11.62) gives indeed $n = 13$ (which is suggestive though, of course, it may be a coincidence).

It must be noted that non-uniformities in the junction will affect the resonance more and more as the resonant wavelength decreases corresponding to higher and higher values of n. According to Eck et al. [400] the individual modes can no longer be distinguished above a certain value of n, and there is a transition then to the behaviour depicted in Fig. 11.13.

We have done here the calculations for a thin film geometry which is more susceptible to analysis but it should be emphasised that self-excited current steps have been observed in other geometries too, namely by Shapiro [385] in solder drop junctions and by Dayem and Grimes [403] in point contact junctions coupled to a resonant cavity. We also mention here the measurements of Matisoo [404] on thin film junctions in which current steps appeared in the absence of a magnetic field. Since no travelling waves can be excited without a magnetic field the probable explanation for the steps in the presence of barrier inhomogeneities which serve as local scattering centres and thus excite resonances.

12. Vortex solutions

12.1 Vortex structure

Certain aspects of the relationship between current and magnetic field have already been discussed in Section 10.1. We obtained a formula (Equation (10.5)) for the critical current I_J as a function of applied magnetic field, correct when the magnetic field created by the current is negligible. We did also get a solution with the self-magnetic fields included (Equation (10.11)) in which ϕ (and consequently the magnetic field) declined exponentially in the interior of the junction. But this was again only an approximate solution valid for small values of ϕ.

In the general case we need to solve the differential equation (Equation (10.9))

$$\frac{d^2\phi}{dy^2} = \lambda_J^{-2} \sin \phi. \tag{12.1}$$

Assuming a semi-infinite junction (from $y = 0$ to ∞) a solution which satisfies the boundary condition, $\phi(\infty) = 0$ was given by Ferrel and Prange [330] in the form

$$\phi = 2 \sin^{-1} \operatorname{sech} \frac{y}{\lambda_J} \tag{12.2}$$

yielding for the magnetic field (Equation (9.16))

$$B_x = -\frac{\hbar}{qd} \frac{d\phi}{dy}$$

$$= \frac{2\hbar}{qd\lambda_J} \operatorname{sech} \frac{y}{\lambda_J} \tag{12.3}$$

and for the current density

$$j = 2j_J \tanh \frac{y}{\lambda_J} \operatorname{sech} \frac{y}{\lambda_J}. \tag{12.4}$$

Plotting Equations (12.3) and (12.4) in Fig. 12.1 as a function of y/λ_J it may be seen that both the magnetic field and the current density are confined to the edge of the junction. It may be noted that the solution is not identical with that obtained for bulk superconductors. There the current density is proportional to the gradient of the phase of the wavefunction and consequently it has its maximum at the surface. In the case of a Josephson junction the current density is proportional to the sine of the phase difference ϕ so the peak occurs where ϕ has dropped to $\pi/2$.

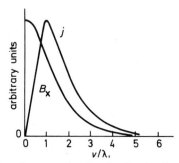

Fig. 12.1 Magnetic field and current density distribution for a semi-infinite junction.

Equation (12.2) is only a particular solution. The general solution was obtained by Owen and Scalapino [333] in the form of elliptic functions.

Multiplying both sides of Equation (12.1) by $d\phi/dy$ and integrating, we obtain

$$\frac{1}{2}\left(\frac{d\phi}{dy}\right)^2 = \lambda_J^{-2}(C - \cos\phi) \tag{12.5}$$

where C is an integration constant. Introducing

$$C = \frac{2 - k^2}{k^2} \tag{12.6}$$

Equation (12.5) reduces to

$$\frac{d\phi}{dy} = \frac{2}{k\lambda_J}\left(1 - k^2\cos^2\frac{\phi}{2}\right)^{1/2}. \tag{12.7}$$

Noting that the elliptic integral of the first kind has the form [405]

$$u = \int_0^\phi \frac{d\theta}{\sqrt{(1 - k^2\sin^2\theta)}}, \tag{12.8}$$

the inverse relation is

$$\phi = \text{am } u \tag{12.9}$$

and the Jacobian elliptic functions are defined as

$$\text{sn } u = \sin \phi$$

$$\text{cn } u = \cos \phi \qquad (12.10)$$

$$\text{dn } u = (1 - k^2 \sin^2 \phi)^{1/2}$$

we may solve Equation (12.7) and get for* $k \leq 1$

$$\sin \tfrac{1}{2}\phi = \text{cn}\left(\frac{y - y_0}{k\lambda_J}\right) \qquad (12.11)$$

$$\cos \tfrac{1}{2}\phi = -\text{sn}\left(\frac{y - y_0}{k\lambda_J}\right) \qquad (12.12)$$

which gives immediately for the current density

$$j = j_J \sin \phi$$

$$= -2j_J \, \text{sn}\left(\frac{y - y_0}{k\lambda_J}\right) \text{cn}\left(\frac{y - y_0}{k\lambda_J}\right) \qquad (12.13)$$

The physical model is shown in Fig. 12.2. Two semi-infinite superconductors

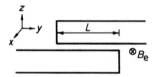

Fig. 12.2 The assumed junction geometry. Two semi-infinite superconductors overlap for a length L.

overlap for a length L and the magnetic field is applied in the $-x$ direction. We may then assume that the current flows in the z direction in the range $0 < y < L$, and it is zero otherwise. Since nothing changes in the x direction we shall use the current

$$I = \int j(y) \, dy \qquad (12.14)$$

flowing per unit x length.

The boundary conditions may be obtained by integrating Equation (9.29)

$$B_x(0) - B_x(L) = \mu_0 I$$

$$= \frac{\hbar}{qd}\left[\frac{d\phi}{dy}\bigg|_L - \frac{d\phi}{dy}\bigg|_0\right] \qquad (12.15)$$

* Similar expressions may be derived [333] for $k > 1$.

and by equating the tangential components of the magnetic field at the edges of the junction

$$B_x(L) + B_x(0) = -2B_e$$

$$= -\frac{\hbar}{qd}\left[\frac{d\phi}{dy}\Big|_L + \frac{d\phi}{dy}\Big|_0\right] \tag{12.16}$$

where B_e is the externally applied magnetic field.

We still need to express $d\phi/dy$ in terms of the tabulated [405] functions of Equation (12.10). Differentiating Equation (12.12) with respect to y we get

$$-\frac{1}{2}\sin\frac{\phi}{2}\frac{d\phi}{dy} = -\frac{d}{dy}\operatorname{sn}\left(\frac{y-y_0}{k\lambda_J}\right). \tag{12.17}$$

Hence

$$\frac{d\phi}{dy} = 2\frac{(d/dy)\operatorname{sn}[(y-y_0)/k\lambda_J]}{\operatorname{cn}[(y-y_0)/k\lambda_J]} \tag{12.18}$$

which, using a further relationship between the Jacobian elliptic functions [405], reduces to

$$\frac{d\phi}{dy} = \frac{2}{ky_J}\operatorname{dn}\left(\frac{y-y_0}{k\lambda_J}\right). \tag{12.19}$$

Fixing B_e and I we now have two equations for determining the two constants k and y_0. In the method used by Owen and Scalapino [333] k and y_0 were related to each other for given B_e and L by Equation (12.16). Substituting them into Equation (12.15) the possible values of I were obtained of which the maximum (or local maxima if there was more than one maximum) was noted.

The results for $B_e = 0$ are shown in Fig. 12.3 for junction lengths $2\lambda_J$, $5\lambda_J$ and $15\lambda_J$. It may be seen that the current density is nearly constant when the length is just two penetration depths but as the length of the junction increases the current density becomes increasingly non-uniform and most of the current is confined to the edge of the junction. When $L = 15\lambda_J$ the solution is very similar to that of the semi-infinite junction shown in Fig. 12.1.

For small values of L/λ_J the total current increases linearly with length because the current distribution is uniform, but for L/λ_J greater than 4 the current rapidly saturates at the value $4\lambda_J j_J$ as shown in Fig. 12.4.

So far there is hardly anything new. The current distribution is somewhat different but the conclusions agree with those drawn in Section 10.1. We get, however, a radically different picture if we reinterpret Equation (12.2) for an infinite (in contrast to a semi-infinite) junction. Then the current density is negative for negative y leading to the following conclusions:

(i) the total current is zero,

(ii) the current peaks are inside the junction,

(iii) the enclosed flux (from Equation (9.10)) is

$$\Phi = \Phi_0 \frac{\phi(\infty) - \phi(-\infty)}{2\pi}$$

$$= \Phi_0 \qquad\qquad (12.20)$$

that is exactly one flux quantum. Thus Equation (12.2) represents a single quantised flux line which we may just as well call a single quantised vortex in analogy with Type II superconductors. The analogy can be pursued further;

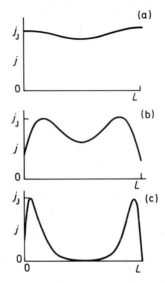

Fig. 12.3 Current density distribution in a junction when the current is maximum and there is no applied magnetic field; (a) $L = 2\lambda_J$, (b) $L = 5\lambda_J$, (c) $L = 15\lambda_J$. After Owen and Scalapino [333].

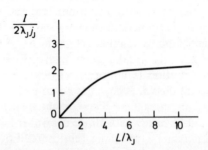

Fig. 12.4 Normalised current as a function of normalised length.

it is possible to have an array of vortices and one may define a lower critical field as well. The latter problem will be discussed in the next section; for the moment we shall return to the numerical results of Owen and Scalapino [333].

For a junction $L = 10\lambda_J$ long the maximum supercurrent as a function of applied magnetic field is shown in Fig. 12.5. It is strikingly different from the

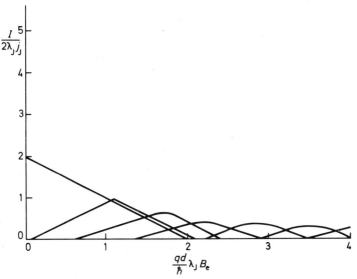

Fig. 12.5 The maximum current through a junction of length $L = 10\lambda_J$ as a function of applied magnetic field. After Owen and Scalapino [333].

interference pattern of Fig. 10.1 measured for a small junction. It is found that for small values of the reduced field

$$\mathscr{B} = \frac{qd}{\hbar}\lambda_J B_e \tag{12.21}$$

the maximum supercurrent reduces linearly but already for $\mathscr{B} = 0.06$ a new mode starts. It is true in general that a number of different modes may exist for the same applied field. In the terminology adopted one may talk about the 'n to $n+1$ vortex mode' when the junction contains more than n but less than $n+1$ vortices. A plot of the maximum supercurrent against B_e for the n to $n+1$ mode is shown in Fig. 12.6. At a point a the junction contains exactly n vortices and at point c exactly $n+1$. The curve for the $n+1$ to $n+2$ mode is of a similar shape but it starts at a point d and overlaps part of the n to $n+1$ mode. A result of this overlapping is that for every $\mathscr{B} > 0$ we can find some mode of operation for the junction which allows $I > 0$. It may be expected that the current distribution switches automatically to the mode which for a given B_e carries the largest supercurrent. Thus measurements of maximum supercurrent against B_e should yield the envelope of Fig. 12.5. Later experimental work [406, 407] did indeed prove this conjecture as may be clearly seen in Fig. 12.7 where Matisoo's results [407] are plotted for a junction length $L \cong 6\lambda_J$.

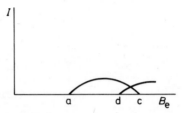

Fig. 12.6 The general shape of the maximum current as a function of applied magnetic field for a mode containing between n and $n+1$ vortices. After Owen and Scalapino [333].

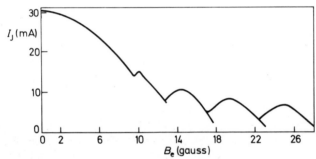

Fig. 12.7 Experimental values of maximum current as a function of applied magnetic field for a junction of length $L \cong 6\lambda_J$. After Matisoo [407].

The shape of the vortices may also be calculated from our equations and are shown for the 0 to 1 vortex mode in Fig. 12.8 for a junction length $L = 10\lambda_J$ for four values of $\mathcal{B}$. It may be seen how the total current decreases to zero as $\mathcal{B}$ increases to 2. The current density vanishes at both ends of the junction for integral number of vortices (0 for $\mathcal{B} = 0$ and 1 for $\mathcal{B} = 2$) but not otherwise.

As it was mentioned before the vortex configuration is dependent upon the fact whether it is part of an $n-1$ to n or an n to $n+1$ mode. In Fig. 12.9 (a) there is exactly one vortex fitted in the junction just as in Fig. 12.8 (d) but the shape of the current distribution is different.

Solutions in the 1 to 2 and 2 to 3 vortex modes are shown in Fig. 12.9 (b) and 12.9 (c) respectively. It may be seen that as more vortices fit into the junction they not only have a smaller dimension but they become more sinusoidal in shape as well. The variation of the magnetic field is also less pronounced as B_e increases because the relative influence of the self-magnetic fields decreases.

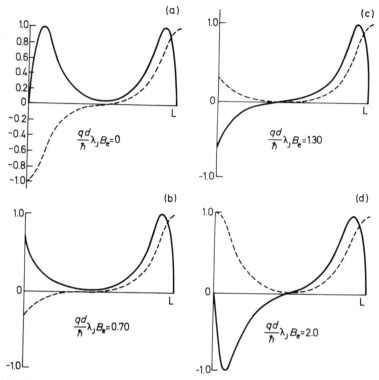

Fig. 12.8 The current density and the local magnetic field for a junction of length $L = 10\lambda_J$. The solid line is j/j_1; the dashed line is $(qd\lambda_J/\hbar)B_x$. (a), (b), (c) and (d) represent four solutions in the 0 to 1 vortex mode. After Owen and Scalapino [333].

12.2 Barrier energy and lower critical field

The discussion of Josephson junctions may very well start with the concept of interaction energy as done by Anderson [329]. Assuming the tunnelling Hamiltonian of Equation (2.29) he calculated this energy by perturbation calculus and hence derived the relative phase and the maximum supercurrent. The physical reasoning is that the relative phase must adjust itself so as to minimise the barrier energy. Since we already know the equations governing the relative phase, we may invert the problem and ask: what expression for energy will reproduce our equations? Considering the one-dimensional, time-independent case the equation to be derived is (from Equation (12.1))

$$\frac{d^2\phi}{dy^2} - \lambda_J^{-2}\sin\phi = 0. \tag{12.22}$$

The relative phase ϕ and its second derivative appear in the above equation

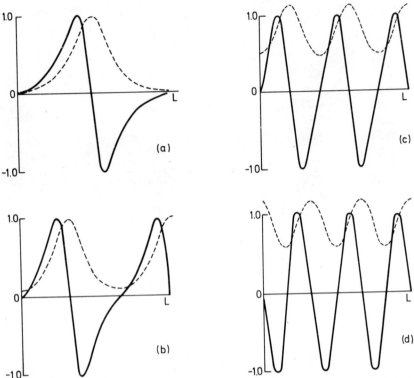

Fig. 12.9 Current density j/j_1 (solid line) and magnetic field $(qd\lambda_1/\hbar)B_x$ (dashed line) for a junction of length $L = 10\lambda_1$. (a) at $\mathscr{B} = 0.06$, the one vortex solution at its lowest possible value of $\mathscr{B}$, (b) at $\mathscr{B} = 1.08$ when the current is at its maximum for the 1 to 2 vortex mode, (c) at $\mathscr{B} = 1.72$ when the current is at its maximum for the 2 to 3 vortex mode, (d) at $\mathscr{B} = 2.44$, the highest value of $\mathscr{B}$ for the 2 to 3 vortex mode. After Owen and Scalapino [333].

so it seems reasonable to assume that the energy function will contain ϕ and its first derivative. So the expression to be minimised is

$$F = \int f(\phi(y), d\phi/dy)\, dy. \tag{12.23}$$

But it is known from variational calculus [408] that the function $\phi(y)$ which minimises F is given by the Euler differential equation

$$\frac{\partial f}{\partial \phi} - \frac{d}{dy}\frac{\partial f}{\partial(d\phi/dy)} = 0. \tag{12.24}$$

It may be seen from the above equation that to reproduce Equation (12.22) f must contain a term proportional to $\cos\phi$ and another term proportional to

$(d\phi/dy)^2$. Adding a further constant for generality, f must be of the form

$$f = C_1 + C_2 \cos\phi + C_3\left(\frac{d\phi}{dy}\right)^2 \tag{12.25}$$

which, in view of Equation (12.24), leads to

$$-C_2 \sin\phi - 2C_3\frac{d^2\phi}{dy^2} = 0. \tag{12.26}$$

Comparing Equation (12.26) with (12.22) we obtain

$$\frac{C_2}{2C_3} = -\lambda_J^{-2}. \tag{12.27}$$

We have now reproduced the differential equation (12.22) but have not yet found all the constants of Equation (12.25). For that we shall use a simple thermodynamical approach due to Josephson [366].

Assume that the free energy and current density depend on the *local* value of ϕ then it is sufficient to consider a barrier so small that the spatial variations of ϕ over it may be neglected. Consider then two systems A and B, one containing a barrier the other not, and both connected to current generators producing a current I. If I is changing, voltages V_A, V_B will appear across A and B. Equating the free energy changes to the work done by the generators

$$dF_i = V_iI\, dt \qquad (i = \text{A or B}) \tag{12.28}$$

yielding

$$d(F_A - F_B) = (V_A - V_B)I\, dt \tag{12.29}$$

Since A and B are identical except for the barrier, $F_A - F_B$ must be the free energy associated with the barrier and $V_A - V_B$ is equal to V, the voltage across the barrier. Hence

$$dF = VI\, dt = \frac{\hbar}{q}\frac{\partial\phi}{\partial t}I\, dt = \frac{\hbar}{q}I\, d\phi \tag{12.30}$$

where Equation (9.17) has been used.

Expressing the current from Equation (12.30) we get

$$I = \frac{q}{\hbar}\frac{\partial F}{\partial\phi} \tag{12.31}$$

or

$$j_z = \frac{q}{\hbar}\frac{\partial f}{\partial\phi} \tag{12.32}$$

if written for the current density. We know that the current density is expressed as

$$j_z = j_{\text{J}} \sin \phi \tag{12.33}$$

so f must be of the form

$$f = -\frac{\hbar}{q} j_{\text{J}} \cos \phi + C_1. \tag{12.34}$$

The value of the remaining constant may be obtained from the consideration that the barrier energy should vanish when no current flows, that is for $\phi = 0$. Hence

$$C_1 = (\hbar/q) j_{\text{J}}. \tag{12.35}$$

Using Equation (12.27) and comparing Equation (12.34) with (12.25), all our constants are determined giving the final form

$$f = \frac{\hbar}{q} j_{\text{J}} \left[(1 - \cos \phi) + \tfrac{1}{2} \lambda_{\text{J}}^2 \left(\frac{d\phi}{dy} \right)^2 \right]. \tag{12.36}$$

The lower critical field B_{c1} may be calculated in analogy with Abrikosov's calculations [26] for Type II superconductors. Transition to the mixed state occurs when

$$B_{c1} = (\mu_0 F_{\text{f}})/\Phi_0 \tag{12.37}$$

where F_{f} is the free energy of an isolated flux line. When the barrier contains no flux line, $\phi = 0$ and $F = 0$. When there is a single flux line in the barrier, $\phi = \phi_{\text{f}}$ is given by Equation (12.2). Thus the additional energy due to a flux line is

$$
\begin{aligned}
F_{\text{f}} &= \frac{\hbar}{q} j_{\text{J}} \int_{-\infty}^{\infty} \left[1 - \cos \phi_{\text{f}} + \tfrac{1}{2} \lambda_{\text{J}}^2 \left(\frac{d\phi}{dy} \right)^2 \right] dy \\
&= \frac{\hbar}{q} j_{\text{J}} \int_{-\infty}^{\infty} \left(2 \operatorname{sech}^2 \frac{y}{\lambda_{\text{J}}} + 2 \operatorname{sech}^2 \frac{y}{\lambda_{\text{J}}} \right) dy \\
&= \frac{4\hbar}{q} \lambda_{\text{J}} j_{\text{J}} \int_{-\infty}^{\infty} \operatorname{sech}^2 y \, dy \\
&= \frac{8\hbar}{q} \lambda_{\text{J}} j_{\text{J}} = 8 \left(\frac{\hbar^3 j_{\text{J}}}{q^3 \mu_0 d} \right)^{1/2}.
\end{aligned}
\tag{12.38}
$$

Equation (12.38) combined with Equation (12.37) yields

$$B_{c1} = \frac{4}{\pi} \left(\frac{\mu_0 \hbar j_{\text{J}}}{qd} \right)^{1/2}. \tag{12.39}$$

This is the field at which it becomes energetically favourable for a single quantised flux line to enter the barrier. Interestingly, our first solution (Equation

(12.3)) showing the exclusion of magnetic field gives for its value at $y = 0$,

$$B_x = \frac{2\hbar}{qd\lambda_J} = 2\left(\frac{\mu_0 \hbar j_J}{qd}\right)^{1/2} \tag{12.40}$$

that is $\pi/2$ times the critical field. This is a metastable solution which may exist (in analogy with Type II superconductors [409]) because of the attractive force between a flux line and the edge of the junction. There is a potential barrier to be overcome before the flux line can get into the interior of the junction. Conversely, metastable solutions may exist in magnetic fields below B_{c1} right down to zero magnetic field.

It may be finally mentioned that there is no direct analogue of the upper critical field of Type II superconductors. The vortex solutions in the junction may exist as long as the materials comprising the junction remain super-conducting, that is up to the bulk critical field.

12.3 Time-dependent solutions

In the two previous sections we discussed time-independent vortex solutions and derived the energy of a single vortex line. Here we shall permit time variation and investigate (following Lebwohl and Stephen [334]) two different types of solutions, namely (i) uniform motion of vortices and (ii) small oscillations of vortices.

Retaining the one-dimensional picture the equation to be solved (from Equation (9.35)) is

$$\left[\frac{\partial^2}{\partial y^2} - \frac{1}{v^2}\frac{\partial^2}{\partial t^2} - \frac{\beta}{v^2}\frac{\partial}{\partial t}\right]\phi = \lambda_J^{-2}\sin\phi. \tag{12.41}$$

In the absence of losses ($\beta = 0$) the solution can be easily found. If $\phi(y/\lambda_J)$ is a solution of the time-independent Equation (12.1) then it may be shown by direct substitution that

$$\phi[(y - v't)/\lambda_J'] \tag{12.42}$$

is a solution of Equation (12.41) where $v' < v$ is an arbitrary velocity and

$$\lambda_J' = \lambda_J\left(1 - \frac{v'^2}{v^2}\right)^{1/2}. \tag{12.43}$$

The general solution will again be obtained in terms of Jacobian elliptic functions replacing y and λ_J by $y - v't$ and λ_J' respectively. A single flux line (Equation (12.2)) in motion is given by*

$$\phi = 2\sin^{-1}\operatorname{sech}\frac{y - v't}{\lambda_J'}. \tag{12.44}$$

*An equivalent representation quoted in the literature [334, 349] is
$$\tan\tfrac{1}{4}\phi = \exp[-(y - v't)/\lambda_J'].$$

Its effective mass per unit length may be defined with the aid of the rest energy as

$$m_{eff} = \frac{F_f}{v^2}. \tag{12.45}$$

Using Equation (12.38) for F_f, Equation (9.32) for v and the numerical data $\lambda_J = 84$ μm, $j_J = 0.5 \times 10^6$ A/m^2, $d = 80$ nm, $w = 1.5$ nm, we get $m_{eff} = 10.8 \times 10^{-27}$ kg which is about 10^4 times as large as the mass of an electron. This is not far from the inertial mass found by Suhl [410] for a flux line in a Type II superconductor.

If we take into account the finite (normal) conductivity of the barrier then we have an additional term $g_0 V$ in the expression for the current density (Equation (9.34)) and the rate of dissipation is

$$\frac{\partial F}{\partial t} = -g_0 \int V^2 \, dy$$

$$= -g_0 \frac{q}{\hbar} \int \left(\frac{\partial \phi}{\partial t} \right)^2 dy. \tag{12.46}$$

If g_0 is small we may use Equation (12.44), that is the dissipationless solution, for ϕ, which leads to

$$\frac{\partial F}{\partial t} = -8g_0 \frac{\hbar^2 v'^2}{q^2 \lambda_J}. \tag{12.47}$$

We may now define a viscosity coefficient, η_0, by equating the rate of dissipation with $-\eta_0 v'^2$. Then from Equation (12.47)

$$\eta_0 = \frac{8g_0 \hbar^2}{q^2 \lambda_J} = \Phi_0 g_0 d B_{c1} \tag{12.48}$$

The above equation for the viscosity coefficient is identical to that found (partly on the basis of empirical relationships) by Kim et al. [411] for Type II superconductors with the only difference that B_{c2} (the upper critical field) replaces B_{c1}.

The solutions in terms of elliptic functions (the equivalents of Equations (12.11) and (12.12)) give moving vortex arrays. Further solutions may be constructed by superposing the solutions obtained, by a mathematical technique called Backlund transformation [412]. One such solution represents vortex lines moving in opposite directions which just pass through each other.*

Another set of solutions for $v'' > v$ may be obtained by noting that if

$$\phi(vt/\lambda_J) \tag{12.49}$$

---

*This is possible because in contrast to Type II superconductors the vortices in Josephson junctions have no normal core.

is a solution of the differential equation

$$-\frac{1}{v^2}\frac{\partial^2\phi}{\partial t^2} = \lambda_J^{-2}\sin\phi \qquad (12.50)$$

then

$$\phi[(v''t-y)/\lambda_J''] \qquad (12.51)$$

is a solution ($\beta = 0$) of Equation (12.41) where

$$\lambda_J'' = \lambda_J\left(\frac{v''^2}{v^2}-1\right)^{1/2} \qquad (12.52)$$

These solutions are referred to as E types and represent a generalisation of those Equations (10.24) and (10.25) to nonsinusoidal waveforms.

The small oscillations of a static vortex array may be investigated by assuming a solution of the form

$$\phi(y_1 t) = \phi_0(y)+u(y)\exp i\omega t \qquad (12.53)$$

where $\phi_0(y)$ is a solution of Equation (12.1). Substituting Equation (12.53) into Equation (12.41) ($\beta = 0$) and linearising with respect to u we obtain

$$\frac{\partial^2}{\partial y^2}+\frac{\omega^2}{v^2}\,u(y) = \lambda_J^{-2}u(y)\cos\phi_0(y) \qquad (12.54)$$

which is not unlike Equation (10.23). An exact solution of the above differential equation exists [334, 413] in terms of eta and theta functions. The dispersion relation (as a function of a parameter p which plays the role of the wavevector) is shown in Fig. 12.10. It may be noted that the upper branch is very similar to that plotted in Fig. 10.4.

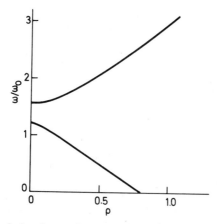

Fig. 12.10 Dispersion relation for small oscillations of a static vortex array. After Lebwohl and Stephen [334].

13. Josephson junctions in parallel; the Mercereau effect

13.1 The maximum supercurrent

The combination of two Josephson junctions in parallel was first investigated experimentally by Mercereau and his coworkers* at the Ford Scientific Laboratory. It was essentially a generalisation of the d.c. Josephson effect for more than one junction.

To derive the theoretical formulae we need first the phase difference between two points connected by a superconducting path. This may be obtained from Equation (9.5) by integrating along the path

$$v_a - v_b = \frac{q}{\hbar} \int_a^b \left(\mathbf{A} + \frac{m}{q^2 \rho} \mathbf{j} \right) ds \tag{13.1}$$

We may now apply this formula to the geometry of Fig. 13.1. Denoting the phase differences across the junctions by ϕ_1 and ϕ_2, and noting that the phase difference between points a and b must be the same independently of the path

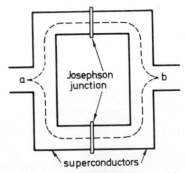

Fig. 13.1 Two Josephson junctions in parallel, connected by a superconducting path.

*There is such a profusion of phenomena associated with Josephson's name (Josephson tunnelling, Josephson junction, d.c. Josephson effect, a.c. Josephson effect, Josephson radiation, Josephson plasma resonance, etc.) that it would actually clarify the situation if some of the related effects bore some other names – preferably the names of their first investigators. It was suggested (more or less on this basis) by Anderson [349] that the quantum interference effects obtained by connecting Josephson junctions in parallel should be named after Mercereau.

taken, we may write

$$\frac{q}{\hbar}\int_{\text{path 1}}\left(\mathbf{A}+\frac{m}{q^2\rho}\mathbf{j}\right)ds+\phi_1 = \frac{q}{\hbar}\int_{\text{path 2}}\left(\mathbf{A}+\frac{m}{q^2\rho}\mathbf{j}\right)ds+\phi_2 \qquad (13.2)$$

leading to

$$\phi_2-\phi_1 = \frac{q}{\hbar}\oint_{\text{loop}}\left(\mathbf{A}+\frac{m}{q^2\rho}\mathbf{j}\right)ds+2k\pi \qquad (13.3)$$

The line integral of the vector potential gives the flux enclosed which we shall call the internal flux and denote it by Φ_{int}. The line integral of the current density may be taken zero when the cross-section of the superconductor is large in comparison with the penetration depth. This condition is nearly always satisfied in the experimental set-up (an exception will be discussed in the next section). Assuming further that the junctions are identical (so that their maximum supercurrents are identical too) we may write for the total current

$$I = I_1+I_2 = I_J(\sin\phi_1+\sin\phi_2)$$

$$= I_J\left[\sin\phi_1+\sin\left(\phi_1+\frac{q}{\hbar}\Phi_{\text{int}}\right)\right]$$

$$= 2I_J\cos\left(\frac{q}{2\hbar}\Phi_{\text{int}}\right)\sin\left(\phi_1+\frac{q}{2\hbar}\Phi_{\text{int}}\right). \qquad (13.4)$$

The maximum supercurrent of the two junctions combined may be obtained by maximising the above expression with respect to ϕ_1 at a given value of Φ_{int}. It may be seen from Equation (13.4) that for any value of Φ_{int} it is possible to choose ϕ_1 in such a way as to make the sine term equal to unity and then

$$I_{\text{max}} = 2I_J\left|\cos\frac{q}{2\hbar}\Phi_{\text{int}}\right|. \qquad (13.5)$$

Remembering that the maximum supercurrent depends also on the magnetic field threading the individual junctions we may use Equation (10.7) to get

$$I_{\text{max}} = 2I_{J0}\left|\frac{\sin(\pi\Phi_j/\Phi_0)}{\pi\Phi_j/\Phi_0}\right|\left|\cos\frac{\pi\Phi_{\text{int}}}{\Phi_0}\right| \qquad (13.6)$$

where Φ_j is the magnetic flux enclosed by each junction.

We have now the product of two interference patterns (analogous to those of two discrete antennas as mentioned in the Introduction). Since the area enclosed by the loop is usually much larger than that of the individual junctions, the second expression in Equation (13.6) varies much faster than the first one. Concentrating for the moment on this fast variation only we may say that it is fully modulated, meaning by this that I_{max} may be zero. We may also add that it is a periodic function, the period being equal to one flux quantum.

However, all the above discussion is true only when the inductance of the loop, L, is negligibly small. For finite inductance the figure of interest is LI_J/Φ_0. When $LI_J/\Phi_0 \ll 1$, that is, the flux due to the self-inductance of the loop is much smaller than the flux quantum, we may expect that the conclusions reached above remain valid. When LI_J is comparable with the flux quantum an appreciable current may circulate* in the loop and the externally applied flux, Φ_{ext} is no longer equal to Φ_{int} the actual flux crossing the loop. The effect of finite inductance is to reduce the modulation of the maximum supercurrent. According to a simple argument advanced by Zimmerman and Silver [341] the modulation of the interference pattern, I_{mod} is equal to $2I_J$ or Φ_0/L which ever is smaller. This is because Φ_0/L is the value of circulating current necessary to contain one flux quantum within the ring. As we shall see presently this simple argument gives good approximation in the two extremes, $LI_J/\Phi_0 \ll 1$ or $LI_J/\Phi_0 \gg 1$ but it is not very accurate in between.

We may now refine our previous model by noting that the external and internal fluxes are related by

$$\Phi_{int} = \Phi_{ext} + LI_{circ} \qquad (13.7)$$

where I_{circ} is the current circulating in the loop. For a perfectly symmetric geometry the circulating current is given by

$$I_{circ} = \tfrac{1}{2}(I_2 - I_1). \qquad (13.8)$$

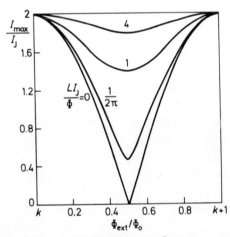

Fig. 13.2 The maximum supercurrent which may flow across a parallel combination of two junctions as a function of magnetic flux.

*This circulating current is analogous to that set-up in ordinary superconducting rings which keep the enclosed flux an integral multiple of the flux quantum.

In view of Equations (13.7) and (13.8) the relationship between the two angles (Equation (13.3)) modifies to

$$\phi_2 = \phi_1 + \frac{2\pi}{\Phi_0}\left[\Phi_{ext} + \frac{L}{2}(I_2 - I_1)\right].$$ (13.9)

Noting further that for identical junctions the maximum supercurrent is

$$I_{max} = I_J[\max(\sin\phi_1 + \sin\phi_2)]$$ (13.10)

we get the following transcendental equation

$$\frac{I_{max}}{I_J} = \max\left\{\sin\phi_1 + \sin\left\{\phi_1 + 2\pi\left[\frac{\Phi_{ext}}{\Phi_0} - \frac{LI_J}{\Phi_0}\left(\sin\phi_1 - \frac{I_{max}}{2I_J}\right)\right]\right\}\right\}$$ (13.11)

where the maximisation is in respect of ϕ_1. This is a rather awkward equation to solve.* It must be done numerically by solving the implicit equation in I_{max} for each value of ϕ_1 and then choose the ϕ_1 which gives the highest I_{max}.

I_{max}/I_J satisfying the above relationship is plotted in Fig. 13.2 against Φ_{ext}/Φ_0 with LI_J/Φ_0 as a parameter. It may be seen that even for the rather small $LI_J/\Phi_0 = 1/2\pi$ the current modulation is still well below its maximum value of $2I_J$. All the curves follow the general behaviour of the $LI_J/\Phi_0 = 0$ curve; the minima of I_{max}/I_J are at half integral multiples and the maxima at integral multiples of the flux quantum. The amount of current modulation is given by

$$\frac{I_{mod}}{I_J} = 2 - \left(\frac{I_{max}}{I_J}\right)_{min}$$ (13.12)

This is shown in Fig. 13.3 against LI_J/Φ_0. It may be seen that the asymptotes are correctly given by the Zimmerman and Silver condition.

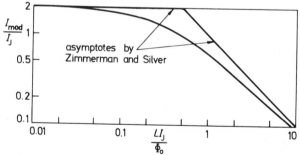

Fig. 13.3 Modulation of the maximum supercurrent as a function of normalised inductance.

*A solution by Clarke [338] yielded the result that the Zimmerman and Silver [341] estimate is optimistic. He derived an inequality showing that the current modulation will always be less than Φ_0/L.

Approximate analytic and numerical solutions were obtained by de Waele and de Bruyn Ouboter [370] both for symmetric and asymmetric geometries. A theory permitting a more general $I(\phi)$ relationship was formulated by Fulton [376].

It is of further interest (relevant to the use of the double junction as a magneto-meter) to determine the relationship between the external and internal fluxes under the condition when maximum supercurrent flows (the point which can be experimentally easily determined). For given Φ_{ext} and LI_J/Φ_0 we obtained the optimum ϕ_1 from Equation (13.11); that gives then ϕ_2, the circulating current and finally the internal flux. The graph obtained is shown in Fig. 13.4. Whenever Φ_{ext} is an integral multiple of the flux quantum the internal flux is the same as the external flux. At half integral multiples of the flux quantum there is a discontinuity, otherwise Φ_{int} is a smoothly varying function of Φ_{ext}.

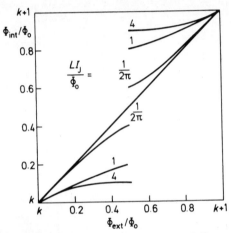

Fig. 13.4 The relationship between internal and external flux under conditions when maximum supercurrent flows.

Next we shall consider the situation when a magnetic field is present and the current exceeds the value possible with zero voltage. Then the voltage is finite and varies periodically with the applied magnetic field. In order to derive this effect we shall take account of the normal current as well, as done in Section 11.2. Note that Equation (13.4) is still valid for the d.c. supercurrent flowing in the circuit only ϕ_1 is now related to the voltage. Adding the normal current GV we may write for the total current

$$I = 2I_J \cos\left(\frac{q}{2\hbar}\Phi_{int}\right)\sin\left(\phi_1 + \frac{q}{2\hbar}\Phi_{int}\right) + GV. \tag{13.13}$$

This is essentially (the difference is a constant term in the argument of the sine function) the same as Equation (11.9) with $\beta_c = 0$. Assuming that the circuit is fed by a constant current generator and that the self-induced flux is negligible the solution is the same as that shown in Equation (11.15). The d.c. voltage across the double junction is

$$V = \begin{cases} G^{-1}(I^2 - I_{max}^2)^{1/2} & I > I_{max} \\ 0 & I < I_{max} \end{cases} \tag{13.14}$$

where I_{max} is given by Equation (13.5).

It follows clearly from Equation (13.14) that the d.c. voltage as a function of magnetic field is in anti-phase with I_{max}. Whenever I_{max} is a minimum V_{dc} is maximum and vice versa. The normalised voltage GV_{dc}/I_J is plotted in Fig. 13.5 against $\Phi_{ext}(= \Phi_{int})$ for $I/I_J = 3, 2$ and $2/3$.

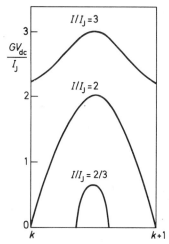

Fig. 13.5 Normalised voltage across the junction as a function of external flux.

The effect of the self-induced flux and lack of symmetry ($I_{J1} \neq I_{J2}$) have been considered in References [370, 415, 416] and reviewed by de Bruyn Ouboter and de Waele [414].

13.2 Experiments on tunnel junctions

The experimental results of Jaklevic et al. [348] show clearly the double inter-ference pattern predicted by Equation (13.6). It may be seen from curves (a) and (b) of Fig. 13.6 (representing experiments on two different sets of junctions) that there is the larger period due to the magnetic flux in the junction, and the shorter period due to the magnetic flux crossing the loop. Curve (a) is nearer to the theoretical prediction because the maximum supercurrent takes periodi-cally zero values. The lack of zeros in curve (b) indicates that the two junctions were not identical. A further reason for departure from ideal behaviour in this case is that $LI_{J_0} \approx \Phi_0$.

The dependence of the maximum supercurrent on the applied magnetic field may be used for the measurement of small magnetic fields or for the accurate measurement of higher magnetic fields. We shall return to this problem in Chapter 18; here we shall mention two further types of interference measurements performed by Jaklevic *et al.* [324, 348].

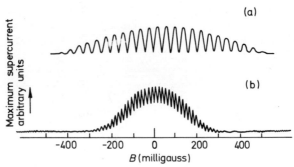

Fig. 13.6 Maximum supercurrent as a function of magnetic field for two thin film junctions in parallel. (a) and (b) represent measurements on two separate pairs of junctions. After Jaklevic *et al.* [348].

The first one is concerned with the interesting problem that the phase of the Cooper-pairs is affected by the vector potential even in the absence of magnetic fields. The experimental arrangement is shown schematically in Fig. 13.7. The circle represents a solenoid constructed by closely winding a fine insulated copper wire around a beryllium–copper core with the core providing the return path. The smallest solenoid made had an overall diameter of 56 μm. The measurement of maximum supercurrent gave again a distinct interference pattern.

Finally, Jaklevic *et al.* [348, 417] investigated the effect of a drift current on the maximum supercurrent. The experimental arrangement is shown schematically in Fig. 13.8. The drift current is carried by the strips b and b′

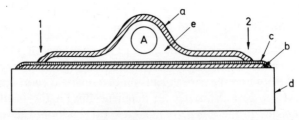

Fig. 13.7 Cross-section of a Josephson junction pair vacuum-deposited on a quartz substrate (d). A thin oxide layer (c) separates thin (∼ 100 nm) tin films (a and b). The junctions (1) and (2) are connected in parallel by superconducting thin film links enclosing the solenoid (A) embedded in Formvar (e). Current flow is measured between films a and b. After Jaklevic *et al.* [324].

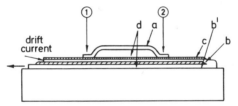

Fig. 13.8 Schematic representation of a junction pair (1) and (2) similar to Fig. 13.7 where the base film b carries a drift current which is returned beneath itself by a second base film b′ designed to keep the field due to the drift current from the area enclosed by the junction loop. The insulating layers d are of Formvar. After Jaklevic *et al.* [348].

designed in such a way as to keep the magnetic field due to the drift current away from the loop.

The theoretical formula for this case may be obtained by retaining the current term in Equation (13.3). Then Equation (13.6) becomes modified, a drift current term being added both to Φ and to Φ_{int}. Thus we may again expect an interference pattern as a function of drift current.* This is borne out indeed by the experimental results shown in Fig. 13.9.

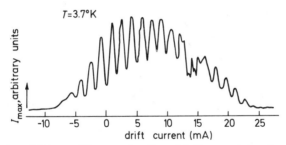

Fig. 13.9 Experimental trace of I_{max} as a function of drift current showing interference and diffraction effects. The zero offset is due to a static applied field. Maximum current is 1·5 mA. After Jaklevic *et al.* [348].

13.3 Experiments on solder drop junctions

The solder junction, a drop of solder on a piece of (usually niobium) wire, was mentioned in the Introduction and shown schematically in Fig. 8.6. A somewhat more accurate representation, showing a longitudinal section, may be seen in Fig. 13.10. The heavy lines represent the niobium oxide and the dotted lines enclose the region into which the flux penetrates. It turns out (the reasons are not clear) that the device usually behaves as if there were two junctions at A

---

* It was noted by Ferrel [418] that one cannot talk of a 'drift current' independently of the gauge used. He shows that in some other gauge the drift current term of Jaklevic *et al.* [348] is absent; he points out further that the origin of phase difference is not qualitatively different in this case than in the previous two.

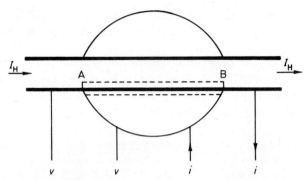

Fig. 13.10 Longitudinal section through specimen made with uninsulated Nb wire. The heavy lines represent the niobium oxide and the dotted lines enclose the region into which the flux penetrates. After Clarke [338].

and B. The magnetic field in the area between the junctions is created by passing a current, I_H, through the niobium wire. The value of this magnetic field is

$$B_H = \mu_0 \frac{I_H}{2\pi r} \tag{13.15}$$

and the corresponding flux

$$\Phi = \mu_0 \frac{I_H}{2\pi r} l(\lambda_{Nb} + \lambda_s) \tag{13.16}$$

where λ_{Nb} and λ_s are the penetration depths in niobium and solder respectively and l is the distance AB. The amount of current needed to cause a change in flux equal to the flux quantum is

$$\Delta I_H = \frac{\Phi_0 2\pi r}{\mu_0 l(\lambda_{Nb} + \lambda_s)}. \tag{13.17}$$

The maximum supercurrent as a function of I_H was first measured by Clarke [338] as shown in Fig. 13.11. The finer periodicity is due to the double junction

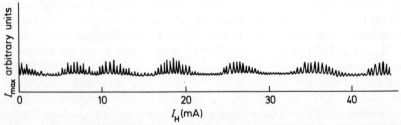

Fig. 13.11 Maximum supercurrent as a function of I_H for a typical specimen. After Clarke [338].

(the value of current needed to cause an oscillation agrees well with that calculated from Equation (13.17)) but the cause of the modulation envelope is not clear.

The same type of junction in an externally applied magnetic field was investigated shortly afterwards by de Bruyn Ouboter et al. [419]. They found the periodic behaviour both in I_{max} and V_{dc}. A measured result for the latter is shown in Fig. 13.12 displaying again a modulation envelope.

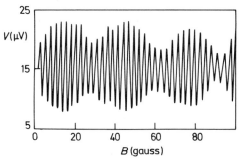

Fig. 13.12 The d.c. voltage measured across a solder drop junction as a function of applied magnetic field. After de Bruyn Ouboter et al. [419].

A modified form of this junction (shown in Fig. 15.4 and discussed in Section 15.4) was developed by Omar and de Bruyn Ouboter [420], and de Bruyn Ouboter et al. [421]. The measured voltage as a function of magnetic field for various values of constant feeding current is shown in Fig. 13.13.

13.4 Experiments with point contact junctions

Point contacts, like solder junctions, can be easily made but are not clearly defined. They may be realised in a number of forms which will be discussed in Section 15.3 concerned with fabrication. In this section we shall briefly return to the double junction mentioned in the Introduction and shown in Fig. 8.6. It consists of two superconducting screws with lock nuts, set one on each side of the centre hole. The screws can be operated with the aid of a mechanical linkage while the device is in the cryostat. With a centre hole of 1 cm in diameter the period in the maximum supercurrent was found to be $\sim 2 \times 10^{-11}$ T. Owing to this high sensitivity Zimmerman and Silver [341] used three Mumetal cylinders around the cryostat and a superconducting lead cylinder in the liquid helium to shield against stray fluctuations. It was also essential to guard against any vibration of the components relative to each other, and to use heavy rf filtering in all leads going into the cryostat. The interference pattern obtained is shown in Fig. 13.14. The maximum supercurrent of the individual contacts was

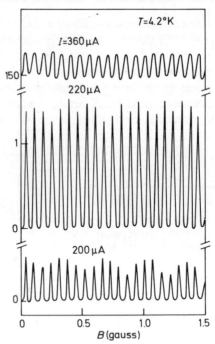

Fig. 13.13 The d.c. voltage measured across a modified solder drop junction (see Fig. 15.4) as a function of applied magnetic field. $I > I_{max}$ is the current forced through the junction. After Omar and de Bruyn Ouboter [420].

Fig. 13.14 Interference pattern obtained with the double point contact shown in Fig. 8.6(*d*). After Zimmerman and Silver [341].

varied (by adjusting the screws) between a few tenths and a few tens of micro-amperes without significantly affecting the interference pattern. The amplitude of the pattern corresponds to 0·2 μA. The calculated inductance of the structure was $L \sim 5 \times 10^{-9}$ H, giving $\Phi_0/L \sim 0.4$ μA.

Double point contacts were also investigated by the Leiden group [370, 415, 422, 423]. A particularly clear set of results is shown in Fig. 13.15 where both I_{max} and V_{dc} are plotted as functions of the applied magnetic field. It may be seen that minima of I_{max} coincide with maxima of V_{dc} in accordance with the predictions of Equation (13.14). The lack of perfect periodicity is attributed [370] to the different critical currents of the junctions.

For the total current of 175 μA the voltage is zero for part of the period and the same effect can just be seen for 300 μA. For higher currents the magnitude of V_{dc} increases but the amplitude of the oscillation decreases. At a current of 1200 μA the voltage is essentially independent of the magnetic field. This is because as the voltage increases the normal current component, unaffected by the magnetic field, increases at the expense of the supercurrent.

More than two junctions in parallel are expected to yield sharper interference patterns. The analogy with point radiators still holds, the general formula for N identical junctions being

$$ I_{max} = I_J \left| \frac{\sin N \frac{q}{2\hbar} \Phi_{int}}{\sin \frac{q}{2\hbar} \Phi_{int}} \right|. \tag{13.18}$$

The effect was first shown qualitatively by Zimmerman and Silver [341] and later studied systematically by de Waele et al. [424]. Their results are shown in Fig. 13.16; they measured, in fact, the d.c. voltage but, as we have mentioned before, it is in close relationship with I_{max}. They conclude that due to experimental difficulties of making all contacts identical devices containing more than six junctions have little chance of being successful.

The behaviour of double point contacts in coaxial cavities was investigated by de Bruyn Ouboter et al. [422]. They found steps in the I–V characteristics at voltages corresponding to the resonant frequency of the cavity.* The steps disappeared when the magnetic flux enclosed by the circuit was equal to a half integral multiple of the flux quantum. This may be explained by reference to the relative phase of the two radiating junctions. When the enclosed flux is

$$ \Phi = (n+\tfrac{1}{2})\Phi_0 \tag{13.19}$$

the phase difference between the junctions (from Equation (13.3)) is 180°. Since the distance of the two junctions is very much less than the free space wavelength the radiation from the junctions cancels and the step structure disappears.

* This is the same phenomenon as that noted in Section 11.8.

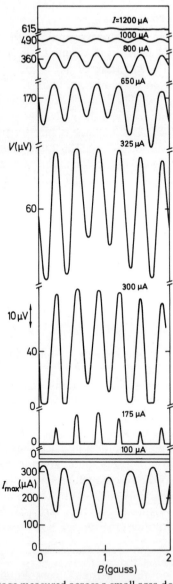

Fig. 13.15 The d.c. voltage measured across a small area double point-contact junction as a function of applied magnetic field. Comparison with theory yields Josephson currents $I_{J1} = 230\ \mu\text{A}$ and $I_{J2} = 90\ \mu\text{A}$ for the first and second junction respectively. I is the total current forced through the two junctions in parallel. (b) I_{max} as a function of applied magnetic field. Note that I_{max} is in anti-phase with V_{dc}. After de Waele and de Bruyn Ouboter [370].

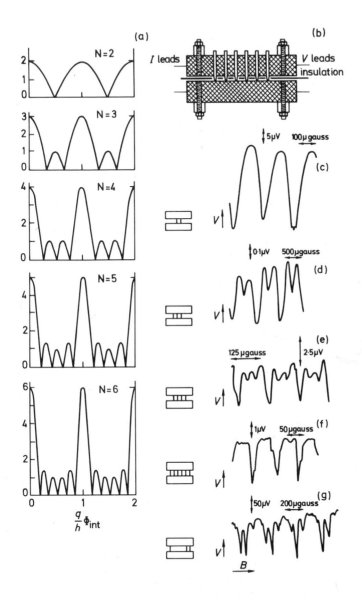

Fig. 13.16 (a) Theoretical interference patterns (in the limit of zero inductance) for multiple point-contact junctions. (b) Practical realisation of the junctions (c)–(d) Point-contact configurations and the corresponding interference patterns. After de Waele *et al.* [424].

13.5 Experiments on thin film bridges

The maximum supercurrent as a function of a perpendicularly applied magnetic field for thin film tin bridges in parallel was studied by Seraphim [425]. He found fairly good periodicity of a few gauss for small enclosed area but failed to observe the theoretical period of about 13 mgauss for a considerably larger loop. A common characteristic of all the measured I_{max} versus B curves was that once I_{max} reduced to zero it never reappeared.

A neat set of experimental results were recently obtained by Fulton and Dynes [377] for roughly rectangular loops of dimensions 3–10 μm wide and 20–100 μm long, the bridges being separated by the larger dimension. A typical set of results is shown in Fig. 13.17. The interference pattern for one period and

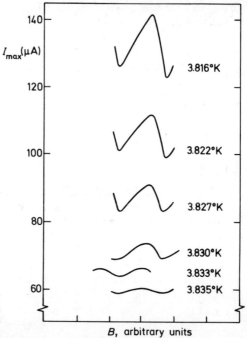

Fig. 13.17 I_{max} as a function of applied magnetic field for thin film bridges in parallel over one period of the interference pattern. After Fulton and Dynes [377].

its dependence on temperature may be clearly seen. Exact comparisons with theory were not attempted because of trapping of flux (due to the small critical perpendicular field) and the resulting hysteresis.

13.6 Coherence in the absence of a zero-voltage current

The existence of a zero-voltage current indicates that the coupling energy between the superconductors is capable of stabilising the relative phases of the wavefunctions. If the coupling is not sufficiently strong it may be destroyed by noise and then no zero-voltage current is observed. It is interesting to note that phase coherence can be established* even under such noisy conditions, as it was conclusively proven by Vant-Hull and Mercereau [426, 427].

The basic idea is to make use of the current steps obtained in resonant structures (as discussed in Section 11.8). The experimental set-up may be seen in Fig. 13.18. The superconducting strip line of length 1·33 m has a dual role

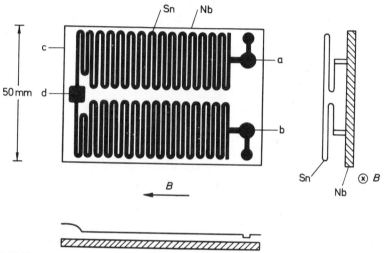

Fig. 13.18 Superconducting interferometer fabricated by evaporating a folded tin film (~ 130 nm) over a Formvar-coated niobium sheet ground plane. The resulting super-conducting strip line (length 1·44 m) connects Josephson junctions at a and b. Voltage is measured between points c and d. After Vant-Hull [427].

here, namely to give a d.c. connection between junctions a and b, and to provide an electromagnetic resonator. The a.c. current of the junctions will then excite the resonators, and the nonlinear interaction between the a.c. currents and fields will produce a finite d.c. current. The excess current (superimposed on the 'resistive' current) may be clearly seen in Fig. 13.19. The peaks correspond to the resonant frequencies of the strip line resonator.

At a given voltage the excess current is a function of magnetic flux. The amplitude modulation of a single peak showed the expected dependence on flux and it was also possible to obtain interference effects. It is significant that this excess current also shows interference properties (not only the zero-voltage current) and that phase coherence still exists for a distance over 1 m.

---

* This was the basic assumption in one of the models trying to account for subharmonic structure (Section 5.4); the a.c. Josephson radiation exists even in the absence of a zero-voltage current.

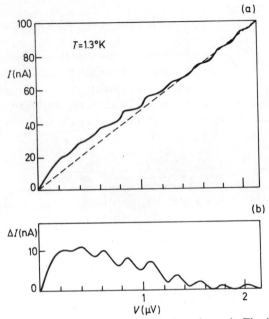

Fig. 13.19 (a) I–V characteristic of the interferometer shown in Fig. 13.18. Magnetic field was chosen to maximise this mode structure. (b) Excess current as a function of voltage. The dependence of this excess current on magnetic flux follows that found for the zero-voltage current in usual double junction interferometers. After Vant-Hull and Mercereau [426].

14. Phenomena in rings containing one junction

14.1 Introduction

It might seem illogical that the problem of two junctions is treated first and the apparently simpler one-junction problem only later. The main reason is that with two junctions in parallel there is an interesting d.c. effect, the Mercereau effect, whereas one single junction in a ring responds only to a.c. excitation so the interesting effects are time dependent. We shall investigate the two basic configurations shown in Fig. 14.1 (a) and (b) where the ring is excited by a magnetic field and by a current respectively. We shall study the arising effects in somewhat more detail partly because of their intrinsic interest and partly because they comprise the basis of a number of very sensitive detectors. The excitation of a persistent current in such a ring was first reported by Smith [428] but the bulk of the work was carried out by Zimmerman and Silver [339, 429–438] who realised the junctions with the aid of point contacts. In fact, in their first experiments they used exactly the same devices as those designed for double point-contact experiments (Fig. 8.6); by adjusting one of the screws to a proper short circuit they were left with a ring containing one junction.

In this chapter we shall concentrate on the basic properties of such rings; their applications as detectors will be discussed in Chapter 18 and some further a.c. properties in Chapter 17.

The linear treatment of the junction is given in Section 14.2 while Section 14.3 is concerned with the solution of Josephson's equations for this particular configuration. In Sections 14.4 and 14.5 static and dynamic experiments are described; in Section 14.6 the dependence of the inductance upon the amplitude of the applied magnetic field is investigated. Finally, in Section 14.7 we analyse the behaviour of a ring modified by the insertion of a piece of resistive material.

We shall use the term 'junction' to describe the point-contact in general; when the nonlinear properties are essential we shall refer to it as a 'Josephson junction'. Theoretically, only the linear and 'Josephson' limits will be treated but the experiments will show the whole range of 'weak' behaviour.

14.2 Linear theory

The simplest approach to the problem is to assume that the junction is linear, i.e. it behaves in the same way as the rest of the ring up to a certain critical current I_c. Obviously, the critical current of the ring will be determined by the junction because the cross-section is the smallest there and so the current density is the highest. Denoting the cross-section of the junction by A_w and the critical current density by j_c, the critical current of the ring is $I_c = j_c A_w$.

We shall start with the equation of fluxoid quantisation (Equation (1.40))

$$\Phi_{int} + \frac{m}{q^2} \oint \frac{\mathbf{j}}{\rho} d\mathbf{s} = n\Phi_0. \tag{14.1}$$

Usually the cross-section of the superconductor is large in comparison with the penetration depth so the line integral of the current may be neglected. The same condition may be applied to the ring with the exception of the junction. Since its cross-section is small the current density may be assumed to be uniform and thus the line integral of the current density over the junction does *not* vanish. Equation (14.1) takes then the form

$$\int_{junction} \frac{m}{q^2 \rho} j\, d\mathbf{s} + \Phi_{int} = n\Phi_0 \tag{14.2}$$

and after integration

$$\gamma L I_{circ} + \Phi_{int} = n\Phi_0 \tag{14.3}$$

where

$$\gamma = \frac{m}{pq} \frac{w}{A_w L} \tag{14.4}$$

is usually small in comparison with unity and w is the length of the junction.

We shall first investigate the arrangement shown in Fig. 14.1 (*a*) where the ring is excited by an external flux. The relationship between the external and internal fluxes is given again by Equation (13.7). Since the external flux is the independent variable we shall express with its aid both the circulating current

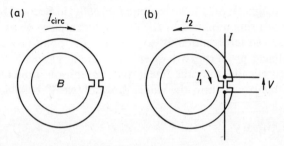

Fig. 14.1 Weakly connected superconducting ring excited by (*a*) a magnetic field, (*b*) an impressed current.

and the internal flux. From Equations (13.7) and (14.3) we get

$$\Phi_{int} = \frac{1}{1+\gamma}(n\Phi_0 + \gamma\Phi_{ext}) \tag{14.5}$$

and

$$LI_{circ} = \frac{1}{1+\gamma}(n\Phi_0 - \Phi_{ext}). \tag{14.6}$$

When $\Phi_{ext} = n\Phi_0$ the circulating current is zero and $\Phi_{ext} = \Phi_{int}$ just as we have seen for the double Josephson junction circuit in Chapter 13. As Φ_{ext} is increasing, the circulating current is decreasing until it reaches the critical current $-I_c$ at an external flux

$$(\Phi_{ext})_c = n\Phi_0 + (1+\gamma)LI_c. \tag{14.7}$$

At this point the ring is no longer capable to carry a supercurrent with the same value of n. The question is whether it will jump directly to the next state with $n+1$ or turn normal and lose its circulating current. To decide which is the case let us work out the value of I_{circ} at $(\Phi_{ext})_c$ and quantum number $n+1$. We get from Equations (14.6) and (14.7)

$$LI_{circ} = \frac{\Phi_0}{1+\gamma} - LI_c. \tag{14.8}$$

Introducing a new parameter α with the relation

$$I_c = \alpha\frac{\Phi_0}{2(1+\gamma)L} \tag{14.9}$$

Equation (14.8) modifies to

$$LI_{circ} = \frac{2-\alpha}{\alpha}LI_c. \tag{14.10}$$

Now, clearly, if $LI_{circ} < LI_c$ the transition to the $n+1$ state is possible. If $LI_{circ} > LI_c$ at this particular value of the external flux then the transition to the $n+1$ state is *not* possible and the junction becomes normal. The boundary between the two cases occurs when $(2-\alpha)/\alpha = 1$, that is when $\alpha = 1$.

The variation of Φ_{int}/Φ_0 and LI_{circ}/Φ_0 for the three cases $\alpha < 1$, $\alpha = 1$ and $\alpha > 1$ is shown* in Fig. 14.2. When $\alpha < 1$ the ring becomes normal when $(\Phi_{ext})_c$ is reached. The normal ring is non-magnetic so $\Phi_{int} = \Phi_{ext}$ until the ring may make a transition to the next state. When $\alpha = 1$ there are no longer normal regions but the transitions are still reversible. For $\alpha > 1$ the transitions are irreversible.

---

*It must be noted that γ, L and I_c are not independent of each other because $I_c \sim A_w^{-1}$ and $\gamma \sim (A_w L)^{-1}$.

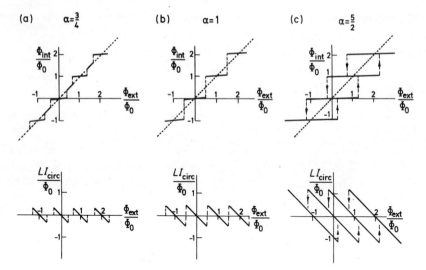

Fig. 14.2 Internal magnetic flux and circulating current of a weakly connected ring as functions of the external flux in the linear approximation for (a) $\alpha = 3/4$, (b) $\alpha = 1$, (c) $\alpha = 5/2$ where α is defined by Equation (14.9). After Silver and Zimmerman [438].

Let us look now at the case of direct excitation by a current I (Fig. 14.1 (b)). Then

$$I = I_1 + I_2 \tag{14.11}$$

and in analogy with Equation (14.3)

$$\gamma L I_1 + \Phi_{int} = n\Phi_0. \tag{14.12}$$

The internal flux is caused now by the current I_2 (the inductance of the link itself is neglected here) thus

$$\Phi_{int} = -LI_2. \tag{14.13}$$

Expressing LI_1 and Φ_{int} with the aid of I we get

$$LI_1 = \frac{1}{1+\gamma}(LI + n\Phi_0) \tag{14.14}$$

and

$$\Phi_{int} = \frac{1}{1+\gamma}(n\Phi_0 - \gamma LI). \tag{14.15}$$

It may be seen that Equations (14.14) and (14.15) are identical with Equations (14.5) and (14.6) under the transformations

$$LI \rightarrow -\Phi_{ext}, \qquad LI_1 \rightarrow LI_{circ}. \tag{14.16}$$

The magnetic behaviour of the ring is apparently independent of the way of excitation. Hence the graphs of Fig. 14.2 (and the conclusions drawn from them) are valid for current excitation as well.

14.3 Nonlinear theory

If the weak link is a proper Josephson tunnel junction then the integration over the weak link is simply represented by a phase angle and we can make use of the formula (Equation (13.3)) derived for the double junction. Dropping one of the phase angles (as we have only one junction now) and performing the integrations we get as our first equation

$$\frac{\hbar}{q}\phi + \Phi_{int} = n\Phi_0. \qquad (14.17)$$

Our second equation is Equation (13.7) which is of course still valid, and our third equation is the nonlinear relationship between phase and current we have met so many times before*

$$I_{circ} = I_c \sin \phi. \qquad (14.18)$$

We may now express from these three equations the internal flux and the

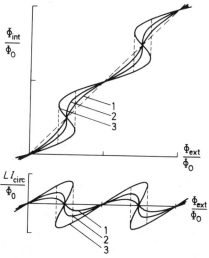

Fig. 14.3 Internal magnetic flux and circulating current of a superconducting ring containing one Josephson tunnel junction as functions of the external flux for 1: $I_c = \Phi_0/\pi L$, 2: $I_c = \Phi_0/2\pi L$, 3: $I_c = \Phi_0/4\pi L$. After Silver and Zimmerman [438].

---

*Since the maximum Josephson current plays the same role here as the critical current in the previous Section we shall use here the notation I_c instead of our usual I_j.

circulating current with the aid of the external flux as follows

$$\Phi_{int} - LI_c \sin\left[\frac{q}{\hbar}(\Phi_{int} - n\Phi_0)\right] = \Phi_{ext} \tag{14.19}$$

and

$$LI_{circ} = -LI_c \sin\left[\frac{q}{\hbar}(LI_{circ} + \Phi_{ext} - n\Phi_0)\right]. \tag{14.20}$$

Graphical solutions for Φ_{int} are shown in Fig. 14.3 for three values of I_c. For $I_c > \Phi_0/2\pi L$ the functions are multivalued and the transitions (denoted by dotted lines) irreversible similarly to the predictions of the linear model. However, when $I_c < \Phi_0/2\pi L$ the nonlinear model gives qualitatively different conclusions; Φ_{int} and LI_{circ} are then continuous single valued functions of Φ_{ext}. There are no sudden transitions, the ring can pass continuously from one state to the next.

14.4 Static magnetisation experiments

The internal flux as a function of external flux was measured by Silver and Zimmerman [438] in an experimental set-up shown in Fig. 14.4. The superconducting ring is realised with the aid of two point contacts. One of which forms the junction and the other is just adjusted to a short circuit. As shown in Fig. 14.4 the magnetometer utilises the same type of ring. The principle of its operation will be explained in Section 18.5 among the other magnetometers; for the time being we may look at it as a measuring instrument.

The ring under study has a field coil closely wound on it to provide Φ_{ext}. This applied flux is slowly cycled over several flux quanta with a very low frequency (VLF) generator. The detector ring experiences an applied flux from two sources; a fraction of the flux in the sample ring $x_1\Phi_{int}$ and a fraction of the applied flux $x_2\Phi_{ext}$. Thus the output from the magnetometer is proportional to $x_1\Phi_{int} + x_2\Phi_{ext}$ and this controls the vertical input to the X–Y plotter.

The internal flux as a function of external flux was observed for a ring 2 cm long and having an internal diameter of 1 mm. The measured curves (a, b, c, d) are shown in Fig. 14.5 for $I_c \cong 3.25$, 0.75, 0.5 and 0.4 times Φ_0/L respectively. Each run represents a cyclic variation of Φ_{ext} over the range shown. The vertical position was arbitrarily adjusted for each scan but each division is one unit of flux. The coordinate axes are not orthogonal in order to subtract the linear term $x_2\Phi_{ext}$ from the vertical scale.

In cases a and b the ring is almost perfectly diamagnetic until the critical current is reached; at these points the flux changes discontinuously by Φ_0. As Φ_{ext} varies in one direction irreversible steps occur with a spacing Φ_0. Upon reversing $d\Phi_{ext}/dt$ there is hysteresis representing irreversible work. When I_c

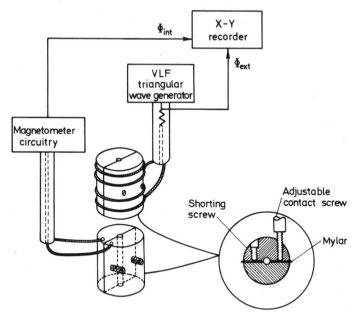

Fig. 14.4 Experimental set-up used to determine the stationary magnetic behaviour of a weakly connected superconducting ring. The upper ring wound with a field coil and connected to the very-low-frequency (VLF) generator is the sample and the lower ring serves as the magnetometer detector. After Silver and Zimmerman [438].

becomes just less than $\Phi_0/2L$ as in case c the magnetic behaviour of the ring becomes continuous and reversible. The flux changes rapidly near the half-quantum points but still reversibly.

Accepting the linear theory in the vicinity of $\Phi_{ext} = n\Phi_0$ we may determine the value of γ from the derivatives of the measured curves. Differentiating Equation (14.5) we get

$$\frac{d\Phi_{int}}{d\Phi_{ext}} = \frac{\gamma}{1+\gamma} \qquad (14.21)$$

which gives $\gamma = 0.17$ and 0.56 for curves c and d respectively.

It may be seen from the experimental results shown in Fig. 14.5 that the selection rule $\Delta n = \pm 1$ is well obeyed. Curves a and b are well described by the linear theory but c and d look more like the results for Josephson tunnel junctions. This is quite reasonable; one would expect that for smaller I_c the weak link becomes 'weaker' resembling more the properties of a Josephson junction.

When $\alpha \gg 1$ jumps of several flux quanta may also occur. In fact transitions $|\Delta n| = 2$ and 3 are more frequent in that case than $|\Delta n| = 1$ as shown in Fig. 14.6.

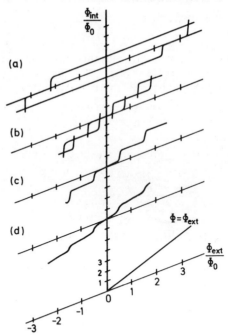

Fig. 14.5 Internal magnetic flux as a function of external magnetic flux for (a) $\alpha = 13/4$, (b) $\alpha = 3/4$, (c) $\alpha = 1/2$ and (d) $\alpha = 2/5$. After Silver and Zimmerman [438].

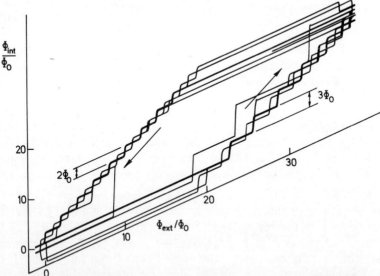

Fig. 14.6 An example of the magnetic behaviour of a weakly connected ring when α is of the order of 10. The recording shows nine cycles of the external flux. After Silver and Zimmerman [438].

14.5 Time-varying solutions

The simplest way to consider the effect of a time-varying magnetic flux is to adopt a quasi-stationary approach or in other words assume the validity of the curves obtained for the stationary case. Then $\Phi_{int}(t)$ as a function of $\Phi_{ext}(t)$ may be obtained from Figs. 14.2, 14.3 or 14.5. The model, however, needs some modification at the points where the internal flux changes suddenly (we shall assume again that only transitions $\Delta n = \pm 1$ occur). The change cannot be infinitely fast because the circulating current can change only as fast as the time constant of the ring permits. Denoting the time constant by $\tau (= L/R$, where R is the normal state resistance) the emf induced in the ring is

$$V_i = -\frac{d\Phi_{int}}{dt} \cong \frac{\Phi_0}{\tau}. \tag{14.22}$$

An approximate formula derived by Silver and Zimmerman [341, 438] for the transition time is

$$\tau \approx eLI_c/\Delta \tag{14.23}$$

which for a choice $LI_c \sim \Phi_0$ reduces to $\tau = e\Phi_0/\Delta \approx 2 \times 10^{-12}$ sec for a gap of 1 mV. The associated emf is roughly Δ/e and it depends [438] little on Φ_0/L.

A case of practical interest (because it provides the basis for the operation of a class of very sensitive magnetometers and voltmeters) is the applied flux of the form

$$\Phi_{ext} = \Phi_x^0 + \Phi_x^1 \sin \omega t \tag{14.24}$$

The emf is then

$$V_i(t) = -\frac{d\Phi_{int}}{d\Phi_{ext}}\frac{d\Phi_{ext}}{dt} = -\omega\Phi_x^1\frac{d\Phi_{int}}{d\Phi_{ext}}\cos \omega t. \tag{14.25}$$

If $\gamma \neq 0$ then $d\Phi_{int}/d\Phi_{ext}$ is finite, thus a time-varying flux means a time-varying emf in the circuit. In addition to this we have the fundamental component of the emf pulses associated with the sudden changes in the internal flux. This fundamental component v_{i1} depends both on Φ_x^0 and Φ_x^1. Hence by measuring the variation of V_{i1} as a function of Φ_x^1 we may determine the unknown Φ_x^0.

The corresponding experiment is to keep Φ_x^0 fixed and measure the variation of V_{i1} with Φ_x^1. The measured curve would contain a strong linear component which should be eliminated in order to reveal the fine structure. This may be done with the aid of the circuit [438] shown in Fig. 14.7. Φ_x^1 (or rather the equivalent rf current) is varied by changing the output of the rf generator which is amplitude modulated at an audio rate. The rectified rf voltage is split into two channels; the horizontal sweep is fed directly whereas a rejection network tuned to the audio frequency is inserted into the vertical channel. Thus the horizontal axis is proportional to Φ_x^1 but the vertical deflection is caused by the nonlinear terms only. The experimental results are shown in Fig. 14.8. It may be seen that for two values of Φ_x^0 differing by as little as $\Phi_0/2$ the rf current-voltage characteristics are noticeably different. The pattern is a periodic function with a monotonically decreasing amplitude; the entire pattern repeats when the d.c. current changes by Φ_0/L. the first peak corresponds to $\Phi_x^1 = LI_c$ and the period is Φ_0.

The theoretical formulae may be derived (see Appendix 7) without too much difficulty for the nonlinear case when $I_c \ll \Phi_0/L$. The resulting expression is

$$V_{i1} = -\omega\Phi_x^1 + 2\omega L I_c \cos\frac{2\pi\Phi_x^0}{\Phi_0} J_1\left(\frac{2\pi\Phi_x^1}{\Phi_0}\right). \tag{14.26}$$

General expressions for the linear case and arbitrary I_c may be found in Reference [438].

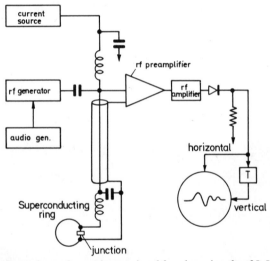

Fig. 14.7 Block diagram of experimental set-up used for observing the rf I–V characteristics of the weakly connected ring. The network 'T' in series with the vertical oscilloscope input is tuned to the frequency of the amplitude modulation. After Silver and Zimmerman [438].

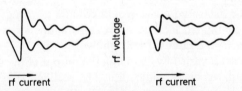

Fig. 14.8 Oscilloscope photographs of the rf V–I curves derived from the experimental arrangement of Fig. 14.7. The two curves are for two values of magnetic field differing by $\Phi_0/2$. After Silver and Zimmerman [438].

The complete experimental behaviour of the ring under the application of a 30 MHz current is summarised in Fig. 14.9 showing multiple exposure photographs of oscilloscope traces [429]. The output is rectified 30 MHz signal displayed as a function of the flux Φ_x^0 (swept sinusoidally at an audio rate of about 300 Hz giving the fm wave trains in the horizontal direction). Φ_x^1 is set at a number of discrete values from zero (at the bottom of the scope) upwards. The four sets of curves a–d were obtained by increasing the contact area and by this the critical current from $I_c \ll \Phi_0/L$ in Fig. 14.9 (a) to $I_c \sim 10\Phi_0/L$ in Fig. 14.9 (d).

For $I_c \ll \Phi_0/L$ we may assume that Equation (14.26) is valid which is roughly born out by the experimental results. Whenever $J_1(2\pi\Phi_x^1/\Phi_0) = 0$, the fundamental component V_{i1} is independent of Φ_x^0. Between the zeros of the Bessel function V_{i1} varies periodically as a function of Φ_x^0, but the dependence is more and more reduced as Φ_x^1 increases. It may also be seen that for the two strongest

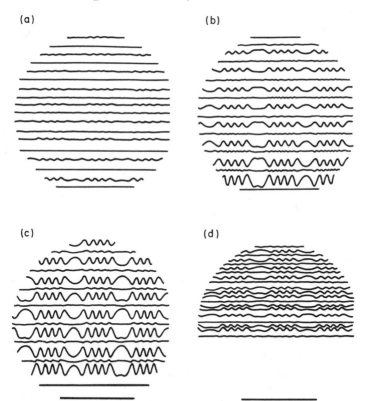

(a)   (b)

(c)   (d)

Fig. 14.9 Rectified response of a weakly connected ring. The rf voltage at the generator (proportional to Φ_x^1) is increased from zero in 16 steps. The abscissa represents Φ_x^0. The critical current I_c is increasing from (a) to (d) from $I_c \ll \Phi_0/L$ to $I_c \sim 10\Phi_0/L$. After Zimmerman [429].

coupling (Figs. 14.9(c) and (d)) the effect starts at higher values of Φ_x^1. This is because the point contact is linear until the critical current is closely approached and the pattern appears with maximum amplitude when Φ_{ext} becomes greater than LI_c.

Similar experiments on a similar geometry were performed by Nisenoff [343]. The ring was realised in the form of a thin film cylinder and the weak link was a Dayem or Notarys bridge as shown in Fig. 14.10. The rectified output (in a similar experimental set-up) was measured as a function of Φ and Φ_x^1 (corresponding only to the second term in Equation (14.26)). The results (as reported by Mercereau [364]) are shown in Fig. 14.11. Both th

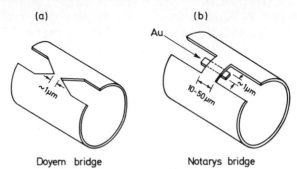

(a)   (b)

Doyem bridge   Notarys bridge

Fig. 14.10 Weakly connected rings using (a) a Dayem bridge, (b) a Notarys bridge. In the latter case the weakness is induced by the proximity effect. After Nisenoff [343].

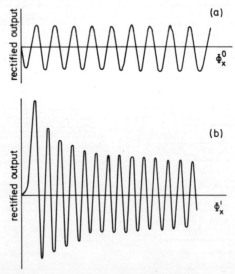

Fig. 14.11 Rectified output as a function of (a) Φ_x^0 while Φ_x' is kept constant, (b) Φ_x' while Φ_x^0 is kept constant. Note that the two scales are different; both Φ_x^0 and Φ_x' are periodic in Φ_0. After Mercereau [364].

harmonic dependence on Φ_x^0 and the Bessel function dependence on Φ_x^1 can be clearly seen. The periodicity in Φ_x^0 was examined to a fraction of a percent but the periodicity (for larger values of Φ_x^1 the zeros of the Bessel function tend to be equally spaced) in Φ_x^1 was found only to an accuracy of a few percent. The period is Φ_0 in both cases (note that Figs. 14.11 (a) and 14.11 (b) have different horizontal scales).

14.6 Parametric inductance

We have met this problem in Section 11.7 where we determined the effective inductance of a Josephson tunnel junction. The problem for the ring may be posed in a similar manner. If we feed the ring of Fig. 14.1 (b) by an alternating current I and measure across the junction a voltage V then an inductance may be defined by the relation

$$L = \frac{V}{dI/dt}. \tag{14.27}$$

Expanding V into a Fourier series and substituting the fundamental component into Equation (14.27) we may define an effective inductance L_1. It is obvious that L_1 will not be a constant and will be dependent both on the d.c. and a.c. part of I (or on Φ_x^0 and Φ_x^1 if an equivalent flux is applied). In principle it is easy to calculate L_1 but since the algebraic and numerical calculations required are rather extensive we shall make here only a few qualitative comments and present some experimental results.

When $\alpha < 1$ the fundamental component of voltage is in phase with dI/dt and hence L_1 is real. This means that an effective parametric inductance may be defined. When $\alpha > 1$ the internal flux Φ_{int} is a multivalued function of Φ_{ext}

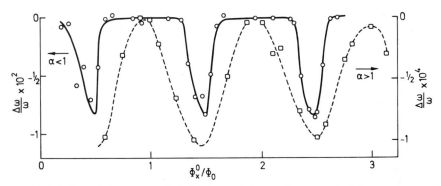

Fig. 14.12 Effect of the parametric inductance of the weakly connected ring on the resonance frequency of the composite LC circuit. The variation in frequency is shown for $\alpha < 1$ and $\alpha > 1$ as a function of applied flux Φ_x^0. Note that the two ordinate scales differ by 10^2 for the two cases. After Silver and Zimmerman [438].

leading to flux jumps and hysteresis losses. Hence L_1 will turn out to have both in-phase and out-of-phase components or in other words we need to define a parametric resistance as well.

Experimentally [438] it is relatively easy to measure L_1 in a set-up similar to that of Fig. 11.11. We replace the single junction by a junction in the ring and we can deduce the inductance of the element (ring+junction) from the measured resonant frequency of the circuit.

The change in resonant frequency as a function of Φ_x^0 is shown in Fig. 14.12 both for $\alpha < 1$ and for $\alpha > 1$. It may be seen that the variation is larger by about two orders of magnitude for the $\alpha < 1$ case. This is because we are sampling the actual derivative $d\Phi_{int}/d\Phi_{ext}$ rather than averaging over a hysteresis loop. Similar results were found by increasing the rf flux. The effective inductance increases and the resonant frequency decreases as more transitions occur per cycle.

14.7 Resistive rings

The rings investigated so far were entirely superconducting; a voltage appeared across the junction only when the critical current was exceeded. It is, however, difficult to achieve an accurately defined small d.c. voltage by this method. There are no low voltage, low impedance sources either which could be put in series with a junction. An elegant solution [339, 430–432] is to insert a small resistance into the ring and feed it by a current source as shown in Fig. 14.13 (a).

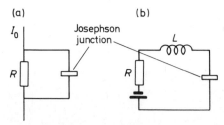

Fig. 14.13 Josephson junction biased with the aid of a resistance in parallel. (b) Equivalent circuit.

This was an easily controllable voltage $V' = I_0 R$ may be realised in the circuit. The voltage across the junction may be calculated from the equivalent circuit of Fig. 14.13 (b) by writing up Kirchof's loop equation as follows

$$IR + L\frac{dI}{dt} + V_J = V'. \tag{14.28}$$

But the current and voltage across the junction must obey the Josephson relationships

$$I = I_c \sin \phi \quad \text{and} \quad V_J = \frac{\hbar \, d\phi}{q \, dt} \tag{14.29}$$

leading to the differential equation

$$\omega'(\eta \sin \phi - 1) + (\xi \cos \phi + 1)\frac{d\phi}{dt} = 0 \tag{14.30}$$

where

$$\eta = \frac{RI_c}{V'} < 1 \quad \text{and} \quad \xi = \frac{2\pi LI_c}{\Phi_0} \tag{14.31}$$

are dimensionless parameters and

$$\omega' = \frac{q}{\hbar} V'. \tag{14.32}$$

The solution of Equation (14.30) is (see Appendix 8)

$$2 \tan^{-1} \frac{\tan \frac{\phi}{2} - \eta}{(1 - \eta^2)^{1/2}} - \frac{\xi}{\eta}(1 - \eta^2)^{1/2} \ln (1 - \eta \sin \phi) = \omega_0 t \tag{14.33}$$

where

$$\omega_0 = \frac{q}{\hbar} V_0 = \frac{q}{\hbar}\sqrt{(V'^2 - I_c^2 R^2)}. \tag{14.34}$$

We may determine ϕ as a function of time from Equation (14.33) and from that we can plot the current. When both η and ξ are zero Equation (14.33) reduces to $\phi = \omega_0 t$ and the current is a sinusoidal function of time. For finite values of η and ξ the sinusoidal is distorted as may be seen in Fig. 14.14 where I/I_c is plotted against $\omega_0 t$ for $\eta = 0, 0.25$ and $\xi = 0.5, 1, 2$. As may be expected (in analogy with the solution in Section 14.3), for higher values of ξ the current becomes multivalued which in practice means sudden jumps in current along the dotted lines. The boundary between the singlevalued and multivalued solutions occurs at $\xi = 1$ when $\eta = 0$ and remains roughly at the same value for finite η.

An alternative way of determining the current is to rewrite the solution in the form [431, 439] (Appendix 8)

$$\frac{I - \eta I_c}{I_c - \eta I} = \sin\left\{\omega_0\left[t + \frac{L}{R} \ln\left(1 - \eta\frac{I}{I_c}\right)\right]\right\} \tag{14.35}$$

from which the $I(t)$ function can be directly obtained.

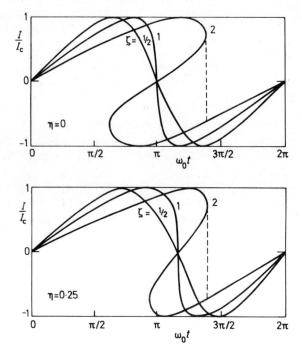

Fig. 14.14 Temporal variation of normalised current for the parameters (defined by Equation (14.31)) $\eta = 0$, $0\cdot25$ and $\xi = \frac{1}{2}, 1, 2$.

Note that for $L = 0$ the problem tackled here is identical to that in Section 11.2 when $C = 0$. Thus the approximate calculation of Sullivan *et al.* [439] for subharmonic steps may be taken as further confirmation of the conclusions of Clarke *et al.* [396] and Waldram *et al.* [397] discussed in Section 11.5.

We shall turn now to the problem how the flux relationships modify for the resistive ring. Equation (13.7) is still valid but not Equation (14.17) because a wavefunction does not exist in the resistive region. Denoting, however, by ϕ_R the difference between the phases on each side of the resistive section we may write instead of Equation (14.17)

$$\frac{\hbar}{q}\phi_R + \frac{\hbar}{q}\phi + \Phi_{int} = n\Phi_0. \tag{14.36}$$

Here ϕ_R is not uniquely specified since it depends on the initial magnetic flux, but we know that it is related to the voltage across the resistive section by the formula $d\phi_R/dt = (q/\hbar)V$. This voltage (in the circuit of Fig. 14.13) depends on R. For sufficiently small resistance the voltage is small and we may argue that everything is the same as for the fully superconducting ring* with the exception of

*A somewhat more rigorous derivation may be found in Reference [440].

the voltage dependence of ϕ_R. This means that Equation (14.19) remains valid if we replace $(q/\hbar)\Phi_{int}$ in the argument of the sine function by

$$\frac{q}{\hbar}\Phi_{int} - \phi_R = \frac{q}{\hbar}\Phi_{int} - \frac{q}{\hbar}Vt \cong \frac{q}{\hbar}\Phi_{int} - \omega_0 t. \qquad (14.37)$$

Hence our magnetisation curve Φ_{int} versus Φ_{ext} (shown in Fig. 14.3) moves bodily along the diagonal south-west or north-east with a speed determined by the Josephson frequency ω_0. It follows that for a given applied external flux (i.e. Φ_{ext} held fixed) the internal flux will vary in time; its actual value may be read on the sliding curve.

The rf I–V characteristics will show little change. Replacing the super-conducting ring by a resistive one in the set-up of Fig. 14.7 and repeating the experiment [432] the results (Fig. 14.15) look very similar to those of Fig. 14.8.

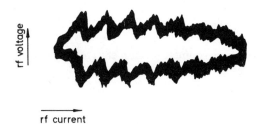

rf voltage

rf current

Fig. 14.15 Experimental rf V–I curves obtained with the experimental set-up of Fig. 14.7 by replacing the fully superconducting ring by a partly resistive ring. After Silver [432].

The critical current of the junction may again be deduced from the first break. Each succeeding signal is at a position where the input current is increased by Φ_0/L. Thus in spite of the fact that the ring contains a resistive portion flux quantisation can still manifest itself if observed fast enough. These experiments by Silver [432] were performed at 30 MHz while the time constant of the ring was about 10^{-4} sec.

15. Fabrication

15.1 Introduction

Most of the mathematical models treat the junction as consisting of t
superconductors separated by a thin, flat barrier. This 'ideal' geometry v
indeed used in many experiments and it differs very little from that used
normal electron tunnelling. Apart from the barrier which needs to be a li
thinner, the only difference is that for many applications a few more evapora
layers are required. This will be discussed in Section 15.2 under the name
tunnel junctions.

The other type of junction geometries, namely point contacts, solder joi
(beads) and superconducting bridges will be discussed in Sections 15.3 to 1

15.2 Tunnel junctions

Most of the initial experiments designed to prove the existence of Josephs
tunnelling were done on thin film junctions. The technology is well descri
by Langenberg *et al.* [379, 441] whose method is essentially the same as t
discussed under thermal oxidation in Section 4.9. The glow discharge techni
was also described in Section 4.9; in fact, the Pb–I–Pb junctions of Schrc
[115] mentioned there were meant for Josephson tunnelling.

The field in which the large-scale production of tunnel junctions may
foreseen is that of computer memories. For that purpose an extra electrod
needed for a control current and possibly also a ground plane to reduce ind
tance. A cross-section of one of Matisoo's [442] junctions may be seen in F
15.1. The fabrication procedure is as follows.

First a 500 nm lead ground plane is evaporated upon an epoxy or gl
substrate through a mask. Then the slide is removed from the vacuum syst

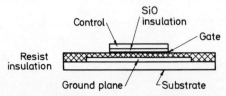

Fig. 15.1 Cross-section of a tunnelling cryotron. The gate contains a Josephson juncti
After Matisoo [442]. See also Fig. 16.2 (b).

and coated with photoresist (1 to 2 μm thick) which serves to insulate the ground plane from the remainder of the structure. Next the electrodes are evaporated and then a strip of tin of thickness 300 nm. The vacuum chamber is then opened to atmosphere for 30 to 60 minutes for oxidising the tin. The second layer of tin is deposited afterwards, then a layer of SiO (300 nm) for insulation and finally lead (500 nm) again. The junction sizes were $(0.25 \text{ mm})^2$ and $(0.125 \text{ mm})^2$.

Junctions of some other dimensions (increasing up to several times the Josephson penetration length) were also made by Matisoo [407] for measuring the maximum zero-voltage current as a function of magnetic field. The procedure was similar to that outlined above with the exceptions that SiO insulation was used instead of the photoresist and the oxidation was carried out in a flow type furnace at gas temperatures less than 75°C using either O_2 or air. The value of j_J was varied by adjusting the gas temperature and the exposure time. Typical current density values obtained by oxidising in air at 50°C were 10^6 A/m^2 for two hours and 2×10^4 A/m^2 for 40 hours.

Matisoo [407] claimed repeatability as good as one part in a thousand for measuring the maximum zero-voltage current. It should also be noted that his samples could be recycled many times between liquid helium and room temperature.

Junctions for the same purpose were made by Schroen [115] by combining the glow-discharge technique of Miles and Smith [114] with the photoresist technique of Pritchard et al. [443–446]. The switching device shown in Fig. 15.2 was made by the following process. First lead is evaporated on the substrate, then a layer of photoresist is deposited. The photoresist is illuminated by ultraviolet light through a mask at the position of the required junction (more about this technique in Section 15.5). The photoresist polymerised by the light may be washed away and the exposed region oxidised by the glow discharge technique. After that there are further depositions of lead, photoresist, lead and often a final layer of photoresist to protect the device. The electrical properties of these junctions were tested; no change was found in the course of eight months' storage at room temperature.

Junctions between niobium and lead have also been fabricated. Nordman [447] used getter-sputtering for the niobium films and heated the sample in oxygen for oxidisation. He found that the conduction of the devices increased between a factor of 1.1 and 20 over several months while the devices were stored at room temperature.

Mullen and Sullivan [448] evaporated niobium on single crystal sapphire substrates. For oxidisation by the glow discharge technique the Nb film was cooled to about 80°K. After termination of the discharge the system was re-evacuated and the substrates warmed to about 240°K. The final barrier thickness was a function of (besides the parameters of the glow discharge) the temperature to which the substrates were warmed and of the duration of the warming period indicating that the oxide layer was not produced during the

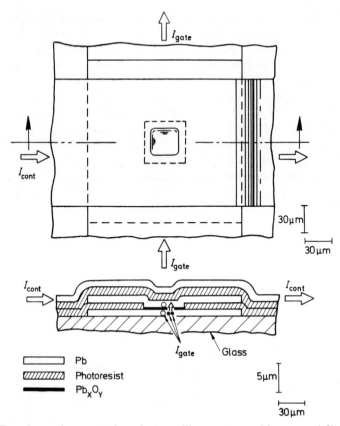

Fig. 15.2 Top view and cross-section of a tunnelling cryotron with a crossed film control. After Schroen [115].

discharge but after the deposition of the lead layer which makes it very likely that adsorbed oxygen is responsible for the reaction. For more details see Kamper, Mullen and Sullivan [449] who also report on life tests. Junctions made by this technique survived both storage at room temperature (for four months) and repeated (about 400 times) thermal cycling between liquid nitrogen and room temperature.

15.3 Point contact junctions

Using a point contact upon the surface of a bulk material is a convenient way of realising a junction whose properties range from those of tunnel junctions to thin film bridges. In the former case there is an insulator between the two materials (one of them is a superconductor, the other one may be a super-conductor, a normal metal or a semiconductor) comprising the point contact whereas in the latter case the two materials are in direct contact over a small area. Applications for normal electron tunnelling were initiated by Levinstein and Kunzler [122] for metal tips and by von Molnar *et al.* [450] for semiconductor tips. Josephson type of applications were considered somewhat earlier by Zimmerman and Silver who have developed various kinds of point contacts. In their first attempt two mutually perpendicular 0·005 in. niobium wires were held in contact with a force of a few grams at the crossing point. Their first double junctions (for interference measurements) were made in the following manner:

(*i*) A niobium ribbon 0·2 mm wide is crossed by a bent wire which contacts the ribbon only near the two edges (Fig. 15.3 (*a*)).

(*ii*) A pair of wires bent in the form of hairpins are held together by a drop of epoxy cement [341] (Fig. 15.3 (*b*)).

The disadvantage of all these devices was the unpredictability of their performance. Clearly, a more useful device can be obtained by allowing some adjustment in the pressure by which the two materials contact. This aim was realised by using niobium screws [341, 437] (as mentioned in Section 8.3 and shown in Fig. 8.6 (*d*)) adjustable by suitable linkage from outside the cryostat.

For some applications (see Section 14.7) the point contact must be in parallel with a resistive section. A solution growing out of the double point contact version of Fig. 8.6 (*d*) is shown in Fig. 15.3 (*c*) and an improved construction in Fig. 15.3 (*d*). The latter one is much more stable mechanically and the symmetric arrangement of the resistive section prevents distortion due to differential thermal expansion.

Even better stability can be reached when the holes are symmetrically arranged with respect to the point contact. These junctions (resistive version shown in Fig. 15.3 (*f*), fully superconductive in Fig. 15.3 (*e*)) can withstand severe mechanical and thermal shock as well as repeated thermal cycling without any change in their electrical properties.

A considerable advantage of these point contact junctions is the facility with which they couple to electromagnetic waves (they act as a short dipole antenna). So they are preferably used in electromagnetic cavities when the aim is the production of microwave power. Dayem and Grimes [403], for example, coupled the junction to the cavity by pressing a tantalum probe against the cavity wall made also of tantalum (Fig. 17.2).

Point contacts were also used by Grimes *et al.* [345, 354] and by Grimes and Shapiro [327] in their studies of infrared detection and frequency mixing respectively. The contacts were made by pressing together the ends of two pieces of wire (0·03 in. diameter), one of which was flattened, and the other sharpened by filing and cutting with a razor blade or by chemical etching. Niobium and tantalum wires were directly used for making up the junctions for these metals whereas copper wire tinned with indium or lead was used for junctions made from the latter materials.

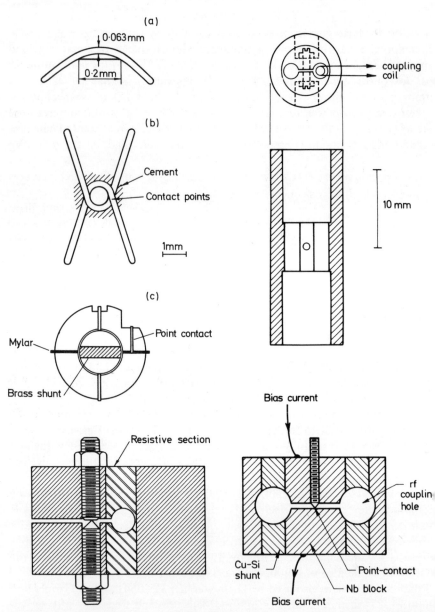

Fig. 15.3 Various realisations of point contact junctions: (*a*) ribbon in contact with bent wire, (*b*) pair of bent wires, (*c*) superconducting ring with a brass shunt; contact made through the Mylar layer, (*d*) stable mechanical arrangement compensated for differential thermal expansion (after Zimmerman *et al.* [435], (*e*) stable symmetric structure (after Zimmerman *et al.* [435]), (*f*) stable symmetric structure with resistive insert (after Kamper and Zimmerman [538]).

Finally, we wish to mention here an experiment by Zimmerman, Thiene and Harding [435] in which a notched screw is broken at the notch at liquid helium temperatures and the contact is then re-established. The electrical performance of this contact was identical with that of ordinary contacts obtained at a certain pressure. Since this special contact may be regarded 'perfectly clean' (in the sense that any impurities present were those in the bulk material rather than on its surface) the experiment proves that a class of point contacts behave as metal bridges.

15.4 Solder junctions

The simplest way of making a Josephson junction is to dip short lengths of wires (made of superconducting materials, usually niobium or tantalum) into molten solder. This is sometimes called a 'solder blob' or 'bead' junction because of the shape of the solder surrounding the wire. The junction is between the two superconductors, the wire and the solder, and the barrier is provided by the oxide layer on the wire.

The first junction of this type was made by Clarke [338] by the following procedure. A short length of 0·004 in. niobium wire was suspended over a semi-circular trough. Molten tin–lead solder to which excess flux had been added was poured into the trough so as to immerse the wire over a length of about 5 mm, the diameter of the solder pellet being about 3 mm. The specimen was allowed to cool and the resistance between the niobium wire and the solder measured in a four-terminal arrangement; specimens with resistances in the range 0·5–1·0 ohms were considered acceptable and the remainder rejected.

Similar junctions were made by Shapiro [326] using indium, lead, tin and a number of solders as the low melting point metal. A particularly good composition was 50% Sn, 32% Pb and 18% Cd which had good wetting properties, low melting point and a fairly high transition temperature (7°K). Sometimes the 'natural' oxide layer on the wires gave satisfactory junctions but when this was not the case the wire was oxidised by heating in air. The heating was accomplished by passing brief surges of current through the wire so as to get it momentarily red hot; or by running the hot tip of a soldering gun along the portion to be oxidised. The room temperature resistance of the specimens varied from a few tenths of an ohm to several tens of ohms and was apparently unrelated to superconducting behaviour. The wire sizes ranged from 0·003 to 0·020 in. without any noticeable change in performance. The size of the solder blob was not critical either though larger sizes tended to show more shorts. A convenient size of junction was achieved when a wire length of about 2 mm was covered by the low melting metal.

For four-probe measurements there is usually another copper wire embedded into the low melting point metal. In a particular realisation [420] shown in

Fig. 15.4 the copper wire is in contact with tin at a and b but insulated from the niobium wire.

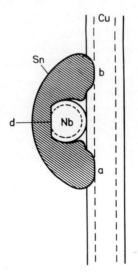

Fig. 15.4 Solder drop junction by Omar and de Bruyn Ouboter [420]. The varnish insulation of a 1 mm diameter Cu wire is removed from two equal spots a and b. The 50 μm Nb wire is stretched at right angles to the Cu wire, its insulation is removed at d.

The main advantage of the solder junction, besides its simplicity, is its ability to withstand repeated thermal cycling between ambient and liquid helium temperatures. Its main disadvantage is that the exact geometry of the junction is unknown. It seems difficult even to predict whether a single, double or multiple junction will be the result. From measured behaviour Clarke [338] concluded that his specimens behaved as if there were two junctions between the wire and the solder located at each end of the bead.

15.5 Superconducting bridges

A superconducting bridge, as the name implies, connects two superconductors by a narrow link. The first bridges were made practically simultaneously by Lambe *et al.* [451], Anderson and Dayem [361] and Parks *et al.* [359].

In the simplest technique a continuous metal film is evaporated and then cut with a sharp point to the required geometry. Bridges of 20 μm [451] and 1–5 μm [377] width were produced by this method.

Another simple technique, developed by Parks *et al.* [359] is as follows. A thin fibre of GE 7031 varnish is stretched over a glass microscopic slide. The fibre is delicately severed with a microknife and metal is then evaporated on to the resulting substrate. The assembly is then soaked in an ultrasonic bath of ethyl alcohol. This step removes the varnish fibre and the metal deposited on the fibre, but not the metal deposited in the juncture where the cut was made. The resulting configuration is shown in Fig. 15.5 (a). Bridges 1–10 μm wide and 1–50 μm long were prepared in this manner.

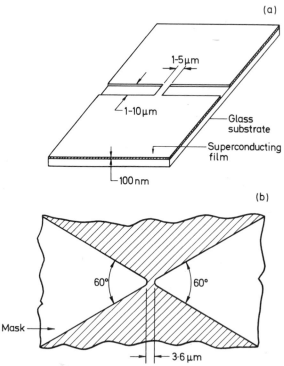

Fig. 15.5 Superconducting bridges by (a) Parks et al. [359], (b) Anderson and Dayem [361].

Similar, rectangular bridges were obtained by Dayem and Wiegand [383] in two distinct steps. A 7 μm tungsten wire was laid across the substrate in one evaporation and a special mask with a 4 μm wide stripe perpendicular to the direction of the wire in the subsequent evaporation.

If the aim is a short bridge then the best method is to evaporate the super-conductor through two V-shaped metal masks [383] yielding the configuration shown in Fig. 15.5 (b). The problems are mainly associated with the finite thickness of the metal masks (they cast some shadow on the substrate) and the design of the holder (for achieving good alignment motion in all three dimensions should be possible). Bridges of 2–6 μm [383], 3–4 μm [452] and 2–12 μm [425] were obtained by this technique. Typical bridge shapes [425] (as found under a transmission microscope) as shown in Fig. 15.6.

Yet another way of making bridges is to use the photoresist technique

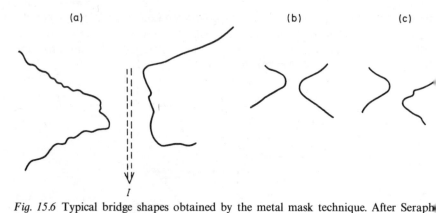

Fig. 15.6 Typical bridge shapes obtained by the metal mask technique. After Seraph [425].

developed for the manufacture of integrated circuits. The main steps in t process [425] are as follows.

(*i*) Daub a drop of photoresist* on a clean substrate and spread it thin spinning at high revolution.

(*ii*) Dry the slides in an oven.

(*iii*) Illuminate the photoresist by ultraviolet light through a photograp mask (Fig. 15.7).

(*iv*) Immerse the slide in the developer, rinse it in deionised water, dry it a flush off dust particles (the photoresist geometry is now the same as th of the emulsion on the photographic mask).

(*v*) Evaporate a thin film (about 100 nm) of the superconductor on the t of the slide (a section of the slide across the bridge region looks now shown in Fig. 15.8).

(*vi*) Place the slide in a beaker of acetone in an ultrasonic cleaner where the unexposed photoresist is dissolved and the metal film on the top shaken off leaving behind the required geometry.

(*vii*) Evaporate contacts.

The Notarys type of bridge based on the proximity effect was shown in F 8.6(*b*). Usually, the normal layer is evaporated first and then the superconduct As far as fabrication is concerned its advantage is that the degree of 'weakne depends on the relative thicknesses of the normal and superconducting laye thus a wide range of properties may be realised with the same set of photograp masks.

The obvious limitation in using photographic masks is light diffracti which becomes a serious obstacle at dimensions approaching the wavelength

---

*The process described here applies to a positive photoresist which has the property that parts exposed to ultraviolet light dissolve on developing. Negative photoresist may also be u but the process then involves chemical etching to obtain the desired geometry.

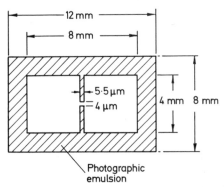

Fig. 15.7 Photographic mask used for fabricating bridges with the photoresist technique [425].

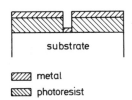

Fig. 15.8 Cross-section across the bridge after stage (v) of the fabrication process.

light. However, this limitation may be removed by going over to a technique where the light source is replaced by an electron beam which has a much smaller wavelength. Some bridges have been made [453] by this method, achieving dimensions as small as 0·1 μm.

Finally, we shall briefly describe here a technique developed by Bondarenko et al. [454, 456] in which the carefully prepared surface of a superconducting plate is oxidised, the oxide layer punctured by a steel needle and then a second superconductor* is evaporated. Bridges of 1 to 20 μm were made in this manner.

---

* SNS junctions may also be made by this method [456]; a normal metal is then evaporated before the second superconductor.

16.   Computer elements

16.1 Introduction

Whenever a physical phenomenon displays two stable states and there
some means for switching from one state to the other one, we have a poten
memory element. A superconducting device (called cryotron) which can perfo
this task was proposed by Buck [457] in 1956. The two stable states are
normal and superconducting states of a superconductor, and the change fr
one into the other state is accomplished with the aid of a small magnetic fie
These cryotrons are fairly fast, may be made very small and produced on a m
scale by vapour deposition techniques. They are not used at the moment
commercial computers partly because of prejudice (that always haunts
unconventional technique) and partly because present computer memories
still not big enough for the full advantages to be realised. A recent comparis
[458] with magnetic memories shows (Fig. 16.1) that above about 100 mill

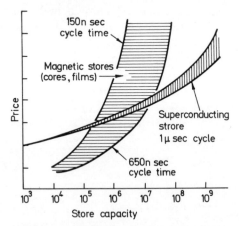

Fig. 16.1 Comparison of the economics of magnetic and cryotron computer memo
After Walker [458].

bits cryotrons represent the cheaper solution, so the chances for future appli
tions may be regarded good.

Josephson junction devices fall into the same category as cryotrons. T can perform the same operations as cryotrons but have several further advantages as discussed in this chapter. Sections 16.2 and 16.3 describe the mechanism of switching in tunnel junctions and thin film bridges respectively while Section 16.4 is concerned with memory elements. The relations for transition and steering time are derived and evaluated in Section 16.5. Some system aspects are briefly discussed in Section 16.6 and conclusions drawn in Section 16.7.

16.2 Switching properties of tunnel junctions

The Josephson tunnel junction as a memory element was first proposed by Rowell [346] and was experimentally investigated by Matisoo [344, 442, 460–462] and Pritchard and Schroen [463].

The similarities and differences between cryotron switching and Josephson junction switching are shown schematically in Fig. 16.2. Assume that both devices are kept below the critical temperature of the superconductors used, that they are fed by a constant current source and there is no current flowing

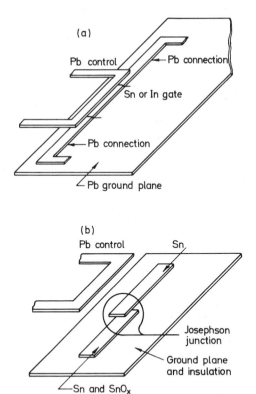

Fig. 16.2 (a) Sketch of an in-line cryotron. After Brennemann [459]. (b) Sketch of an in-line tunnelling cryotron. After Matisoo [442].

in the control electrode. Then there is no output voltage because both g
currents flow in resistance-free media. However, driving sufficiently la
currents through the control electrode (producing a high enough magn
field) an output voltage will appear. This is due to the fact that the tin in
gate circuit of the cryotron (the particular version shown here is called
in-line cryotron because the control electrode is in line with the gate) beco
normal and similarly the Josephson junction changes from Josephson to nor
electron tunnelling.

We may roughly say that the quality of the switch depends on two fact

(i) speed of transition,

(ii) magnitude of the output voltage.

The Josephson junction device is superior to the cryotron on both cou
The speed of transition for the cryotron depends on how fast the piece of
can change from the superconducting into the normal phase. Although
phase-transition is practically instantaneous at a number of discrete point
takes considerable time (of the order of 10 nsec) for the phase boundaries
propagate between these nucleation sites. In contrast, there is no pha
transition in the Josephson junction device, merely the type of tunnel
changes. This can occur very fast, so the transition time will be prima
determined by the capacitance of the junction. Matisoo's measurements [4
showed that it was within the resolving time of the apparatus, about
picosecond.

The output voltage from the Josephson junction device is determined so
by the energy gap of the superconductor used. In the case of tin this is ab
1 mV. It would be difficult to produce a voltage of similar magnitude with
cryotron. The obvious reason is that the resistance of a short piece of tir
liquid helium temperatures is bound to be small. A typical figure for the volt
output may be 200 μV. It may be noted that by using lead instead of tin for
junction material (as done by Pritchard and Schroen [463]) the output volt
is about 2·5 mV, that is one order of magnitude higher than that obtained in
cryotron.

A set of current–voltage characteristics are shown in Fig. 16.3 for the tin-

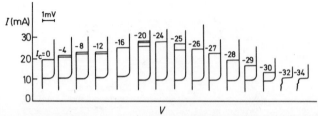

Fig. 16.3 I–V characteristics of gate (junction) taken for various values of control cur
(I_c). The minus sign indicates that the currents are antiparallel. Each curve was tra
three times. After Matisoo [442].

oxide–tin junction of Fig. 16.2. The parameter is I_c, the control current, which varies from zero to -34 mA (the negative sign indicates that the currents are antiparallel).

It may be seen that above a certain gate current (what we have called the maximum supercurrent, I_J) a voltage appears across the junction. Interestingly, this maximum gate current (I_g) depends strongly on the control current and does actually increase with increasing control current. This case is a little different from that worked out theoretically in Section 10.1 because for the junction geometry of Fig. 16.2(b) the self-magnetic field of the gate current must also be taken into account. The dependence of I_g on I_c may be explained by assuming the simple relationship

$$I_g = I_0(1 - |H|/H_0) \tag{16.1}$$

where H_0 is the magnetic field required to reduce the maximum gate current to zero, H is the magnetic field created by the joint action of the gate and control currents and I_0 is the maximum gate current occurring when $H = 0$.

Since the gate and the control are above one another the magnetic field in the junction is

$$|H|\frac{1}{d} = \left|\frac{I_g}{2} + I_c\right| \tag{16.2}$$

where d is the width of the lines. The factor 2 comes in because the current flows next to the ground plane and so the lower superconductor contributes very little to the magnetic field in the junctions. Substituting Equation (16.2) into Equation (16.1) we get

$$I_g = \frac{I_0}{1 + I_0/2dH_0}[1 - I_c/dH_0], \quad I_g \geq 0, \quad I_c \geq -\frac{I_0}{2}$$

$$I_g = \frac{I_0}{1 - I_0/2dH_0}[1 - |I_c|/dH_0], \quad I_g \geq 0, \quad I_c \leq -\frac{I_0}{2} \tag{16.3}$$

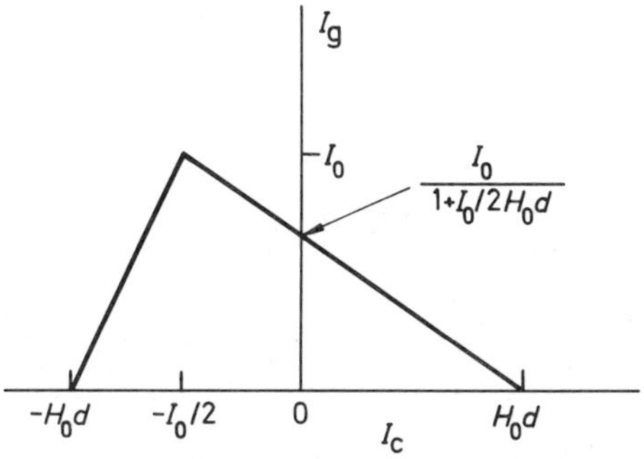

Fig. 16.4 The gate current at which switching occurs as a function of control current according to Equation (16.1). After Matisoo [442].

This is plotted in Fig. 16.4 and compared with the experimental results in Fig. 16.5.

Similar experimental results for five separate elements in an array were obtained by Pritchard and Schroen [463]; the curves are well reproducible as shown in Fig. 16.6. Pritchard and Schroen [463] investigated also the crossed-film arrangement (see Fig. 15.2) in which case I_g is a symmetric function of I_c.

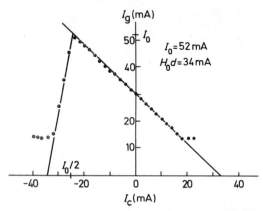

Fig. 16.5 Experimentally found I_g versus I_c curve. The points which lie off the lines are due to the next period in the diffraction pattern. After Matisoo [442].

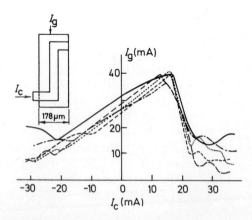

Fig. 16.6 Experimentally found I_g versus I_c curve for five in-line tunnelling cryotrons in an array. After Pritchard and Schroen [463].

16.3 Switching by thin film bridges

A typical current-voltage characteristic of a superconducting thin film bridge is shown in Fig. 16.7. It was found by Seraphim and Solymar [464] that this

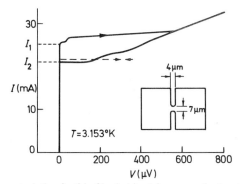

Fig. 16.7 I–V characteristic of a thin film bridge (shown on the inset). Dashed lines show the path taken when a small magnetic field is applied. After Seraphim and Solymar [464].

device may be switched between a zero voltage and finite voltage state by applying a magnetic field of a few gauss. In most of the range the switching went only one way, i.e. it switched to a finite voltage at the application of a magnetic field but did not switch back to zero voltage when the magnetic field was removed. However, when the current bias was in the vicinity of I_2, switching in both directions became possible. This latter type of switching was observed with six out of eight samples having bridge dimensions varying in width from 3 μm to 8 μm and in length from 3 μm to 11 μm. The magnitude of the voltage step also varied from a few hundred microvolts to one millivolt depending on sample geometry and operation temperature.

16.4 Memory elements

A possible way of building a flip-flop is shown in Fig. 16.8. It consists of two Josephson junction devices in parallel, fed from a constant current source. The current flow in one or the other branch represents the 'zero' or 'one' state of the flip-flop.

Consider the case [442] when $I = 40$ mA, $I_1 = 30$ mA and $I_2 = 10$ mA and both junctions exhibit Josephson tunnelling. Apply now a current pulse through the control so that junction 1 changes to normal electron tunnelling. Then a voltage appears across junction 1, the current across junction 1 diminishes and across junction 2 increases. However, when the current through junction 1 is reduced to 10 mA then the junction (according to the experimental evidence of Fig. 16.3) reverts to Josephson tunnelling and the voltage drops to zero. Now the control pulse can go to zero and there will be no further change in the circuit.

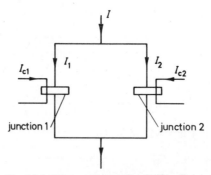

Fig. 16.8 Schematic circuit of a flip-flop.

Thus with the aid of a control pulse we could steer the current from branch 1 to branch 2.

It must be noted that in order to perform logic operations a gain larger than unity is needed, that is the control current causing the transition should be smaller than the current switched. This may be achieved by biasing the control. In this particular case the bias current is -26 mA and a further -5 mA is needed to steer 20 mA current. Hence the gain is

$$G = \frac{20 \text{ mA}}{5 \text{ mA}} = 4 \tag{16.4}$$

A memory element (Fig. 16.9) working on somewhat similar principles was proposed by Anacker [465]. There are two in-line gates B_0 and B_1 for writing and a sense gate S for nondestructive readout. A drive line w splits into two branches A_1 and A_2 and combines again before entering the next memory element. A common line b (called the bit line) overlays and controls the in-line gates B_0 and B_1. The sense gate S is part of an interrogate line i; it is between the ground plane and a part of branch A_2 which acts as a control line.

Information is stored in the ring A_1, A_2 by clockwise and counterclockwise circulating currents I_p representing 1 and 0 respectively.

Let us assume now that a counterclockwise circulating current is stored in the ring (0 state) and we wish to write in 1. We start by supplying currents I_w and I_b as shown in Fig. 16.9. I_w splits into $I_{A1} = I_{A2} = \frac{1}{2}I_w$ because of symmetry and is superimposed on the circulating current I_p so that $I_{A1} = \frac{1}{2}I_w + I_p$.

The in-line gate B_0 is designed so that it switches only if I_b and $I_g = \frac{1}{2}I_w + I_p$ coincide and flow in the same direction. This is what happens now so that B_0 switches to normal tunnelling. As a consequence (in the same manner as in the memory element discussed before) the current in branch A_1 decreases and in branch A_2 increases. As the current decreases in branch A_1 it reaches a value $(I_{g \min})$ at which the gate B_0 switches back to Josephson tunnelling. This distribution of current would remain stable if the current source I_w would be always

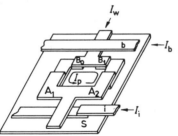

Fig. 16.9 Memory element comprising two gates B_0 and B_1 for writing and one gate S for nondestructive readout. After Anacker [465].

on. If I_w is discontinued then (in order to conserve the magnetic flux) a circulating current is set up but now in the clockwise direction. This was our purpose; 1 is now written in. We may change back to 0 in a similar manner; I_w has the same direction but I_b must be reversed.

Information is read out nondestructively by applying a current I_w and an interrogate current I_i. Gate S is designed to switch only if I_i and $I_{A2} = \frac{1}{2}I_w + I_p$ coincide, that is when 1 is stored in the selected memory element. When S switches the current I_i is rerouted supplying the required information. During reading no current redistribution occurs between branches A_1 and A_2 so that the circulating current I_p is undisturbed.

Another type of memory element was proposed by Clark and Baldwin [466]. It is based on Richards' one-cryotron-per-bit memory element [467] where the cryotron is replaced by a Josephson junction as shown schematically in Fig. 16.10. This is nothing else but the ring containing one Josephson junction

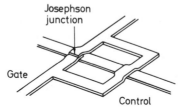

Fig. 16.10 Schematic drawing of a one-cryotron-per-bit memory element. After Clark and Baldwin [466].

discussed at length in Chapter 14. If properly excited a supercurrent may circulate in it storing one bit of information. Readout may be accomplished nondestructively by applying a pulse to the control.

Finally, we wish to mention here the possibility of storing information in the form of n flux quanta in a ring. This was investigated for simple superconducting rings by Dumin and Gibbons [468]. The presence of Josephson junctions in the ring may facilitate the processes of writing in and reading out but no experiments on this possibility have been reported so far.

16.5 Transition time and steering time

As we have seen in the previous section a certain bias current flows across the junction in a practical case. Then the application of a small control current may switch the junction to normal electron tunnelling. A similar but somewhat less favourable (as far as switching speed is concerned) transition was investigated by Stewart [469]. A junction was fed by a current I_J and the rise of the voltage measured (Fig. 16.11). The essential physical limitation is that the capacitance must be charged up before the voltage $V = 2\Delta/e$ may be reached.

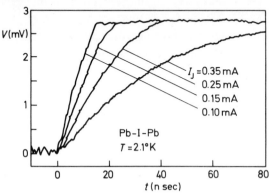

Fig. 16.11 Waveforms of junction voltage at several values of maximum supercurrent, I_J. Curvature is caused by shunt resistance of current and voltage line terminations. After Stewart [469].

The relevant equations describing the variation of d.c. voltage and current were discussed in Section 11.2. If we single out the capacitive current only, then the equation for the voltage is

$$I = C\frac{dV}{dt} \tag{16.5}$$

which with the initial conditions $V = 0$ at $t = 0$ may be solved to give

$$V = \frac{I}{C}t \tag{16.6}$$

If $I < I_J$ there is a d.c. supercurrent flowing across the junction without causing any voltage but for $I > I_J$ we need to resort to the differential equation given by Equation (11.5)

$$I = \frac{\hbar C}{q}\frac{d^2\phi}{dt^2} + \frac{\hbar G}{q}\frac{d\phi}{dt} + I_J \sin\phi \tag{16.7}$$

When the impressed current just exceeds the critical current the voltage has not had time to rise across the resistance so it is sufficient to take into account the first and third terms on the right-hand side of the above equation. This was integrated numerically by Stewart [469] showing that the effect of the d.c. supercurrent term is to introduce a small delay (less than $(\pi C \Phi_0/2I_J)^{1/2}$) and a rapidly diminishing oscillation to the otherwise linear voltage waveform associated with the charging of the capacitance. As the voltage builds up an increasing fraction of the input current is shunted into resistive terminations of the current and voltage lines, thus accounting for the exponential shapes of the waveforms. When the voltage reaches $2\Delta/e$ both the d.c. supercurrent and the capacitive current vanish and the input current divides between the resistive terminations and the normal electron tunnelling current of the junction.

The measured results shown in Fig. 16.11 are consistent with this model. The initial slope of the curves

$$\frac{dV}{dt} = \frac{I}{C} \tag{16.8}$$

lies indeed on a straight line as shown in Fig. 16.12. The capacitance calculated from this is $C = 1{\cdot}7 \times 10^{-9}$ F leading to a transition delay of less than 1% of the

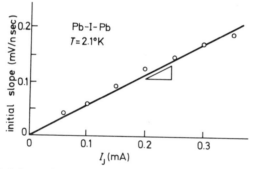

Fig. 16.12 The initial slope of the voltage waveforms as a function of I_J for a junction of area 0·064 mm². After Stewart [469].

total transition time. In the absence of a shunt conductance the transition time from Equation (16.6) is

$$\tau_t = \frac{2\Delta}{e}\frac{C}{I} \tag{16.9}$$

thus much shorter transition times may be achieved for larger currents. Taking $I = 30$ mA, $2\Delta/e = 1{\cdot}1$ mV and the above derived value of the capacitance we get for the transition time 0·24 ns, which in the light of Matisoo's experimental results ($\tau_t \le 0{\cdot}8$ nsec) seems the right order of magnitude.

The steering time may be calculated on the assumption that the steering of the current in the circuit of Fig. 16.8 starts only when the transition is completed. Assuming further that the appearance of a voltage $V_0 = 2\Delta/e$ is equivalent to the insertion of a battery in series with the normal resistance of the junction we get the equivalent circuit of Fig. 16.13. With V_0 taken as a step

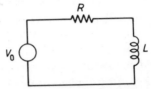

Fig. 16.13 Equivalent circuit for steering of current. R stands for the differential resistance at $V_0 = 2\Delta/e$, typically less than a mohm; L is the inductance of the loop.

function the current is given as

$$I = \frac{V_0}{R}[1 - e^{-(R/L)t}] \qquad (16.10)$$

But in a practical case V_0/R is much larger than the amount of current needed to transfer, so Equation (16.10) may be approximated by

$$I = \frac{V_0}{L}t \qquad (16.11)$$

Hence the steering time is

$$\tau_s = \frac{L\,\partial I}{V_0} \qquad (16.12)$$

where ∂I is the current transferred.

The steering time in the circuit of Fig. 16.8 was measured by Matisoo [460] by two different methods. The total inductance of the ring was measured separately to be $L = 1{\cdot}08 \times 10^{-10}$ H. The expected values for steering time are 1·8 nsec for 18 mA and 2·0 nsec for the transfer of 20 mA whereas the experimental results gave $1{\cdot}4 \le \tau_s \le 3$ nsec and $\tau_s \le 2{\cdot}0$ nsec respectively

The energy dissipation during current transfer can be estimated by assuming that the voltage V_0 remains constant and the current through the gate decreases linearly from $I_{g\,\min} + \frac{1}{2}\partial I$ to $I_{g\,\min}$ during the transfer time τ_s. The dissipated energy is then given by

$$E = V_0(I_{g\,\min} + \tfrac{1}{2}\partial I)\tau_s \qquad (16.13)$$

Taking $I_{g\,\min} + \frac{1}{2}\partial I = 20$ mA, $V_0 = 1$ mV and $\tau_s = 2$ nsec we get $E = 4 \times 10^{-4}$ joule, an extremely low value.

Note that both the transition time and the steering time are about one nanosecond, considerably better than the corresponding figures for other types of memory elements but this is by no means the lower limit which can be achieved by Josephson junctions.

16.6 System aspects

On the basis of the well-established switching properties of Josephson junctions, Anacker [465] designed two random access memory modules, one for 64,000 and the other for 4000 memory elements and peripheral circuits. The dimensions of the former one come to $(6 \text{ in.})^2$ while the latter one needs an area of $(1\cdot8 \text{ in.})^2$. Summing all the transfer times which occur in sequence during read and write operations, and multiplying the results by a factor 2 to account for tolerance margins, time overlap, and necessary recovery times leads finally to read and write cycle time estimates of 46 and 30 nsec, respectively, for the 64 Kbit array and of 12 and 9 nsec, respectively, for the 4 Kbit array.

A notable feature of the design is that the number of interconnections between memory arrays is exceptionally small. About 50 contacts should suffice for the 64 Kbit array or for 16 individual 4 Kbit arrays.

The power requirements would be small because Josephson tunnelling gates do not require standby power; the voltage across the junction is zero except during current transfers as discussed before. The power dissipation and refrigeration requirements were estimated for a memory module of 30 Mbit capacity. Such memory would comprise 512 substrates with individual 64 Kbit arrays or sets of 16, 4 Kbit arrays per substrate; it could be housed in a cryostat of about a cubic foot volume. For parallel access of 256 bits per cycle and one memory access every 50 nsec or every 15 nsec (depending on the array type used) the total power dissipated comes to about 110 mW. The total heat influx via the transmission lines was estimated at 800 mW so that a closed cycle liquid helium refrigerator with 1 W refrigeration capacity is sufficient to keep the memory operable.

16.7 Conclusions

The properties of Josephson junctions are very well suited to practical applications as switches and memory elements. They share with cryotrons the advantages of high density but are superior to them on a number of counts:

(*i*) faster speed,
(*ii*) lower power dissipation,
(*iii*) lower current supply level,
(*iv*) higher output voltage,
(*v*) possibility of working above liquid helium temperature (by using lead junctions).

The weakest point of Josephson junction devices is the need to reproduce in actual mass production thicknesses of a few atomic layers. No large units have so far been made; on a smaller scale the plasma discharge technique gave encouraging results.

17. Generation, mixing and detection of electromagnetic waves

17.1 Radiation from the junction

It was shown in Section 10.2 that a constant voltage, V_0, across the junction leads to an a.c. current of the form

$$j = j_J \sin\left(\alpha + \frac{q}{\hbar} V_0 t\right) \qquad (17.1)$$

In principle, all we need to do is to vary the voltage from zero upwards and we can cover the whole frequency spectrum. In practice the situation is not so rosy because the available a.c. power is rather low and cannot be easily extracted (because of impedance mismatch) from the junction. There is also an upper limit in frequency; not a sharp cut-off at the energy gap as one might naively expect but a gradual decay above the gap extending over several octaves.

The emitted radiation was first detected by Giaever [380] with the aid of another tunnel junction (the experiment is briefly discussed in Section 10.2) and by Yanson *et al.* [325] using an external circuit. They measured about 10^{-14} W; this was improved by Dmitrenko and Yanson [470] by an order of magnitude, and the same authors reported [471] later 3×10^{-10} W with an

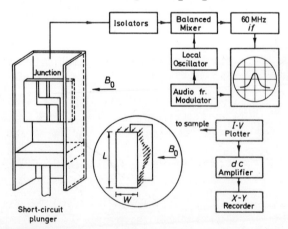

Fig. 17.1 The junction in the waveguide and a block diagram of the superheterodyne detector. The inset shows the electric field distribution in the junction when it is biased on the second mode. After Langenberg *et al.* [472].

efficiency of 2%. Similar measurements were carried out by Langenberg *et al.* [379, 472, 473] with the experimental set-up shown in Fig. 17.1. The local oscillator was frequency modulated using an audio frequency signal which also provided the *X*-axis deflection for the oscilloscope; the *Y*-axis was fed from the video output of the *if* amplifier. The *if* signal resulting from the radiation of the junction was swept in frequency through the bandpass (4 MHz) of the *if* amplifier producing a resonance-like curve on the oscilloscope.

A better coupling of the microwave current to the outside world was achieved by Dayem and Grimes [403] who used a point-contact junction in a coaxial cavity as shown in Fig. 17.2. We may look upon this as an impedance transformer. The junction is placed at a point where the current is maximum (impedance minimum) and the radiation is led away at the part of the cavity where the impedance is high (i.e. it is matched to the waveguide).

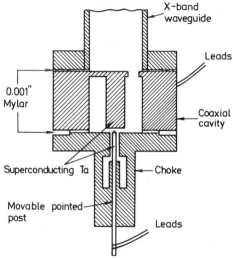

Fig. 17.2 Coaxial cavity with point-contact as part of centre conductor. After Dayem and Grimes [403].

The measured *I–V* characteristics displayed steps but in contrast to thin film tunnel junctions the resonances responsible for the steps were not due to the junction itself but to the external microwave cavity. Dayem and Grimes [403 measured microwave power outputs of 10^{-10} W at the fundamental and 10^{-12} W and 10^{-14} W at the second and third harmonics respectively. Note that only the fundamental resonance of the cavity at ω_r was used; the second and third harmonics were obtained by voltage biasing at steps corresponding to $\omega_r/2$ and $\omega_r/3$.*

---

* In a similar experiment microwave output power was obtained by Krasnopolin and Khaikin [475] when biased at voltages corresponding to ω_r/n down to $n = 10$.

Evidence for radiation in the millimeter wave region was reported by Clark [474] using a novel geometrical arrangement. The junctions were formed by packing closely 1 mm Sn balls into a two-dimensional array of 10×10 mm^2. The advantage of this arrangement is the large number of junctions which, if radiating in phase, can significantly enhance the output power. The balls were placed into a resonant structure consisting of parallel mirrors.

The differential resistance as a function of voltage (at a given mirror spacing) was found to display dips corresponding to harmonics and subharmonics. Keeping the voltage constant and varying the mirror spacing a large number of periodically spaced (period $\sim 1{\cdot}1$ mm) dips were found again in the differential resistance indicating a series of resonances.

It is, of course, of little practical interest to produce small amounts of microwave power by a cryogenic device. The question is whether it is possible to extend the frequency limit further to the submillimeter (far infrared) region where coherent oscillators are scarce. The theoretical predictions are favourable. On the basis of microscopic theory both Riedel [127] and Werthamer [128] came to the conclusion that the amplitude of the a.c. current initially increases with frequency, has a singularity at the frequency corresponding to the energy gap, and decays slowly above the gap as shown in Fig. 17.3. This means that with

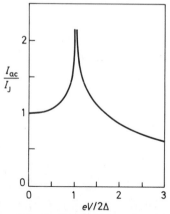

Fig. 17.3 The amplitude of the a.c. current as a function of normalised voltage. After Riedel [127].

a high gap material it should be possible to obtain oscillators down to the wavelengths around 100 μm. There has, as yet, been no report of direct observation of radiation in this range (one of the principal difficulties is the lack of sensitive detection systems). There are, though, indications from the measured steps of the I–V characteristics that such radiation is indeed generated by the junctions; this experimental evidence will be discussed later in this chapter.

Rings. As discussed in Section 14.7 a current biased resistive ring is roughly equivalent to a voltage biased junction. It was also shown in Fig. 14.14 that the current flowing in the circuit is a periodic function of time. If we want resonably sinusoidal current we have to choose ξ less than unity, say $\xi = \frac{1}{2}$ that gives for the amplitude of the current

$$I_{peak} = I_c = \frac{\xi \Phi_0}{2\pi L} = \frac{\Phi_0}{\pi L} \tag{17.2}$$

Since the voltage is $(\hbar/q)\omega_0$ we get for the a.c. power in the ring

$$P = \frac{1}{L} \Phi_0^2 \omega_0 \tag{17.3}$$

and for the impedance

$$Z = \tfrac{1}{2}\omega_0 L \tag{17.4}$$

With inductances of $L = 10^{-10}$ H the available power is about $7 \cdot 5 \times 10^{-12}$ W at 30 MHz and $2 \cdot 3 \times 10^{-9}$ W at 10 GHz.

The a.c. output was measured [430–432] both around 30 MHz and 10 GHz under resonance conditions (realised by an L–C circuit at the lower frequency and by a cavity at the higher frequency). By changing the current bias several resonances of the cavity were observed. The maximum power obtained was about 10^{-11} W.

It is quite remarkable that the same basic structure, a simple resistive ring, is capable to produce oscillations from audio frequencies to microwaves and possibly even further. No conventional oscillator could perform a similar feat.

Note finally the studies (at 30 MHz) of Silver and Zimmerman [476] on two resistive rings connected in parallel. The rf amplitude across the junctions was essentially constant but they observed interference effects due to the variation of the relative phase with the applied flux.

17.2 Basic equations of frequency conversion

We shall determine here the current flowing through the junction in the presence of two applied high frequency signals [327] at frequencies ω_1 and ω_2. This is essentially a simple generalisation of the formulae derived in Section 10.2 where we analysed the effect of a single high frequency signal. We shall assume that the voltage across the junction is

$$V = V_0 + V_1 \cos(\omega_1 t + \alpha_1) + V_2 \cos(\omega_2 t + \alpha_2) \tag{17.5}$$

hence the current (from Equations (9.24) and (9.17)) takes the form

$$j = j_J \sin\left\{ \omega_0 t + \alpha_0 + \frac{qV_1}{\hbar\omega_1} \sin(\omega_1 t + \alpha_1) + \frac{qV_2}{\hbar\omega} \sin(\omega_2 t + \alpha_2) \right\} \tag{17.6}$$

where ω_0 is defined by Equation (10.15) and α_0, α_1 and α_2 are constants. By using standard trigonometric identities and the expansions in terms of Bessel functions [381] we obtain for the current

$$j = j_J \sum_{k=-\infty}^{\infty} \sum_{l=-\infty}^{\infty} J_k(X_1)J_l(X_2)\sin\left\{\omega_0 t + \alpha_0 + k(\omega_1 t + \alpha_1) + l(\omega_2 t + \alpha_2)\right\} \quad (17.7)$$

where

$$X_1 = \frac{q\,V_1}{\hbar\,\omega_1} \quad \text{and} \quad X_2 = \frac{q\,V_2}{\hbar\,\omega_2} \quad (17.8)$$

We may now determine from Equation (17.7) the components of the current at various frequencies. At the difference frequency $\Delta\omega = \omega_1 - \omega_2$ when the junction is biased at zero voltage ($\omega_0 = 0$) we get for example

$$j = -2j_J J_1(X_1)J_1(X_2)\cos\alpha_0. \quad (17.9)$$

We may work out in a similar manner the a.c. current at other frequencies but here we shall calculate only the height of the d.c. steps (mathematical condition $\omega_0 + k\omega_1 + l\omega_2 = 0$) because that can be relatively easily measured. At bias voltages

$$V = \frac{\hbar}{q}\omega_0, \quad (\omega_0 = 0, \omega_1, \omega_2, \omega_1 - \omega_2, \omega_1 + \omega_2, 2\omega_2 - \omega_1, 2\omega_2) \quad (17.10)$$

we find that the current amplitudes are proportional to

$$
\begin{array}{lll}
J_0(X_1)J_0(X_2) & \text{when} & \omega_0 = 0 \\
J_1(X_1)J_0(X_2) & & = \omega_1 \\
J_0(X_1)J_1(X_2) & & = \omega_2 \\
J_1(X_1)J_1(X_2) & & = \omega_1 - \omega_2 \\
J_1(X_1)J_1(X_2) & & = \omega_1 + \omega_2 \\
J_1(X_1)J_2(X_2) & & = 2\omega_2 - \omega_1 \\
J_0(X_1)J_2(X_2) & & = 2\omega_2
\end{array}
\quad (17.11)
$$

17.3 Experimental results on frequency conversion

There have been a number of experiments [327, 352, 384, 432, 477–479] concerned with the mixing properties of tunnel junctions and weak links though in most cases the various components were not directly measured but their presence ascertained from the observed steps in the I–V characteristics. The first measurements were done by Higa [477] and Gaule et al. [478] both of them working in the 10 GHz region.

We shall start here the discussion by describing in somewhat more detail the elegant experiments of Grimes and Shapiro [327]. They realised the junction by pressing together the ends of two pieces of Nb wire, one of which was flattened and the other pointed by a chemical etch. The junction was placed about midway across the narrow dimension of a waveguide as may be seen in Fig. 17.4 which shows the block diagram of the measuring apparatus.

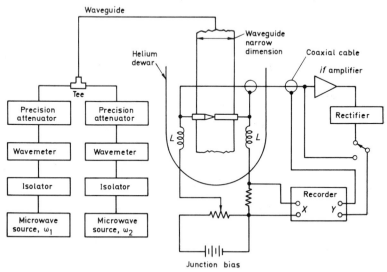

Fig. 17.4 Block diagram of the apparatus for the measurement of frequency conversion. After Grimes and Shapiro [327].

The two microwave sources were isolated from each other and both the amplitude and frequency of each source could be separately adjusted and measured. The two microwave signals incident upon the junction interacted producing (among others) the difference frequency. The *if* circuit (to be matched by this difference frequency) was fixed at about 1 MHz and brought out by a coaxial line connected directly across the voltage terminals of the junction. The *if* output was amplified and after rectification connected to the Y terminal of the recorder through a switch. The junction bias circuitry was isolated from the *if* signal by *rf* chokes labelled by L in Fig. 17.4. The d.c. voltage across the junction was also fed to the Y terminal whereas a signal proportional to current was fed to the X terminal.

A clear summary of the experimental results may be seen in Fig. 17.5. Curve 1 at the bottom shows the I–V characteristic in the absence of microwaves. Applying now a microwave signal at a frequency $\omega_1/2\pi = 72$ GHz, steps appear at intervals of $V = (\hbar/q)\omega_1 = 150$ μV as shown in curve 2 where the steps corresponding to ω_1 and $2\omega_1$ may be clearly discerned. Keeping now the power at 72 GHz fixed and applying a signal at $(\omega_2/2\pi) = 64$ GHz there are now steps (curve 3) at ω_2 and $2\omega_2$ as well. Increasing successively the power of

the 64 GHz signal by 23 db's (curves 4 to 13) it may be seen that steps character-
ising the mixing procedure appear at $\omega_1 - \omega_2$, $\omega_1 + \omega_2$, etc. The amplitude of
these steps as a function of the 64 GHz signal is shown in Fig. 17.6 where it is
compared with the predictions of Equation (17.11). The grouping into a, b and c
is according to the $J_0(X_2)$, $J_1(X_2)$ and $J_2(X_2)$ dependence on microwave voltage.
The height of the steps is only a relative measure, so to decide on the scale of the
ordinate one point can be arbitrarily selected to match the experimental curve
to the theoretical one. This is done at the first maximum of the data for $\omega_0 = \omega_2$

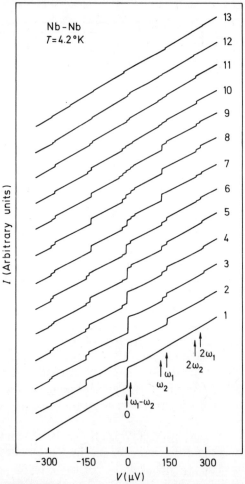

Fig. 17.5 I–V characteristics of a Nb–Nb point-contact junction. Curve 1, no incident
microwaves; curve 2, input at $\omega_1/2\pi = 72$ GHz; curves 3–13, input power at ω_1 is kept
constant while the input power at $\omega_2/2\pi = 64$ GHz is increased. After Grimes and Shapiro
[327].

lenoted by a double circle in Fig. 17.6 (b). In view of the fact that the model dopted in the previous section was rather simple (feeding arrangements and phmic dissipation were disregarded) and that the junction was realised by a point contact (for which the application of the Josephson equations is not always ustified) the agreement between theory and experiment may be regarded atisfactory.

Grimes and Shapiro [327] performed a number of further experiments vith two signals which differed by a frequency of about 1 MHz from each pther and detected directly the *if* output as outlined in Fig. 17.4. In one experinent both signals were around 72 GHz, in another both were around 23 GHz

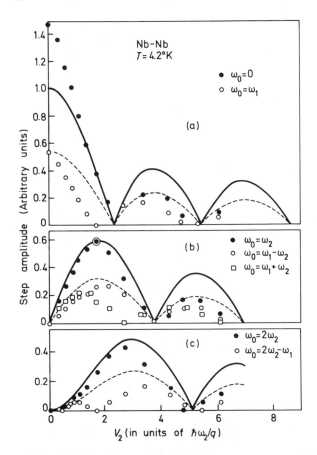

Fig. 17.6 Data of Fig. 17.5 plotted to compare the current steps at constant voltages with the Bessel function dependence expected for a tunnel junction. The full circle data points are to be compared with the solid theoretical curves and the open data points are to be compared with the dashed theoretical curves. Data and theory were fitted at only one point denoted by the double circle. After Grimes and Shapiro [327].

and in a third one the signal at 72 GHz interacted with the third harmor the other signal at 24 GHz to yield the *if* output.

Mixing experiments in the same frequency range but using only one pressed signal at $\omega_1/2\pi = 75$ GHz were performed by Longacre and Sha [384]; the other signal was obtained by exciting a coaxial cavity (resona $\omega_2/2\pi = 20$ GHz) by the d.c. voltage across the junction. The 75 GHz s was fed to the junction with the aid of a waveguide as shown in Fig.

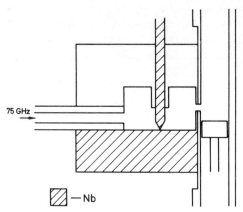

75 GHz

—Nb

Fig. 17.7 Point-contact junction in a 20 GHz cavity exposed to 75 GHz radiation. Longacre and Shapiro [384].

Another difference in measurement technique is that the differential resist dV/dI is measured rather than the I–V characteristic, hence a step in the cu appears as a dip in the differential resistance.

The differential resistance against voltage in the absence of an input si is shown in Fig. 17.8 (*a*). There is a dip corresponding to the resonant frequ of the cavity; a proof that at that particular bias voltage some measur power is excited. When some microwave power at ω_1 is incident upon junction we can see (Fig. 17.8 (*b*)) a pronounced dip at ω_1 and smaller di $2\omega_1$ and at some subharmonic frequencies. Increasing further the microw power (Fig. 17.8 (*c*)) the dips due to mixing at $n\omega_1 \pm \omega_2$ can be observed. mechanism of mixing is not obvious because the signal at ω_2 is now ne impressed across the junction nor excited by a d.c. voltage $V = (\hbar/q)\omega_2$. argument may be put in the following form. When the voltage across the junc is (say) $V = (\hbar/q)(\omega_1 + \omega_2)$ then a microwave signal at the frequency ω_1 is excited. This signal mixes with the incident signal at ω_1 and produces difference frequency that is ω_2. But ω_2 is the resonant frequency of the ca so the signal created at ω_2 can have a reasonable amplitude. The dip at ω_1 can now be simply explained by saying that there are signals present bo ω_1 and at ω_2 so there must be a dip at $\omega_1 + \omega_2$.

Further increase in microwave power led to the appearance of dips at even higher voltages. In Fig. 17.8 (d) the dip corresponding to $5\omega_1 + \omega_2$ can be recognised. This means that microwaves up to a frequency of 395 GHz can be excited with this geometrical arrangement.

In some further experiments Longacre and Shapiro [384] found dips well above 500 GHz with a point contact junction in which the point, due to excessive pressure, bent over. They explained the results by assuming that the distorted point formed a resonant structure coupled strongly to the junction.

Experiments yielding steps at even higher frequencies were reported by McDonald *et al.* [352]. To improve the signal-to-noise ratio for observing the

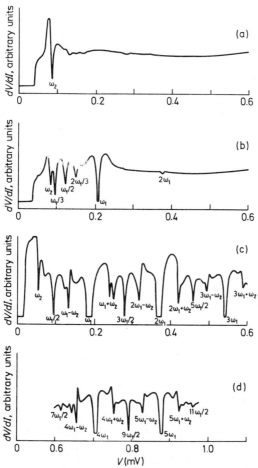

Fig. 17.8 Differential resistance as a function of voltage for the point-contact junction shown in Fig. 17.7 with increasing amounts of 75 GHz radiation: (*a*) no applied radiation, (*b*) 30 db, (*c*) 12 db and (*d*) 6 db. After Longacre and Shapiro [384].

current steps they averaged a large number of I–V characteristics. This was done by employing a 1024 channel digital instrument into which the junction voltage was fed. Depending on the junction voltage one of these channels is addressed whose output is proportional to dwell time of the voltage in that particular channel. When the junction is driven by a linear current ramp the dwell time on a current step is proportional to the amplitude of the step since the voltage remains constant. Hence the current steps appear (in contrast to the dips in the experiments described previously) as peaks in the output of the digital instrument. Using again a niobium–niobium point contact junction and incident radiation at 70 GHz, the amplitude of steps (zero voltage step excluded) as a function of voltage at a given microwave input is shown in Fig. 17.9. It is

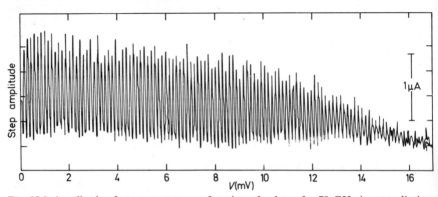

Fig. 17.9 Amplitude of current steps as a function of voltage for 70 GHz input radiation. After McDonald *et al.* [352].

remarkable that steps up to the 103rd harmonic* (that is, up to 7210 GHz) can be distinguished, which is about 12 times higher than the frequency of the photon capable to break a Cooper-pair.

Assuming that the high frequency cut-off of the Josephson effect is determined by the relaxation time of the superconducting order parameter McDonald *et al.* [352] estimate for the cut-off frequency 7500 GHz in good agreement with their experimental results.

In further experiments McDonald *et al.* [352] investigated the current steps produced by an HCN laser. The output from the digital instrument when two lines of the laser ($\omega_1/2\pi = 891$ GHz and $\omega_2/2\pi = 805$ GHz) were incident is shown in Fig. 17.10. Besides the two peaks corresponding to the frequency of the incident radiation it is possible to identify peak 1 at $\omega_1 - \omega_2$, peak 2 at $2(\omega_1 - \omega_2)$, peak 4 at $\omega_1 + \omega_2$ and peak 5 at $2\omega_1$. The peak at 3 is the Riedel

---

*This is an indirect proof of the generation of high harmonics. Not much work has been done on direct observation of harmonic generation. There are Shapiro's [326] results on solder drop junctions and those of Seraphim and McDermott [480] on thin film bridges.

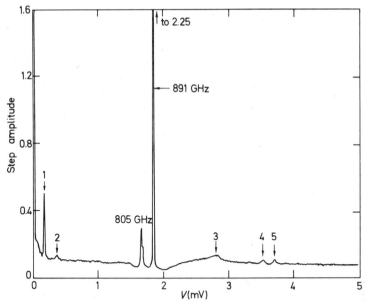

Fig. 17.10 Amplitude of current steps as a function of voltage for 805 GHz and 891 GHz input radiation. The numbered peaks correspond to 1: $\omega_1-\omega_2$, 2: $2(\omega_1-\omega_2)$, 3: Riedel peak at the gap voltage, 4: $\omega_1+\omega_2$, 5: $2\omega_1$. After McDonald *et al.* [352].

peak [127] (going to infinity in Fig. 17.3 but finite, of course, in practice) corresponding to the energy gap of niobium. It appears in the absence of laser input as well.

Rings. Many of the nonlinear properties of Josephson junctions in rings have already been discussed in Chapter 14. We shall describe here a few experiments by Zimmerman [436]. The mechanisms investigated were as follows:

(*i*) mixing two externally applied signals at ω_1 and ω_2 in a fully super-conducting ring; ·

(*ii*) mixing an external signal at ω_1 with the internally generated Josephson oscillation at ω_2 in a partly resistive ring;

(*iii*) mixing the signal at ω_1 generated by one partly resistive ring with a signal at ω_2 generated by a second partly resistive ring, and taking the mixed output from the second ring;

(*iv*) mixing of signals generated in the special double contact partly resistive configuration shown in Fig. 17.11.

In all these experiments the critical current of the contacts was kept low enough, $LI_c < \Phi_0$, to avoid high-order quantum transitions; ω_1 and ω_2 varied over a wide range but $|\omega_1 \pm \omega_2|/2\pi$ was kept fixed at 30 MHz. The first three mechanisms gave the expected result that the maximum mixed power

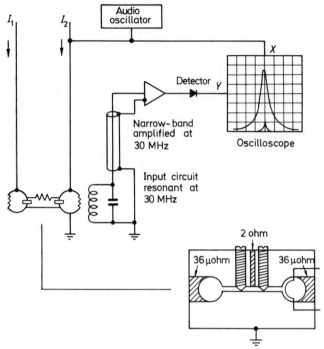

Fig. 17.11 Schematic representation of the experiment for observing heterodyne signal from a pair of closely-coupled point-contacts. An audio sweep signal applied to one of the bias current leads sweeps the heterodyne signal through the amplifier pass band. A typical oscilloscope pattern is shown. After Zimmerman [436].

output is proportional to the mixed frequency and independent of the signal frequency.

The tightly-coupled contacts gave, however, qualitatively different results. Firstly, output at the mixed frequency varied approximately linearly with ω_1 and ω_2 up to a few hundred MHz and irregularly above that. Secondly, the output power greatly exceeded the value given by Equation (17.3). The cause of these effects is not known; apparently the close interaction of the two high-frequency signals is responsible for some sort of amplification at 30 MHz.

17.4 A spectrometer

The mixing of two signals in a ring may be used for realising various instruments. Some of these will be discussed in Chapter 18 among sensitive voltmeters and magnetometers. Here we shall be briefly concerned with the working principles of a spectrometer devised by Silver and Zimmerman [481] (see also Reference 432).

The basic experimental set-up is shown in Fig. 17.12. The resistance R,

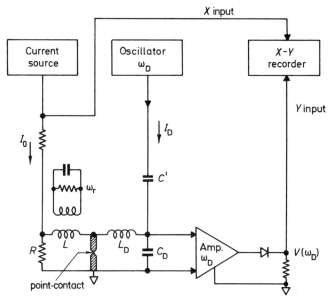

Fig. 17.12 Schematic diagram of the spectrometer. The absorber at ω_r is represented by a lossy parallel resonant circuit inductively coupled to L. After Silver and Zimmerman [481].

inductance L and the point contact comprise one of the sources oscillating at the frequency

$$\omega_0 \approx \frac{q}{\hbar} I_0 R \qquad (17.12)$$

where I_0 is a d.c. current bias. In addition, there is an impressed rf current I_D at the frequency ω_D. The coupling capacitor C' is very small so that I_D is determined by the source voltage. A sensitive amplifier tuned to the frequency ω_D is used to measure the voltage across the resonant circuit L_D, C_D (tuned also to ω_D) as the current I_0 and hence ω_0 is varied.

The resonant circuit (resonant frequency ω_r) coupled to the resistive ring oscillator consists of a small coil at the end of a 1 m open-ended coaxial line. This circuit has a low frequency resonance at 63 MHz from the lumped inductance of the coil and the capacitance of the cable. The lowest frequency mode of the coaxial line is 194 MHz.

The observed rf voltage $V(\omega_D)$ is plotted in Fig. 17.13 as a function of ω_0. It may be seen that there is a response whenever ω_0 is an integral multiple of ω_D and also when the condition

$$\omega_0 = \omega_r \pm \omega_D \qquad (17.13)$$

is satisfied.

At $\omega_0 = \omega_D$ the resistive ring oscillator feeds power directly into the amplifier.

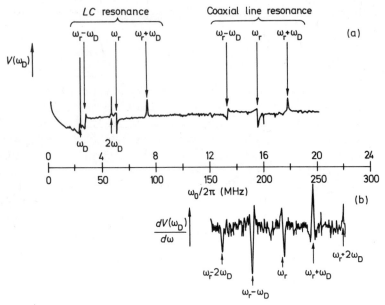

Fig. 17.13 X–Y recording of observed spectra as functions of the biasing current and oscillator frequency. (*a*) Direct measurement of $V(\omega_0)$ showing the lumped constant LC resonance centred at 63 MHz and the coaxial line resonance centred at 194 MHz. Signals at ω_0 and $2\omega_0$ are due to the detection system. (*b*) The Co^{59} nuclear magnetic resonance centred at 218 MHz observed by measuring $dV(\omega_0)/d\omega$. After Silver and Zimmerman [481].

At $\omega_0 = 2\omega_D$ the mixing between ω_0 and ω_D leads to enhanced output. At $\omega_0 = \omega_r$ there is resonant absorption so the effective impedance of the point contact decreases such that the *I–V* characteristic has a peak at $V = (\hbar/q)\omega_r$. The oscillating source ω_D modulates V and the subsequent narrow-band detector signal at ω_D is the derivative of the *I–V* characteristic.

Changes in $V(\omega_D)$ at frequencies satisfying Equation (17.13) may again be explained by the nonlinearity of the contact and the subsequent mixing action. For a qualitative answer, however, it seems easier to formulate the problem in terms of two quantum oscillators capable of absorbing or emitting photons of energies $\hbar\omega_0$ and $\hbar\omega_D$. Thus when $\omega_r = \omega_0 + \omega_D$ both photons are absorbed and there is a dip in the response. When $\omega_r = \omega_0 - \omega_D$ then the absorption of a photon at ω_r entails the absorption of a photon at ω_0 with the simultaneous *emission* of a photon at ω_D. Hence there is a peak in the response.

As far as spectrometric applications are concerned what matters is that any resonant absorption gives a response. Silver and Zimmerman [481] demonstrated this by filling the inductance L by finely powdered cobalt and observing the response shown in Fig. 17.13 (*b*). The experimental arrangement is somewhat more complicated (there is an additional audio modulation) but the principle is the same. The nuclear magnetic resonance of Co^{59} is clearly detected. The measured absorption is estimated at 10^{-16} W so we have indeed a sensitive spectrometer.

17.5 Detection

We have already mentioned the experiments of McDonald *et al.* [352] in which two laser lines incident upon the junction produced several peaks in their measured characteristics (Fig. 17.10). For detection purposes all that matters is that the incident radiation should cause some measurable change somewhere, so we could just as well regard the above mentioned experiments as investigations of infrared detection.

The obvious way of using a Josephson junction as a detector is to look at the *I–V* characteristic in the vicinity of $V = (\hbar/q)\omega$ where ω is the frequency of the incident radiation. If nothing conspicuous occurs then there is no incident radiation; if there is a step or dip or peak (depending on instrumentation) then radiation must be present. One can, of course, expand on this principle when building detectors; we shall return to this a little later. For the time being let us concentrate on the detection of incoherent radiation which obviously cannot lead to distinct structure. The *I–V* characteristic is, however, still different in the absence and in the presence of radiation so it is still suitable for detection.

Grimes, Richards and Shapiro [345, 354] (see also References 482 and 483) rely on the change in the zero-voltage current for indicating the input radiation. Their experimental arrangement is shown schematically in Fig. 17.14. The incoherent radiation from a mercury lamp illuminated a Michelson interferometer, one arm of which was adjustable. The recombined light was chopped at about 90 Hz, focused into a light pipe and led to the point contact junction (realised again by pressing together flat and sharpened wires). The incident radiation diminished the zero-voltage current and with it depressed the *I–V*

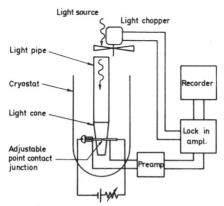

Fig. 17.14 Experimental arrangement for far infrared detection. After Grimes *et al.* [354].

characteristic for low voltages. Choosing a current bias as shown in Fig. 17.15, the incident radiation produced a voltage at the chopper frequency. This voltage suitably amplified and rectified in a lock-in amplifier, was plotted on a chart recorder as a function of the path difference in the interferometer. The spectral response of the junction was obtained by computing the Fourier transform of the interferogram on a digital computer [484].

Results for In–In junctions are shown in Fig. 17.16 where the fine structure due to geometrical resonances is omitted. It may be seen that there is a peak

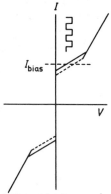

Fig. 17.15 I–V characteristic showing how applied radiation causes a change in zero-voltage current which is converted to a voltage output by constant current external bias. Dashed curve applies in the presence, solid curve in the absence of input radiation.

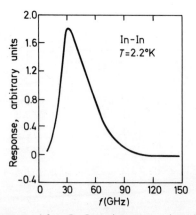

Fig. 17.16 Spectral response of an In–In point-contact junction at low resolution which smooths out structure associated with geometrical resonances. After Grimes et al. [354].

and a high-frequency cut-off (the low-frequency cut-off is due to instrumentation) both of them reminiscent of Riedel's theoretical curve (Fig. 17.3). However, the peak is at the wrong position, about 15% below the gap which corresponded to 216 GHz at the temperature used.

For Nb–Nb junctions the response extends to much higher frequencies in accordance with the higher energy gap but otherwise there is little similarity (Fig. 17.17) with Riedel's theory. Instead of a peak at the gap, there are minima at the gap, at half the gap, at quarter the gap and at one and a half times the gap.

Pb–Pb junctions tended to give sharp response and high frequency cut-offs well below the gap (Fig. 17.18). One of the possible reasons is that all Pb–Pb junctions contained hysteresis regions in their I–V characteristics and the response was measured by a different technique (still relying on the change in

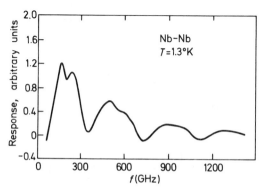

Fig. 17.17 Spectral response of a Nb–Nb point-contact junction. After Grimes et al. [354].

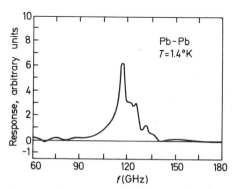

Fig. 17.18 Spectral response of a Pb–Pb point-contact junction. After Grimes et al. [354].

the zero-voltage current). Another reason may be the presence of high Q resonances; thus the zero-voltage current was influenced not only by the incident radiation but also by the radiation excited by the junction itself.

The sensitivity of at least one of these junctions (Nb–Nb) was determined by Grimes *et al.* [354] using microwave techniques at a frequency of 75 GHz. They found a noise equivalent power of 5×10^{-13} WHz$^{1/2}$ with a time constant of 1 sec. This already compares favourably with the performance of other liquid helium temperature detectors even though the frequency of measurement was well below the maximum response of the junction.

Grimes *et al.* [354] investigated the response time of the Nb–Nb junction as well. In one experiment the detector was found to respond to 75 GHz radiation amplitude modulated by 1 MHz. In another experiment a pulse with a rise time of 10 nsec was applied to the detector which was also reproduced. Hence the response time must be below 10 nsec.

We shall now return to the investigation of the step in the I–V characteristic caused by the incident radiation but with the added complication that there is a resonant cavity at the same frequency [355]. The experimental set-up is identical with that shown in Fig. 17.14 apart from details of coupling, and the presence of a coaxial cavity (typically 4 mm in length and diameter). The constant current bias is now on the high differential resistance region connecting the cavity-induced step with the rest of the I–V characteristic. The incident radiation changes the height of the step and the voltage as well, resulting in an output at the chopper frequency.

The problem is here a little more complicated than any of those investigated before because besides the self-excited resonance there is an impressed field present as well at the same frequency. Richards and Sterling [355] attacked the problem numerically by generalising the solution of Werthamer and Shapiro [485]. The essential conclusion of the computation is that the impressed field leads to the sharpening of the resonance.

The resulting bandwidth was better than the resolution limit; from the measured data an upper limit of 300 MHz was placed for the bandwidth at a resonant frequency of 190 GHz. With this assumption the observed noise equivalent power was estimated at 10^{-14} WHz$^{-1/2}$.

The fundamental limit to the noise equivalent power for a square law detector (which the present one probably is) coupled to a single mode of the input radiation with a background temperature $T_B \gg h\nu$ is

$$\text{NEP} \cong kT_B \Delta\nu$$

where $\Delta\nu$ is the geometric mean of the pre- and post-detection bandwidths. For $T_B = 300°$K and 1 Hz post-detection bandwidth, NEP $\cong 3 \times 10^{-16}$ W. The advantage of the present detector over others is the narrowness of the pre-detection bandwidth. Signal–photon shot noise begins to dominate the NEP when $T_B \approx 1°$K.

Finally we shall mention here detection with the aid of magnetometers, in which case the magnetic flux generated by the incident radiation is measured. Recent theoretical calculations by Chiao [486] suggest that a single junction magnetometer (see Section 18.5) should be able to detect single photons.

18. Voltmeters and magnetometers

18.1 Introduction

For the measurement of small voltages and small changes in the magnetic field the devices based on Josephson tunnelling are not only competitive but are superior to any other existing device. The galvanometer developed by Clarke [338, 340, 487, 488] (known as SLUG, Superconducting Low-inductance Undulating Galvanometer) measured voltages as small as 10^{-15} V with a time constant of one second. Similarly impressive results obtained by another type of voltmeter were reported by Zimmerman and Silver [339] and also referred to by Silver *et al.* [489].

Clarke's SLUG converted to magnetometer is capable to detect magnetic fields of the order of 10^{-8} gauss [488, 490, 491]. The various weak link devices of the Ford group [341–343, 364, 438, 492] have given magnetic field resolutions from 10^{-9} gauss to 7×10^{-11} gauss (the best at the time of writing) with a 1 sec time constant. A digital magnetometer was also developed [343, 350, 364, 493] counting increments of 10^{-7} gauss at a rate of 10^4 per second.

Though the operational principles of these devices differ somewhat, it is possible to make the general statement that their sensitivity is a consequence of flux quantisation. They are all based on certain variants of a slightly perturbed superconducting ring which admits flux but retains its periodicity with the flux quantum. If one can resolve n^{th} part of the period then the accuracy of magnetic field measurement is

$$\Delta B = \frac{\Phi_0}{nA} \qquad (18.1)$$

where A is the area of the ring. For $A = 2$ cm^2 and $n = 100$ the resolution is
$\Delta B = 10^{-10}$ gauss.

18.2 Clarke's SLUG

The application of Josephson tunnelling to the measurements of low voltages was first proposed by Clarke [338]. The device is based on the property of the double junction that I_{max} the maximum supercurrent flowing across them is a periodic function of the enclosed flux. As a digital voltmeter the principle of its operation is quite straightforward. Assume that an external primary coil of inductance L_0 (in practice stray inductances will be present as well but we shall disregard them in this discussion) is tightly coupled to the double junction and has in series with it a resistance R_0 (Fig. 18.1). The current in the primary

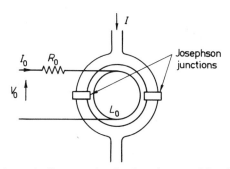

Fig. 18.1 Schematic diagram of a simple voltmeter. After Clarke [487].

required to create one flux quantum in the double junction and thus one oscillation in critical current is $I_0 = \Phi_0/L$. The voltage which will cause such a current to flow is

$$V_0 = I_0 R_0 = \frac{\Phi_0 R_0}{L_0} = \frac{\Phi_0}{\tau_0} \tag{18.2}$$

where τ_0 is the time constant of the primary circuit. A voltage which is n times larger than V_0 will cause n oscillations in the maximum supercurrent. Thus we measure voltage by counting the number of oscillations. The minimum voltage V_0 for a time constant of 1 sec is 2×10^{-15} V from Equation (18.1). If we can detect say 1% of one oscillation then the voltage sensitivity goes up to 2×10^{-17} V though the digital character of the information is lost.

The practical realisation of this device is on a somewhat different basis. There are not two separate junctions but a solder drop junction, although the latter behaves as if it were a double (or multiple) junction. The sensitivity to magnetic fields appears as sensitivity to current flowing to the junctions.

Clarke's experiments on these solder drop junctions were discussed in Section 13.3 and illustrated in Figs. 13.11 and 13.12. Since I_{max} depends strongly on I_H we can use* this set-up for measuring I_H.

For the measurement of voltage and resistance the circuit was modified as shown in Fig. 18.2. With the currents through the resistors zero, I_H was chosen so that dI_{max}/dI_H is a maximum, that is for maximum sensitivity. The currents in the two resistances were then adjusted so as to maintain the null position.

---

* The variation in I_{max} may in fact be smaller (even at the steepest point of the $I_{max} - I_H$ curve) than the variation in I_H. The reason why we are still better off measuring changes in I_{max} is that it flows in a circuit with a resistance of a few ohms, in contrast to an impedance of about 10^{-8} ohm in the I_H circuit. So we have an impedance transformer of 10^8!

If R_s is a standard resistor whose resistance is known and I_s is measured then the voltage across R_1 is given by $V_1 = I_s R_s$. If I_1 is measured as well then an unknown resistance may also be determined from the balance condition, giving $R_1 = I_s R_s / I_1$.

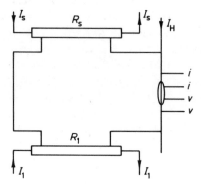

Fig. 18.2 Circuit for measurement of small voltages. The terminals i and v serve for measuring the current and voltage across the solder-drop junction.

The role of the junction is to provide a low inductance null detector and that is where the name of the device comes from, Superconducting Low-inductance Undulating Galvanometer.

The maximum supercurrent was determined in the following way. A sinusoidal current sweep at about 20 kHz was applied to the leads (i) (shown both on Fig. 13.10 and on Fig. 18.2). The sweep was adjusted so as to exceed I_{max} only for a fraction of the period giving rise to a series of voltage pulses across the terminals (v). The pulses were amplified and integrated to give an output related to I_{max}.

Clarke [338, 487, 488] achieved a current sensitivity of 1 μA which for a resistance of $R_1 = 10^{-8}\,\Omega$ corresponds to a voltage sensitivity of 10^{-14} V.

A different method of measurement (using a feedback loop) by McWane *et al.* [494] led to improved current sensitivity of 10^{-7} A and thus to a voltage sensitivity of 10^{-15} V.

Further improvements may be effected [488] by placing the cryostat in a shielded room (the noise picked up by the cryostat leads is apparently the most important one). Other sources of noise [494] (in order of importance) were: imperfect regulation of the device temperature, bias current drifts, amplifier noise and (only at the last place) Johnson noise.

A method using phase sensitive detection has obvious potential though the reported voltage sensitivity [495] (limited probably by the picked up electrical noise) was only 10^{-14} V.

Note that the current resolution may be improved (in principle to an arbitrarily small value) with the aid of a superconducting transformer [487, 496] as shown in Fig. 18.3. The Nb wire through the SLUG is bent into a circle and the ends welded together to form a superconducting ring. A magnetic field applied to this ring (not equal to an integral multiple of the flux quantum) generates a circulating current which is measured by the SLUG. If the field is generated by an ideally coupled primary coil of N turns, the current sensitivity is improved by a factor N. In practice, the coupling is not ideal, the stray inductances associated with the N-turn coil increase the time constant appreciably so the overall gain may not be so much. The design of such a superconducting transformer is discussed by Clarke et al. [496]. The voltage resolution may be improved by several orders of magnitude so that it becomes limited by Johnson noise which is about $8 \cdot 10^{-16}$ V at $4°$K for $R = 10^{-8}$ ohm and a 1 sec time constant.

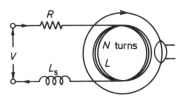

Fig. 18.3 Schematic diagram showing a superconducting transformer coupled to a SLUG. After Clarke [487].

Magnetometer. By joining the two ends of the Nb wire and forming the superconducting ring we have, in fact, made a magnetometer. With a current resolution ΔI_H the minimum flux change which can be detected is

$$\Delta\Phi = L\Delta I_H \qquad (18.3)$$

It should be further noted that the accuracy of this device for measuring a uniform magnetic field may be indefinitely increased by simply increasing the area of the superconducting ring. This is because the inductance is approximately proportional to r, the radius of the ring yielding for the magnetic field resolution

$$\Delta B = \frac{\Delta\Phi}{\pi r^2} = \frac{L\Delta I_H}{\pi r^2} \sim \frac{\Delta I_H}{r} \qquad (18.4)$$

The smallest value reported [488, 490–491] for ΔB is 10^{-8} gauss for a loop of 1·5 cm^2 with a 1 sec time constant.

18.3 A resistive ring voltmeter

A schematic diagram of the device is shown in Fig. 18.4. The voltage to be measured is the one developed across the resistance R. If the critical current of the ring (and so the circulating current) is small in comparison with the d.c. current then this voltage is essentially $I_0 R$ and it is independent of the state of the ring. The voltage creates an a.c. current in the ring oscillating at $\omega_0 = (q/\hbar)V$ as discussed in Section 14.7. Impressing a further a.c. current at ω_D upon the ring, a signal at the frequency $\omega_D \pm \omega_0$ will appear. This is due to the mixing properties of the junction as amply discussed in Chapter 17. The sign is then amplified (ω_0 is assumed small so that the sidebands are within the bandwidth of the amplifier centred around ω_D) and demodulated, and the resulting ω_0 is displayed on an oscilloscope. Measuring ω_0 we can deduce

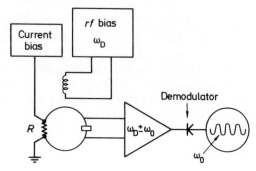

Fig. 18.4 Schematic diagram of a resistive ring voltmeter. After Zimmerman and Silver [339].

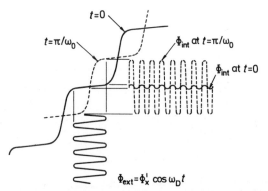

Fig. 18.5 Graphical derivation of the response. After Zimmerman and Silver [339].

The principle of operation is simple indeed. We shall, though, give here an alternative explanation [339] as well because it throws some light on the depth of modulation. Going back again to Section 14.7 we came to the conclusion there that in the presence of a small impressed voltage the static Φ_{int} versus Φ curve moves diagonally with a speed dependent on voltage. Two positions of this curve, separated by a time interval π/ω_0 are shown in Fig. 18.5 by solid and dotted lines respectively. Since $\omega_0 \ll \omega_D$ we may assume that the curve is stationary for a few periods of the signal at ω_D. A sinusoidal input will then lead to outputs of quite different amplitudes at the two specified instances of time. In other words the output is modulated at the low frequency ω_0 and the amplitude modulation is nearly 100%.

The sensitivity of this voltmeter will be limited by the noise in the resistor which is

$$\Delta V = \langle V_n^2 \rangle^{1/2} = \left(\frac{2}{\pi} kTR\Delta\omega \right)^{1/2} \tag{18.5}$$

here $\Delta\omega$ is the noise bandwidth to which the system responds. It may be argued [489] that this bandwidth is just that created by the noise voltage across the junction, i.e.

$$\Delta\omega = q\langle V_n^2 \rangle^{1/2}/\hbar \tag{18.6}$$

hen from Equations (18.5) and (18.6) we get for the voltage sensitivity

$$\Delta V = \frac{4kTR}{\Phi_0}. \tag{18.7}$$

In a practical realisation [339] the actual values were $\omega_D/2\pi = 30$ MHz, $= 10^{-10}\,\Omega$, $I_0 = 10^{-4}\,$A, $I_c = 10^{-5}\,$A, $L = 10^{-10}\,$H. Hence the voltage cross the resistor was $\sim I_0 R \approx 10^{-14}\,$V and the modulating frequency $_0/2\pi = 20$ Hz. The corresponding voltage sensitivity from Equation (18.7) $\Delta V \sim 10^{-17}\,$V.

8.4 Double junction magnetometers

We have already discussed the properties of double junctions in Chapter 13, and a special use of them earlier in this chapter in connection with Clarke's SLUG. A more straightforward application of double junctions is to sense the flux enclosed by the ring containing the junctions. We know that both I_{max} and the voltage across the junctions (as shown in Fig. 13.15) are periodic functions of the magnetic field. Hence by measuring the change in voltage we gain information about the change in magnetic field. Such a magnetometer was developed by Zimmerman and Silver [341] with the instrumentation provided by Forgacs and Warnick [493].

A block diagram of the device is shown in Fig. 18.6. With the control loop open the effect of magnetic field modulation for three different values of the

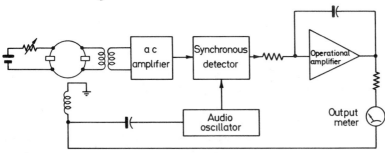

Fig. 18.6 Schematic digram of a double-junction lock-on magnetometer. After Forgacs and Warnick [493].

ambient magnetic field is shown in Fig. 18.7 where V is the output signal from the double junction. At B_1 and B_2 the output signal is at the modulation frequency but in opposite phase with each other. At B_0 the output signal is essentially sinusoidal at twice the modulation frequency so it contains no fundamental component.

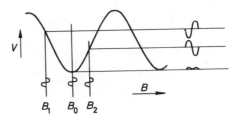

Fig. 18.7 Voltage output from the double-junction as a function of biasing magnetic field for small signal modulation. After Forgacs and Warnick [493].

The output at the fundamental frequency is converted to d.c. control voltage by a synchronous detector. With the control loop closed and the net flux in the ring at a value other than B_0 a correction voltage will be applied to the integrator and the counter flux will change until a net flux equal to B_0 is attained. Each time the external flux is changed, the system comes to equilibrium with a new value of counter flux to keep the net flux constant at B_0. The current producing the counter flux is then a direct measure of the change in external flux.

The best resolution obtained by this device [364] (1970) with a 1 sec time constant is 7×10^{-11} gauss which beats by more than an order of magnitude its nearest rival the rubidium vapour magnetometer (see for comparison Reference 490).

A further advantage of this device is that larger magnetic fields may be measured by counting the number of flux quanta moving in or out of the ring. A circuit capable to do so was designed by Forgacs and Warnick [492]. It is based on a modulation similar to that shown in Figs. 18.6 and 18.7 but the second harmonic signal is used as well. The maximum counting rate reported was about 2000 period/sec.

In principle this device is capable to measure high magnetic fields but in practice the perfect periodicity extends only to a limited range. Due to some sort of irregularity (usually the presence of multiple contacts very close to each other) there is a further modulation envelope necessitating changes in the bias current.

Small changes in high magnetic fields were measured by Beasley and Webb [351] who were able to detect changes of less than 10^{-7} gauss with a 1 sec time constant in an applied field of 2500 gauss; a magnetic field resolution better than 1 part in 10^{10}. The main difficulty is to eliminate flux motion in the superconductor itself because it has the same effect upon the device as the change of magnetic field in the ring. The problem was solved by putting a d.c. super-conducting transformer between the high field region and the measuring device. The transmitted flux change may then be measured in the conventional manner.

18.5 Single junction magnetometers

The physical phenomena occurring in fully superconducting rings containing one junction have been discussed in Chapter 14. For the purpose of measuring magnetic fields the most important relationship is Equation (14.26) which gives the emf as a function of the rf and ambient magnetic fields. It may be seen that the variation with Φ_x^0 is harmonic so the measurement (whether analogue or digital) may proceed in the same manner as for the double junction device. The essential advantage of the rf driven single junction device is that the output is proportional to the frequency used. Hence further improvement in sensitivity may be expected when the rf frequency is raised substantially above the 30 MHz used at present.

There are two types developed; the point contact junction of Silver and Zimmerman [438] and the bridge device (shown in Fig. 14.10) of Mercereau and Nisenoff [342, 343, 364]. There are no new principles involved; the details of instrumentation may be found in the respective papers. The best resolution reported is $\sim 10^{-10}$ gauss with a 1 sec time constant and counting rate $\sim 10^4$ period/sec.

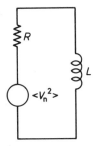

Fig. 18.8 Circuit for calculating the noise current. $\langle V_n^2 \rangle$ represents thermal noise.

A rough indication of the noise properties of this device may be deduced from the circuit of Fig. 18.8. The ring contains an inductance, a resistance and a noise generator corresponding to the Johnson noise of the resistance. We get for the noise current

$$\langle I_n^2(\omega) \rangle = \frac{2}{\pi} \frac{RkT \, d\omega}{R^2 + \omega^2 L^2} \tag{18.8}$$

and integrating for the whole bandwidth

$$\langle I_n^2 \rangle = \frac{2}{\pi} \int_0^\infty \frac{RkT\, d\omega}{R^2 + \omega^2 L^2} = \frac{kT}{L} \tag{18.9}$$

The total flux noise $\langle \Phi_n^2 \rangle$ in a single turn is $L^2 I_n^2 = kTL$ and is independent of R.

We interpret here the resistance as the one developing in the ring when the induced current exceeds the critical current. As the resistance drops the entire noise is confined to a decreasing bandwidth R/L and eventually when R disappears LkT represents the uncertainty in the flux. When this uncertainty is larger than the square of the flux quantum the periodicity disappears and quantum effects are no longer observable. Hence the inductance must obey the inequality

$$L \le \frac{\Phi_0^2}{kT}. \tag{18.10}$$

Thus at $T = 4°K$ the inductance should be smaller than 10^{-7} H. Since the inductance for a cylindrical ring of area A and length l is

$$L = \frac{\mu A}{l}, \tag{18.11}$$

for a length of 1 mm we get a maximum permitted diameter of ~ 1 cm.

The noise may be reduced by putting the signal through a resonant circuit. For a time constant τ the reduction is $(\tau \omega_r / Q)^{1/2}$ where ω_r and Q are the resonant frequency and Q factor of the resonant circuit respectively.

18.6 Gradiometers

A gradiometer measures the *variation* of the magnetic field. The basic set-up is shown in Fig. 18.9. There are two coils wound in opposite directions and

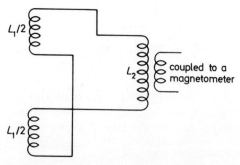

Fig. 18.9 Schematic drawing of a gradiometer. The two primary coils ($L_1/2$) are wound in opposite directions.

connected by superconducting leads. A uniform field change will induce opposing currents in the two coils which cancel, resulting in zero circulating current. Thus the Earth's magnetic field and its fluctuations as well are automatically cancelled. However, a spatially varying magnetic field does induce a net circulating current which creates a magnetic field in L_2. Measuring this magnetic field by any of the magnetometers discussed in this chapter we have a measure of $\partial H/\partial z$. A particular design using the symmetric point contact of Fig. 15.3(e) is described by Zimmerman [497]. The obtained resolution was $2\cdot3 \times 10^{-11}$ gauss/cm.

19.   Determination of *e/h*

19.1 Choice of method

As mentioned in Chapter 8 and shown (not very rigorously*) in Chapter 9 the voltage across a Josephson junction is related to the emitted radiation frequencies by

$$qV = n\hbar\omega \tag{19.1}$$

and the same relationship is obeyed by the voltage at the microwave-induced current steps. Hence the following two** methods readily offer themselves for measuring *e/h*:

  (*i*) bias the junction at one of its self-induced steps, measure the voltage and measure the frequency of emitted radiation,
  (*ii*) irradiate the junction by microwaves, measure the frequency of the microwaves, and measure the voltage of a selected induced step.

Method (*i*) has several disadvantages. Firstly, the power radiated by the junction is small, of the order of 10^{-10} W, not enough for a comfortable frequency measurement. Secondly, the highest frequency at which self-induced steps have been observed (in a rather special circuit [384]) is about 500 GHz corresponding to about 1 mV. This is not too low for accurate voltage measurement but still an order of magnitude below that achievable by microwave-induced steps. In the third place the steepness of the steps is determined by the *Q* of the cavity resulting in a step less vertical than that induced by radiation. In the fourth place higher order self-induced steps were found to emit radiation at frequencies *not* satisfying the Josephson frequency–voltage relationship and showing strong dependence on the applied magnetic field [499]. In spite of these draw-backs the accuracy achieved by this method is very good (3·4 part per million (ppm) according to Parker *et al.* [500]) but still below that of method (*ii*) which is now universally accepted [353, 500–504] as the better one.

*A rigorous derivation would yield the difference of the chemical potentials $\Delta\mu$ and not the electrostatic potential qV. In general, $\Delta\mu$ may contain contributions from the Bernoulli effect and from temperature, stress and gravitational potential gradients. However, any physical voltmeter measures $\Delta\mu$ rather than qV, so we are entitled to use Equation (19.1) provided *V* refers to the measured voltage.
** In fact there is a third method as well. This is based on the work of Langenberg *et al.* [498] in which a microwave input yielded d.c. output voltages (in the absence of a d.c. feed) obeying Equation (19.1). Preliminary measurements [328] yielded quite good values of *e/h* but the voltages (especially for higher values of *n*) were rather unstable so the method was soon abandoned.

19.2 The independence of the voltage–frequency ratio

According to Equation (19.1) the voltage–frequency relation is independent of the type of junction, of junction material, of the order of steps – in fact of everything. Two questions immediately arise: (i) whether the relationship is exact according to our present theories and not only a good approximation, and (ii) whether it can be shown to be valid experimentally. We shall return to the first question in Section 19.3 and investigate the second one here.

For testing the relationship experimentally Clarke [505, 506] used junctions made of Pb, Sn and In, and the barrier was an approximately 1 μm layer of Cu or Ag. A normal metal was chosen as the barrier because it provides good vertical steps. Clarke showed that exposing the junctions to radiation of 500 kHz (first induced step about 1 nV) the current step was vertical to 1 part in 10^8.

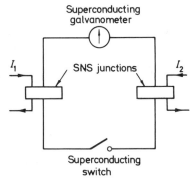

Fig. 19.1 Schematic diagram of the circuit used for comparing the voltages developed across two junctions made from different superconductors. The currents I_1 and I_2 are adjusted to bias each junction on the same order current step. After Clarke [506].

The experimental set-up in which two junctions made from different superconductors were compared is shown schematically in Fig. 19.1. The bias currents I_1 and I_2 were adjusted so that both junctions were biased at an (usually first) induced step. The superconducting galvanometer (the SLUG discussed in Section 18.2) could detect current as low as 3×10^{-7} A. Now if at $t = 0$ the switch is closed, a current satisfying the equation

$$L\frac{dI}{dt} = \Delta V \qquad (19.2)$$

would start to flow in the circuit, where ΔV is the difference between the junction potentials (more correctly chemical potentials) and L is the inductance of the circuit. After 30 minutes in each case (for any combination of the junctions) the current was still below the resolution of the galvanometer. Hence for an inductance of $L = 10^{-7}$ h the upper limit may be set as

$$\Delta V < L\frac{\Delta I}{\Delta t} = 10^{-7}\frac{3 \times 10^{-7}}{1 \cdot 8 \times 10^3} \cong 1 \cdot 7 \times 10^{-17} \text{ V.} \qquad (19.3)$$

Since the voltage of the induced step is about 1 nV this means that the voltages on the two junctions are identical to an accuracy of at least 1 part in 10^8. And this result was found to be independent of the material of the junction or of the barrier. In addition, e/h was found to be independent of temperature (from 1·2 to 2·2°K), barrier thickness, rf power (over a factor of 5), rf frequency (from 100 kHz to 1 MHz), magnetic field (up to ± 1 gauss) and direction of bias current.

A comparison involving induced steps of different order in different types of junctions was made by Finnegan et al. [507] in the experimental arrangement shown in Fig. 19.2. First the Nb$_3$Sn point contact junction was irradiated by a laser of 891 GHz frequency and biased to the first order step ($n = 1$) with a

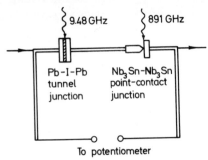

Fig. 19.2 Schematic diagram of the experimental set-up used for comparing two different types of junctions biased at different steps. Double lines represent superconducting portions of the circuit. After Finnegan et al. [507].

current which was below the maximum supercurrent of the Pb–PbO–Pb tunnel junction. Hence the voltage across the potentiometer corresponded to the laser-induced step amounting to about 1·84 mV. Next the point contact was adjusted to increase its zero-voltage current above the value required for the proper bias (on the $n = 94$ step for an input radiation of about 9·48 GHz) of the tunnel junction. By varying the frequency of the X-band source and choosing the appropriate step, the step voltage could be made nominally equal to the previously measured laser-induced step. The potentiometer therefore functioned only as a stable standard and determination of its absolute calibration was unnecessary.

The voltage–frequency ratio was calculated from both measurements and found to be different by 0·4 $\pm$ 1·5 ppm. This experiment may be taken as proving the independence of the frequency–voltage relation (at least down to the 1 ppm level) on the order of the induced step.

The type of junction might possibly have an influence but again only at the 1 ppm level. This was proved by Parker et al. [500] who made measurements both on tunnel junctions and point contacts, and by Petley and Morris [501, 502] who used solder drop junctions.

19.3 Frequency pulling effects

It was proposed by Scully and Lee [508] that the frequency of Josephson radiation may be slightly pulled from the value qV/h. They used a technique developed for the treatment of laser oscillators where it is shown that the radiation frequency emitted by an atom is influenced by a detuned cavity. The analogue of the detuned cavity is the finite lifetime of the pairs in the Josephson junction problem. Their result is a frequency pulling of the order of 1 part in 10^8.

A more general calculation by Stephen [509, 510] which treats the problem of noise in general, showed also the presence of a small shift. It was, however, pointed out by McCumber [511] that the interaction with the reservoir is already taken into account when the correct value of the chemical potential is established. Hence Josephson's formula in terms of the difference of the chemical potentials is exact.

Bloch [512, 513] arrived at the same conclusions by considering a super-conducting ring containing a Josephson junction. His derivation contains none of the approximations inherent in both microscopic and phenomenological theories; it is based on no more than some fundamental tenets of electro-dynamics and quantum mechanics. He proves that the Josephson frequency–voltage relation follows directly from the simple induction law

$$V = -\frac{d\Phi}{dt} \qquad (19.4)$$

in conjunction with the theorem (proven from first principles) that the free energy of the system must be a periodic function of magnetic flux. The transition from the ring to an open-ended geometry may then be made (this is of course where some approximation does creep in) by assuming that the conditions at some distance from the junction are immaterial to the manifestation of the Josephson effects.

Obviously, one cannot exclude the possibility that some future theories will predict a slight shift in frequency but for the present we seem to be entitled to take the frequency–voltage relation as being exact.

19.4 Effect of noise

We have already mentioned in Section 18.3 that the effect of noise is to broaden the radiation from a Josephson junction, and derived an approximate formula for the bandwidth. This finite bandwidth means one more source of error (in locating the centre frequency) in the method relying on emitted radiation.

One may feel intuitively that the effect of noise will be smaller for microwave-induced current steps. There are in that case two oscillators locked in phase (as discussed in Section 10.2) so a larger amount of noise is needed to break up this relationship.

A microscopic study of noise in a driven Josephson oscillator is given by Stephen [514]. We shall follow here the treatment of Kose and Sullivan [515] who, without enquiring into the origin of the noise, investigated its effect upon the slope of the current steps.

Let us first look at the case when besides the applied voltage at ω_m there is a purely sinusoidal noise voltage at ω_s present as well. The corresponding current we have already worked out in Section 17.2 concerned with frequency conversion. The d.c. component of the current is given by

$$I(V) = I_J(-1)^{k+l} J_k\left(\frac{qV_m}{\hbar\omega_m}\right) J_l\left(\frac{qV_s}{\hbar\omega_s}\right) \sin(\alpha_0 - l\alpha_s) \qquad (19.5)$$

where α_s is the initial phase difference between the two voltages and V satisfies the relationship

$$V = \frac{\hbar}{q}(k\omega_m + l\omega_s). \qquad (19.6)$$

For a given value of $k(= K)$ we have steps for each discrete value of l. If $\omega_s \ll \omega_m$ (the case of interest) then these steps (each one adding to the previous one) lie so close to each other that they appear smeared out in the I–V characteristic. Taking the current zero in the middle of the step its value at the voltage

$$V = \frac{\hbar}{q}(K\omega_m + L\omega_s) \qquad (19.7)$$

is given by

$$I(V) = I_J \left| J_k\left(\frac{qV_m}{\hbar\omega_m}\right) \right| \left\{ \left| J_0\left(\frac{qV_s}{\hbar\omega_s}\right) \right| + 2\sum_{l=1}^{L} \left| J_l\left(\frac{qV_s}{\hbar\omega_s}\right) \right| \right\}. \qquad (19.8)$$

Since

$$|J_l(x)| = |J_l(-x)|, \qquad (19.9)$$

the modified step is antisymmetric in the chosen coordinate system. Equation (19.8) gives an idea of the modified shape of the current step and in fact taking a realistic noise spectrum makes little qualitative difference.

Mathematically, a general noise voltage leads to the formula for the current

$$I = I_J \sin\left(\omega_0 t + \alpha_0 + \frac{qV_m}{\hbar\omega_m}\sin\omega_m t + \frac{q}{\hbar}\int_0^t V_n(t')\,dt'\right) \qquad (19.10)$$

which can be expanded to yield

$$I = I_J \sum_{k=-\infty}^{\infty} (-1)^k J_k\left(\frac{qV_m}{\hbar\omega_m}\right) \sin\left(\omega_0 t + \alpha_0 - k\omega_m t + \frac{q}{\hbar}\int_0^t V_n(t')\,dt'\right). \qquad (19.11)$$

For noise input of this form (and assuming white noise) the output noise spectrum was calculated by Stewart [516] and Parker [517]. It was found that depending on whether

$$D = \frac{q}{\hbar} V_{rms} \tag{19.12}$$

is large or small in comparison with the assumed noise bandwidth, the output spectrum has Gaussian or Lorentzian form. For a Gaussian distribution the d.c. current in the vicinity of the Kth step is found to be

$$I = I_J \left| J_k \left(\frac{qV_m}{\hbar \omega_m} \right) \right| \frac{\int_0^\omega \exp\left[-(\omega')^2/4D^2\right] d\omega'}{(4\pi D^2)^{1/2}} \tag{19.13}$$

Experiments. Kose and Sullivan [515] used an external noise source (a Zener diode biased into the breakdown region followed by amplification) to study the effect of noise upon a current step. They made a distinction between noise voltage (fluctuations generated by a low internal impedance voltage source) and noise current (fluctuations generated by a high internal impedance current source) and could experimentally simulate both. A very neat result is shown in Fig. 19.3 where the d.c. current is plotted in the absence and in the presence

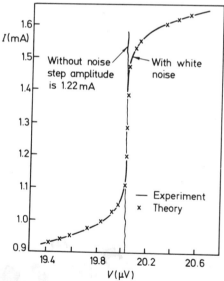

Fig. 19.3 Effect of noise current on the first current step of a Nb–Nb point-contact junction. After Kose and Sullivan [515].

of current noise. The agreement between the above outlined phenomenological theory and experiments is remarkable.

It is concluded that voltage noise broadens the step whereas current noise affects only the amplitude of the steps. As far as the determination of e/h is concerned the most valuable result of this study is that fluctuations do not break the antisymmetry of a current step and so the centre point is unaffected by noise.

19.5 Measurements and numerical results

There are only two quantities to be measured: the voltage of the chosen current step and the frequency of the impressed radiation (or of the emitted radiation if self-induced steps are used). There is no difficulty in measuring frequency to high accuracy (provided the power is not too small) and it is also relatively easy to stabilise a microwave oscillator (e.g. with the aid of a quartz crystal) to a few parts in 10^8. Hence the accuracy of determining e/h depends on the accuracy of the voltage measurement. The main factors are: uncertainty in the calibration and linearity of the potentiometer, resolution of the null detector and stability and absolute calibration of the voltage standard. This is described in great detail by Parker et al. [500] who did advance the art of high-accuracy low-level d.c. measurements by about an order of magnitude. The main steps in their measuring procedure were as follows:

(i) observe the I–V characteristics of the junction;
(ii) adjust the amplitude of the steps by varying parameters like microwave power, magnetic field, etc.;
(iii) stabilise the microwave source;
(iv) measure the frequency;
(v) bias the junction to the chosen current step;
(vi) measure the voltage with the specially developed potentiometer;
(vii) reverse the junction current (adjust if necessary) and measure the voltage again.

Taking the average of the two voltage measurements served to eliminate the thermoelectric voltages in the cryostat leads.

In most measurements the microwave frequency was around 10 GHz and the step order high, up to $n = 70$.

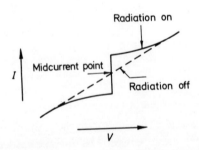

Fig. 19.4 Part of the measured I–V characteristic showing that the midcurrent point on a current step remains unaltered when the radiation is switched off.

The current steps were not always vertical, in which case the voltage at the middle of the step was taken as the correct value. This is well justified by the Kose–Sullivan theory [515] though at the time the decision was taken on experimental grounds. It was found (for point contacts) that the midcurrent point remained undisturbed in the presence of microwave radiation as shown in Fig. 19.9. Considerable part of the noise affecting the slope of the current steps was probably due to external factors. Finnegan et al. [507] observed that the steps became more vertical (in fact perfectly vertical within the resolution of their instruments) when the cryostat was placed in a shielded room.

The sources of uncertainty in the measurements of Parker et al. [500] are summarised in Table 19.1. It gives an idea of the importance of the various factors involved.

Table 19.1 Contributions to the uncertainty in the measured value of e/h (in units of ppm). All uncertainties are intended to be 1 standard deviation.

I.	Frequency		0·01
II.	Voltage		
	(a) Random error	0·4	
	(b) NBS calibration of standard cells	0·6	
	(c) Transportation and ageing of standard cells	0·5	
	(d) Standardising potentiometer	0·5	
	(e) Establishing 1/10 ratio, 0·3 ppm, additive per decade	1·0	
	(f) Lead-resistance correction, 0·3 ppm, additive per decade	1·0	
	(g) Ground-loop currents	1·0	
	(h) Stability of operating current	0·5	
	(i) Self-heating during calibration	0·5	
	(j) Linearity of divider	0·5	
	(k) Temperature drift of output voltbox	1·0	
	rss total		2·4

In terms of the volt of the National Bureau of Standards, as it existed in 1966–67, the result is

$$2e/h = 483{\cdot}5976 \pm 0{\cdot}0012 \text{ MHz}/\mu V_{NBS} \quad (2{\cdot}4 \text{ ppm})$$

The conversion to the absolute volt is not at all straightforward. It is discussed by Taylor, Parker and Langenberg [518] yielding

$$V_{NBS}/V_{\text{absolute}} = 1{\cdot}000\,008\,8 \pm 2{\cdot}6 \text{ ppm}$$

Similar results for e/h were obtained by Petley and Morris [501, 502, 519] using solder drop junctions. Their value is 0·6±3·3 ppm higher than that of Parker et al. [500]. Interestingly, this is exactly the same as that reached by the Pennsylvania group after a reassessment [503]. There is no doubt that such spectacular agreement is no more than a coincidence but it is remarkable to note that two different experiments conducted in different countries and involving different voltage standards (their intercomparison is not a trivial task either) yielded so close results.

Finnegan, Denenstein and Langenberg [353] improved the accuracy of the measurements by using higher steps up to $n = 500$ so the resulting voltage (around 10 mV) was more easily measurable. They eliminated further the adjustable resistance element of the potentiometer and adjusted instead the frequency of the microwaves to get a step at a specified voltage.

Improved accuracy was also achieved by Harvey *et al.* [504] at the National Standards Laboratory in Australia. Their results are compared in Fig. 19.5 with those obtained earlier.

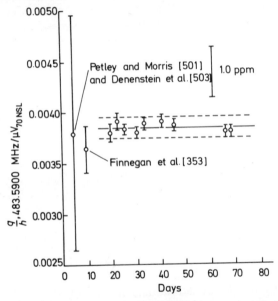

Fig. 19.5 Measurement of q/h by Harvey *et al.* [504] taken over a period of 50 days. The error bars on the individual day's results are the root sum square of all known and estimated errors. The mean value is indicated by the solid line while the dashed lines represent an estimated uncertainty for the NSL to BIPM voltage intercomparison. Values of q/h determined by others are also shown.

19.6 Significance of the measurements

There are very few applied scientists whose heart would beat faster on hearing of a new determination of a fundamental constant. A change in the sixth or seventh significant figure would indeed have little influence on the operation of devices but as far as the whole of physics is concerned their importance should not be underestimated. A new, more consistent set of constants may lead to the discovery of a previously unknown inconsistency or to the removal of a known inconsistency in our description of nature.

The accurate measurements of e/h, for example, made possible an unambiguous comparison between theory and experiments in quantum electrodynamics (QED). Taylor, Parker and Langenberg [518] managed to obtain a value for the fine structure constant α by combining the value of e/h with the measured values of some other constants (Faraday constant, the gyromagnetic ratio of the proton, the magnetic moment of the proton, the ratio of the ampere as maintained by the NBS to the absolute ampere and certain accurately known auxiliary constants) not dependent on QED. Comparisons between this value of α and those obtained from QED led in some cases to large (several times the standard deviation) discrepancies, suggesting either that the experimental errors were underestimated or that the theories need further refinement (or both).

Table 19.2 A comparison of the values of five constants resulting from the *1963* adjustment with those resulting from the *1963* adjustment.

Quantity	Symbol	Units	1963 adjustment	Error ppm	1969 adjustment	Error ppm	Change ppm
Inverse of the fine-constant	α^{-1}		137·0388	4·4	137·03602	1·5	− 20
Electronic charge	e	10^{-19} C	1·60210	12	1·6021917	4·4	+ 57
Planck's constant	h	10^{-34} J sec	6·62559	24	6·626196	7·6	+ 91
Electronic mass	m_e	10^{-31} kg	9·10908	14	9·109558	6·0	+ 52
Avogadro's number	N	10^{26} kmole^{-1}	6·02252	15	6·022169	6·6	− 58

Taylor *et. al.* [518] in fact undertook a much bigger task. Armed with the new value of e/h they could reject a number of data used for the previous adjustment of the fundamental constants [520]. Further sifting of all the data since available, they arrived at a new set of values. This is shown in Table 19.2. The change in some of the constants is remarkably large.

19.7 Voltage standards

The volt is at present maintained as the average emf of a group of Weston–Cadmium cells. This is rather unsatisfactory because the emf's drift with time and so the various national laboratories have to do a triennial pilgrimage to France to compare their cells with those maintained by the Bureau International des Poids et Mesures (BIPM). The usual drifts are several ppm but it is not clear how much of this is due to the drift of the BIPM cells. Thus clearly any method which could measure the change in the maintained volt would be of great value. The gyromagnetic ratio of the proton (γ_p) is suitable for this purpose and was indeed used by the NBS to check on the constancy of the United States legal volt to an accuracy of 1 ppm. However, these measurements are rather complex and their duplication by other laboratories would require environments which are stable in time both magnetically and thermally. The Josephson effect has none of these disadvantages. It has now been experimentally proven that at the 1 ppm level the Josephson frequency–voltage relation is independent of all conceivable parameters and in addition it is a relatively easy and inexpensive measurement to set up. Hence, as first pointed out by Taylor, Parker, Langenberg and Denenstein [521], it is already feasible to maintain and compare the voltage standards of the various national laboratories with the aid of the Josephson effect.

There is also the distinct possibility of using the Josephson effect to *define* the volt in terms of frequency. However, as the SI unit is the ampere that would mean a major readjustment not on the cards at the moment.

20.   Fluctuations

20.1 Introduction

Noise and fluctuations are important both as fundamental concepts and as limitations to the sensitivity of devices. We have mentioned noise a number of times in this book, e.g. in Sections 9.2 and 13.6 (relation to maximum zero-voltage current), Section 18.5 (flux noise), Section 19.4 (deviation of current steps from vertical), but made no attempt to derive the results from a more general formulation. The difficulties are twofold. Firstly, no general theory is available which would be applicable to all the individual problems of interest (e.g. what is the minimum signal detectable by a Josephson junction) and secondly, most of the theories are based on microscopic theory beyond the scope of the present book. The solution adopted here is to try to give some insight into the problem on the basis of a simple macroscopic theory, quote some further theoretical results and discuss experimental results.

20.2 Fluctuations in a thin superconducting hollow cylinder

It is quite feasible to expect that fluctuations, even in a superconducting system, will be related to the normal resistance of the material. So we shall express here following Burgess [522, 523]) the magnetic flux in terms of both normal and superconducting quantities. For a hollow cylinder of inner radius r, length a and wall thickness d, the magnetic inductance for circulating currents is

$$L = \frac{\mu_0 \pi r^2}{a} \qquad (20.1)$$

Assuming that the penetration depth is large in comparison with the wall thickness but much smaller than the radius of the cylinder, the kinetic energies of the normal and superfluids may be expressed in terms of 'kinetic' inductances (see Appendix 9) as

$$L_N = \frac{2\pi r}{ad}\frac{m_N}{\rho_N e^2}, \qquad L_S = \frac{2\pi r}{ad}\frac{m_S}{\rho_S q^2}. \qquad (20.2)$$

The total energy associated with electron motion is

$$E = \tfrac{1}{2}LI^2 + \tfrac{1}{2}L_N I_N^2 + \tfrac{1}{2}L_S I_S^2 \qquad (20.3)$$

where

$$I = I_N + I_S. \tag{20.4}$$

Let us work out now the fluxoid Φ_c for this configuration from Equation (1.40). The current density and charge density may be assumed uniform, leading to

$$\Phi_c = \Phi + \frac{m}{(2e)^2} \int \frac{j_S}{\rho_S} \, ds$$

$$= \Phi + \frac{m}{(2e)^2} \frac{2\pi r j_S}{\rho_S} \tag{20.5}$$

which according to Equation (1.40) must be equal to an integral multiple of Φ_0. With the aid of Equation (20.2) we may then rewrite Equation (20.5) in the form

$$LI = n\Phi_0 - L_S I_S \tag{20.6}$$

whence I_S can be expressed in terms of I_N. At the end we get for the flux

$$\Phi = \frac{n\Phi_0 + L_S I_N}{1 + L_S/L}. \tag{20.7}$$

It may be seen from the above equation that the magnetic flux will fluctuate because of the fluctuations in I_N^*.

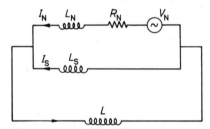

Fig. 20.1 Equivalent circuit of a thin superconducting hollow cylinder.

All what we have said so far (with the exception of fluxoid quantisation) may be incorporated into the equivalent circuit shown in Fig. 20.1. The resistance R_N represents the dissipation in the normal fluid (Appendix 9)

$$R_N = \frac{2\pi r}{ad} \frac{m}{\rho_N e^2 \tau_N} \tag{20.8}$$

where $\tau_N = L_N/R_N$ is the momentum relaxation time of the normal fluid. V_N is an rms noise voltage obtained from the Nyquist formula $\langle V_N^2 \rangle = kTR_N/\pi$ per unit angular frequency.

---

* If the normal fluid is ignored (on the basis that it does not contribute to the net current) we come to the erroneous conclusion that there can be no fluctuation of flux.

In equilibrium the energy is a minimum with respect to I_S and I_N for a given value of the fluxoid. The result is (Appendix 10)

$$I_N = 0 \quad \text{and} \quad I_S = \frac{\Phi_c}{L+L_S} \tag{20.9}$$

but note that both components will fluctuate around these equilibrium values. The fluctuations in I_S occur by an indirect mechanism. The normal fluid is scattered by phonons and hence I_N is fluctuating which causes a fluctuating magnetic flux (Equation (20.7)) whose rate of change gives a voltage accelerating both the normal and the superfluid.

Let us now calculate the fluctuations in magnetic flux. On the basis of the equivalent circuit we get

$$\langle \Phi^2(\omega) \rangle = \frac{A}{1+\omega^2 \tau^2} \tag{20.10}$$

where

$$A = \frac{2kT}{\pi R_N} \left(\frac{LL_S}{L+L_S} \right)^2 \quad \text{and} \quad \tau = \frac{1}{R_N} \left(L_N + \frac{LL_S}{L+L_S} \right). \tag{20.11}$$

The spectrum of fluctuations decreases with increasing frequency. The relaxation time τ has a lower bound τ_N.

We may get the total variance by integrating Equation (20.10) ove r frequency, yielding

$$\langle \Phi^2 \rangle = \int_0^\infty \langle \Phi^2(\omega) \rangle \, d\omega$$

$$= \frac{L^2 L_S^2 kT}{(L+L_S)(L_N L_S + LL_N + LL_S)} \tag{20.12}$$

In a similar fashion we may get the variances of I, I_N and I_S. Note that the sum of the contributions to the mean energy from the fluctuations of the three currents add to $\frac{1}{2}kT$ as may be expected from equipartition considerations. The system has one degree of freedom which also follows from the quadratic denominator in ω (in general a polynomial of degree $2N$ in ω means that the system is N-variate).

20.3 Fluctuations in Josephson tunnelling

We can use again the model of Fig. 11.1(a) and write the relationship for the current (Equation (11.4))

$$I_J \sin \phi + GV + C\frac{dV}{dt} = I_G \tag{20.13}$$

and voltage (Equation (9.17))

$$V = \frac{\hbar}{q}\frac{d\phi}{dt}. \tag{20.14}$$

The only difference is that now I_G is a noise current due to the conductance G, that is

$$\langle I_G^2 \rangle = \frac{2}{\pi}kTG. \tag{20.15}$$

If the phase fluctuations are small we can replace $\sin\phi$ by ϕ and then with the aid of Equations (20.13) and (20.14) we get

$$\langle \phi^2(\omega) \rangle = \frac{2}{\pi}kTG\frac{(q/\hbar C)^2}{(-\omega^2 + qI_J/\hbar C)^2 + (\omega G/C)^2}. \tag{20.16}$$

In this linear approximation (as we have already discussed in Section 11.7) the inductance of the junction is

$$L_J = \frac{\hbar}{qI_J}. \tag{20.17}$$

There is a resonance at

$$\omega_{pJ} = \left(\frac{1}{L_J C}\right)^{1/2} \tag{20.18}$$

and the bandwidth is

$$\Delta\omega = G/C. \tag{20.19}$$

By integration we can get again the variances

$$\langle \phi^2 \rangle = \frac{qkT}{\hbar I_J} \tag{20.20}$$

and

$$\langle V^2 \rangle = \frac{kT}{C}. \tag{20.21}$$

Burgess [523] goes on from here to discuss the arising quantum mechanical formulae (obtained by a simple rewriting of the classical ones) when the assumption $\hbar\omega_{pJ} < kT$ no longer holds; expressions are also given for the case when ϕ cannot be regarded small.

20.4 Fluctuations in a superconducting ring containing a Josephson junction

If a Josephson junction is inserted into the superconducting ring discussed in Section 20.2 then there will be two further terms added to the energy of the ring. Firstly the junction coupling energy*

$$E_J = \frac{\hbar I_J}{q}(1 - \cos \phi) \tag{20.22}$$

and secondly the electrostatic energy associated with the junction voltage V

$$E_e = \tfrac{1}{2}CV^2. \tag{20.23}$$

The technique of handling fluctuations is the same as before. The system is now trivariate, the three statistically independent variables are I_N, V and ϕ (each of them contributing $\tfrac{1}{2}kT$ to the energy in the classical limit). The variances are

$$\langle V^2 \rangle = \frac{kT}{C} \tag{20.24}$$

$$\langle \phi^2 \rangle = \frac{qkT}{\hbar I_J} \frac{L + L_S}{L_J + L + L_S} \tag{20.25}$$

$$\langle I_N^2 \rangle = kT \frac{L + L_S}{LL_N + L_N L_S + L_S L} \tag{20.26}$$

In the limit of strong coupling ($I_J \to \infty$, $L_J \to 0$) the results reduce to those of a simple superconducting ring whereas in the limit of very large ring inductance ($L \to \infty$) we get back the formulae for an isolated junction.

20.5 Notes on the I–V characteristics

Looking at the expression (Equation (20.22)) for the junction coupling energy we can see that the energy is minimum when $\phi = 0$, that is when no current flows across the junction. The coupling energy becomes maximum, equal to

$$E_{J\,max} = \frac{\hbar}{q}I_J \tag{20.27}$$

at $\phi = \pi/2$, that is when the d.c. supercurrent is maximum. If we try to force a current greater than that across the junction then it switches over to single particle tunnelling. We may now use the rough argument that in the presence of

---

*This is the first term of Equation (12.36). The second term is disregarded here because only junctions without spatial variation of phase are considered.

noise the transition will occur when the sum of the coupling energy and thermal energy reaches E_{max}, that is when

$$\frac{\hbar I_J}{q}(1-\cos\phi)+\tfrac{1}{2}kT = \frac{\hbar I_J}{q}. \tag{20.28}$$

Solving Equation (20.28) for $\cos\phi$ and substituting it into our usual expression for the Josephson current we get

$$I_{max} = I_J \sin\phi$$

$$= \left[I_J^2 - \left(\frac{kTq}{2\hbar}\right)^2 \right]^{1/2} \tag{20.29}$$

which is nearly identical to Vant Hull's formula [427] obtained by similar arguments. Note that the observable d.c. supercurrent is always reduced by thermal effects; it becomes unobservable when

$$I_J = \frac{kTq}{2\hbar}. \tag{20.30}$$

We can also make some qualitative comments on the hysteresis region (as shown, for example, in Fig. 11.5) of tunnel junctions. As the current increases above I_{max} there is a transition to single particle tunnelling and the junction capacitance is charged up to an energy $\tfrac{1}{2}CV^2$. When the current is reduced there is no immediate return to Josephson tunnelling because the energy is too high; the characteristic follows the single particle tunnelling curve. The switch back occurs when the voltage (and thus the energy) is sufficiently reduced.

One can, of course, devise more sophisticated theories concerned with the effect of noise upon the I–V characteristic of the junction. We shall mention here a theory by Ambegaokar and Halperin [524] who showed that the equations of the junction ((20.13) and (20.14)) have the same mathematical form as that of a particle in a field of force. Taking then the phase equivalent to the position of the particle the problem may be solved in analogy with Brownian motion. Experiments by Anderson and Goldman [525] and Simmonds and Parker [526] agreed reasonably well with this theory. There is though a basic difficulty in showing the effect of noise on the I–V characteristics. In order to have measurable effect one has to operate very close to the critical temperature of the superconductors with subsequent uncertainties in the measurement. An alternative solution is to impress noise from an external source upon the junction in which case the effect of noise can be separately studied without affecting the other parameters.

In a series of experiments Buckner et al. [392, 393] have measured the I–V characteristics of a small (area = 10^{-10} m²) tunnel junction (c.f. Section 11.5) under conditions when a variable amount of noise was fed into the junction circuit from a noise generator. The measured results are shown in Fig. 20.2.

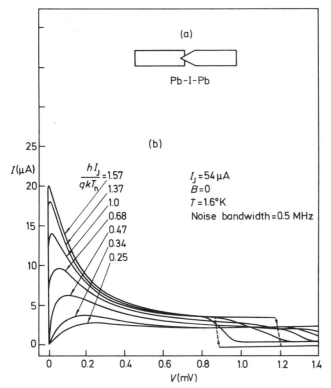

Fig. 20.2 (a) Junction geometry of small tunnel junction. (b) Typical I–V characteristics of a small tunnel junction at various noise levels characterised by the parameter $\hbar I_J/qkT_n$. After Buckner *et al.* [392].

The shape of the characteristics resembles very much the theoretical curves of Ivanchenko and Zilberman worked out for small junctions [394, 395].

For a quantitative comparison a relationship must be derived between the noise voltage of the generator and the effective noise temperature T_n. This was done by assuming the noise temperature in the form

$$T_n = T_{n0} + b\langle V^2\rangle \tag{20.31}$$

where $\langle V^2\rangle$ is the mean square noise voltage of the noise generator, T_{n0} is the noise temperature of the junction in the absence of impressed noise and b is a constant. T_{n0} was determined by comparing the measured zero-voltage current (in the absence of impressed noise) with the theoretical one, and b was obtained by fitting the observed maximum current to the theory for one value of $\langle V^2\rangle$. The maximum current in each curve of Fig. 20.2 is plotted in Fig. 20.3 (another set of results for 5 MHz noise bandwidth are also shown) against the parameter $\hbar I_J/qkT_n$. The agreement with the theoretical curve may be seen to be very good.

Similar experiments on Nb–Nb point-contact junctions were performed by Kanter and Vernon [527]. The junction was placed into a waveguide and irradiated by a gas discharge tube. The amount of noise reaching the junction was controlled by an attenuator.

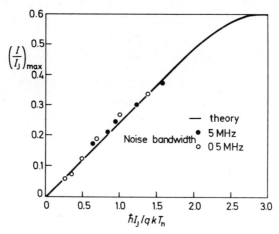

Fig. 20.3 Comparison of the experimental and theoretical values of the current maxima as a function of $\hbar I_J/qkT_n$. The solid curve was calculated from the theory of Ivanchenko and Zilberman [395]. After Buckner *et al.* [392].

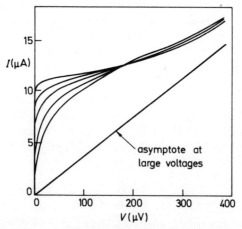

Fig. 20.4 Recorder plots of *I–V* characteristics in the presence of impressed noise. The attenuator setting is ∞, 7, 4, 2 and 0 db going from left to right. After Kanter and Vernon [527].

The effect of noise upon the *I–V* characteristics is shown in Fig. 20.4; it leads again to a reduction of zero-voltage current. There is good qualitative agreement with the theoretical curves of Kurkijärvi and Ambegaokar [528].

20.6 Quantum phase fluctuations

A magnetic flux

$$\Phi_{ext} = \Phi_x^0 + \Phi_x^1 \sin \omega t \tag{20.32}$$

applied to a double junction interferometer will give rise to a component in supercurrent [529]

$$\Delta I = A J_1 \left(\frac{2\pi \Phi_x^1}{\Phi_0} \right) \sin \left(\frac{2\pi \Phi_x^0}{\Phi_0} + \gamma \right) \sin \omega t \tag{20.33}$$

where A is a constant. This formula is similar to Equation (14.26) derived for the a.c. excitation of a single junction ring. The essential difference is that we have introduced here γ, an arbitrary quantum phase to account for possible phase fluctuations. Well below the critical temperature we may expect to find a stable value for ΔI but nearer to the critical temperature this no longer applies. Fluctuations and reductions in the value of ΔI may then be attributed to changes in the quantum phase integral around the ring.

The basic experiment was performed by Ulrich [529] using two Nb–Sn point contact junctions as shown in Fig. 20.5 (a) and $\omega/2\pi = 750$ Hz. The temperature was first raised above the critical temperature of tin and then slowly reduced to observe the onset of quantum phase coherence in bulk tin. The measure of coherence is the value of $\Delta I / (\Delta I)_{max}$ plotted in Fig. 20.5 (a) against temperature for a 3 sec detector output time constant. At each temperature Φ_x^0 was adjusted to give maximum output. At temperatures below about $3.72°$K the quantum phase was found stable for at least several minutes, and above $3.73°$K the fluctuation frequency exceeded 1 kHz. For comparison the Meissner transition and the maximum supercurrent across the junction (persisting up to about $4.0°$K) are shown in Figs. 20.5 (a) and (b). Ulrich's [529] main conclusions are as follows.

(i) The transition to the stable phase coherent superconducting quantum state is not discontinuous but proceeds through a fluctuation state where the fluctuation frequency is strongly temperature-dependent.

(ii) Quantum phase coherence becomes established concurrently with the complete exclusion of the magnetic field.

20.7 Paraconductivity of tunnel junctions

It was predicted by Ferrel [530] that instantaneous Josephson currents will flow in a tunnel junction when one of the electrodes is in the superconducting state and the other is above its critical temperature. This paraconductivity* is strongly temperature-dependent. For thin films it tends to infinity as $(T - T_c)^{-2}$.

---

*Term introduced by Ferrel in analogy with the diverging paramagnetic susceptibility when the Curie point is approached.

When the critical temperature of one film (T_{c1}) is considerably higher than that of the other one (T_{c2}) the junction conductance may decrease by several orders of magnitude below its normal value as the temperature decreases from T_{c1} to just above T_{c2}. Under these circumstances it is relatively easy to detect the effect of instantaneous Josephson currents because the excess conductance needs only be comparable with a rather small conductance.

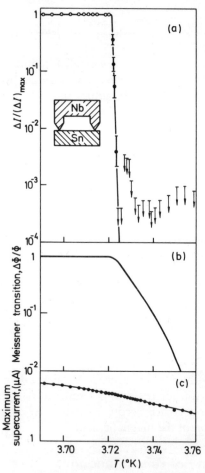

Fig. 20.5 (a) The quantum phase coherence signal $\Delta I/(\Delta I)_{max}$ as a function of temperature for a 3 sec detector output time constant. Arrows denote upper limits due to background noise where no quantum coherence signal was observed with this time constant. (b) Meissner transition in the tin sample measured at 10 Hz. The fraction of flux excluded related to the total flux excluded at 10 Hz is plotted versus temperature. (c) Temperature dependence of the maximum supercurrent through the two junctions in parallel. The junctions become superconducting at a temperature significantly higher than the bulk tin critical temperature. After Ulrich [529].

The variation of the conductance was measured experimentally by Yoshihiro and Kajimura [531] on an Al–I–Pb junction ($T_{c2} = 2 \cdot 16°K$). Their results are shown in Fig. 20.6 where $(G/G_{NN})^{-1/2}$ is plotted against temperature. As may be seen the functional relationship agrees well with Ferrel's prediction but the magnitude of the excess conductance is below the predicted value by a factor of about 40. More detailed measurements in a wider temperature range were also reported by Yoshihiro and Kajimura [532].

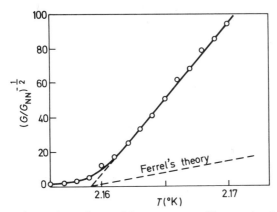

Fig. 20.6 Temperature dependence of the excess tunnelling conductivity. After Yoshihiro and Kajimura [531].

20.8 Shot noise and thermal noise

We have derived formulae earlier in this chapter for the spectrum of current and voltage fluctuations. They are valid in the classical limit in the absence of a d.c. voltage across the junction. In some respects that approach can be easily generalised. For example if we abandon the model of constant conductance and want to take into account single particle tunnelling, we can simply replace G in Equation (20.13) by the zero bias differential conductance. However, to get the full picture one is bound, sooner or later, to turn to microscopic theory. The contribution of single-particle tunnelling to the current noise spectrum was calculated by Scalapino [369, 533] and Dahm et al. [534] in the form

$$\langle I^2(\omega) \rangle = \frac{e}{2\pi} \left[I_{SS}(V + \hbar\omega/e) \coth\left(\frac{eV + \hbar\omega}{2kT}\right) + \right.$$
$$\left. + I_{SS}(V - \hbar\omega/e) \coth\left(\frac{eV - \hbar\omega}{2kT}\right) \right] \tag{20.34}$$

where $I_{SS}(V \pm \hbar\omega/e)$ is the single-particle part of the d.c. current at a d.c. voltage $V \pm \hbar\omega/e$. When $eV \gg \hbar\omega$ and $eV \ll kT$ the above equation reduces to Equation (20.15).

The contribution from Cooper-pair tunnelling was derived by Stephen [510]. In the low frequency limit

$$\langle I^2(\omega)\rangle = \frac{q}{\pi} I_P(V) \coth \frac{qV}{2kT} \tag{20.35}$$

where $I_P(V)$ is the average value of the supercurrent

$$I_P(V) = I_J\langle \sin \phi \rangle + G_1(V)V\langle \cos \phi \rangle. \tag{20.36}$$

Note that the cosine term in the equation for the supercurrent (Equation (9.33)) must also be included.

Adding the contributions from single-particle and Josephson tunnelling we get the total noise spectrum. In the limit of $\hbar\omega \ll eV$ and $kT \ll eV$ the sum of Equations (20.34) and (20.35) reduces to

$$\langle I^2(\omega)\rangle = \frac{1}{\pi}(eI_{SS} + qI_P) \tag{20.37}$$

which is the familiar shot noise formula. It is interesting to see that the contribution of Josephson tunnelling is exactly twice as large, corresponding to the double charge of the Cooper-pairs.

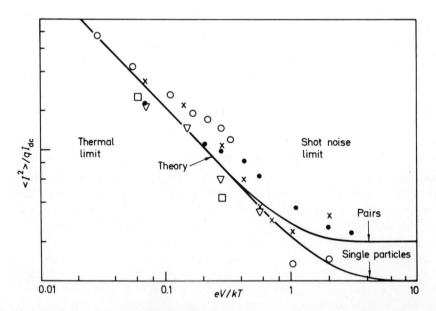

Fig. 20.7 Comparison of measured voltage dependence of normalised mean square noise currents with theory for various samples. $kT/e \approx 360\ \mu V$ ($T = 4\cdot2°K$) and $1\ \mu A \leq I_{dc} < 10\ \mu A$. After Kanter and Vernon [536].

If we look at the low frequency region only and in addition take account of the capacity of the junction (but not of its inductance) we get for the noise voltage

$$\langle V^2(\omega)\rangle = \frac{\langle I^2(\omega)\rangle}{G_D^2(1+\omega^2\tau^2)} \tag{20.38}$$

where $G_D = dI/dV$ is the differential conductance and τ is the time constant.

The voltage fluctuations were directly measured by Kanter and Vernon [535, 536] for point contact junctions realised by pressing a Nb wire tip against a flat oxidised Nb surface. The bandwidth of the measuring apparatus was 30 kHz centred at 150 kHz (chosen high enough to avoid flicker noise). The current noise derived from the measured voltage noise is plotted in Fig. 20.7 against eV/kT. The contribution of single-particles and pairs could be separately assessed by suppressing the pair current by a magnetic field. The agreement with theory (the sum of Equations (20.34) and (20.35)) is good.

20.9 Bandwidth of Josephson radiation and noise thermometry

Since the frequency of Josephson radiation is related to the voltage across the junction it is obvious that voltage fluctuations will modulate the output frequency and lead thereby to a finite bandwidth. Thus the theoretical treatment needs to combine the previous calculations on fluctuations with standard frequency-modulation noise theory [516, 517]. The formula derived by Dahm et al. [534] is as follows

$$\Delta\omega = \left(\frac{q}{\hbar}\right)^2 \frac{1}{G_D^2}\left[eI_{SS}(V)\coth\frac{eV}{2kT} + qI_P(V)\coth\frac{qV}{2kT}\right] \tag{20.39}$$

When $kT \gg eV$ the above formula reduces to

$$\Delta\omega = \frac{q^2}{\hbar^2}\frac{2kT}{G_D^2 V_0}(I_{SS} + I_P). \tag{20.40}$$

Dahm et al. [534] compared this theoretical result with experimental ones obtained by biasing Sn–I–Sn and Pb–I–Pb tunnel junctions on a self-induced current step. The dependence on temperature, current and differential conductance agreed well with the theoretical predictions though the experimental values were in general larger by a factor of about two.

An alternative way of voltage biasing the junction is to use the resistive ring configuration introduced in Section 14.7 and discussed as a voltmeter in Section 18.3. The bandwidth may then be calculated with the aid of Equations (18.5) and (18.6) which we reproduce below

$$\langle V_n^2\rangle^{1/2} = \left(\frac{2}{\pi}kTR\Delta\omega\right)^{1/2}, \qquad \Delta\omega = \frac{q}{\hbar}\langle V_n^2\rangle^{1/2} \tag{20.41}$$

yielding

$$\Delta\omega = C\frac{q^2}{\hbar^2}kRT \qquad (20.42)$$

where $C = 2/\pi$. A similar calculation by Silver *et al.* [489] which takes into account the spectrum of frequency modulation gives $C = 4/\pi$, whereas more rigorous calculations by Burgess [522] lead to $C = 2$.* Taking this latter value the bandwidth measurements of Silver *et al.* [489] show good agreement with the theory (a typical result is $\Delta f(\text{exp}) = 8200$ Hz, $\Delta f(\text{theory}) = 8400$ Hz).

It was suggested by Kamper [537] that the linear relationship between bandwidth and temperature given by Equation (20.42) could be used for thermometry. The feasibility of this proposal was explored by Kamper *et al.* [449] and Kamper and Zimmerman [538].

The experimental set-up needed is practically identical with that shown in Fig. 18.4. The resistance R (about 10^{-5} ohm) is kept at the temperature to be measured and the information is to be derived from the bandwidth of the oscillation at ω_0. Thus the demodulator in Fig. 18.4 must be followed by a device capable of measuring the bandwidth. This may be done either by direct measurement or by the use of a frequency counter which counts the number of cycles in a fixed gate time τ. In this latter method fluctuations in frequency appear directly as fluctuations in the count. A convenient measure of the fluctuations is the mean square deviation

$$\sigma^2 = \langle(\omega - \langle\omega\rangle)^2\rangle \qquad (20.43)$$

which is also proportional to absolute temperature provided only thermal noise is present. Advantages of the method are the automated output (information may be recorded on tape and processed immediately) and possible discrimination against spurious sources of noise as for example flicker noise.

The junction used for the measurement by Kamper and Zimmerman [538] is a point contact in a symmetric resistive circuit as shown in Fig. 15.3 f). The lowest noise temperature measured was 75 mK though according to the authors' analysis of the interfering noise sources the thermometer could work down to μK temperatures.

---

* Note that Equation (20.40) reduces to Equation (20.42) when $C = 2$ and the differential resistance may be taken equal to the d.c. resistance.

21.   Further topics

21.1 Introduction

The aim of this chapter is to discuss a few more topics concerned with various properties of Josephson junctions or weak links. The only common link between them is that they could not be easily fitted into any of the other chapters. We shall include here measurement results on SNS junctions, mechanical and superfluid analogues, an interesting effect found for thin film bridges (enhancement of the supercurrent by the application of a microwave field), various diagnostic and device applications, parametric amplification and relaxation oscillations.

21.2 SNS junctions

The proper explanation of the properties of superconductor–normal metal–superconductor junctions should be based on the proximity effect. The reason why a relatively thick (few tenths of a μm) layer of normal metal can sustain a zero-voltage current is that it acquires some superconducting properties due to the proximity of the superconductors. Conversely, the order parameter of the superconductors will be depressed in the vicinity of the normal metal. The theory for these junctions was formulated by de Gennes and Guyon [539] and de Gennes [540] and it is reviewed in de Gennes' book [367]; a simplified version due to Clarke [362] is also available.

The theory is strongly bound to the proximity effect but if we look at the experimental results it becomes obvious that we are faced with just another variation of the Josephson effect. There is a zero-voltage current up to a certain maximum current and then a gradual transition to ohmic behaviour quite similarly to that found, for example, for thin film bridges. The maximum zero-voltage current is dependent on magnetic field and the analogy holds also for the a.c. effect. Impressing an a.c. signal upon an SNS junction Clarke [506] obtained current steps in the $I-V$ characteristic at voltages corresponding to the Josephson relationship (this was already mentioned in Section 19.2).

The amount of experimental work done is rather limited. Although the first experiments were carried out as early as 1958 (by Meissner [253] on crossed copper-coated tin wires) there is essentially only one detailed set of experiments available (by Clarke [362]) which we shall briefly review here.

The normal metal chosen was copper alloyed by about 3% aluminium in order to shorten the mean-free-path (that makes the physical processes simpler and comparison with theory easier). The superconductors were of pure lead. The junctions were fabricated by evaporating successively on a water-cooled 3×1 inch glass slide a 0.7 μm strip of lead (0.2 mm wide), a disc of Cu/Al alloy (5 mm diameter) and a second strip of lead at right angles to the first as shown in Fig. 21.1.

A typical measured I–V characteristic is shown in Fig. 21.2. Note that the voltage scale is now in nV, about six orders of magnitude below that obtained with SIS junctions or weak links, which is simply due to the fact that the normal resistance is smaller by about the same six orders of magnitude. The shape of the detailed I–V curve (Fig. 21.2(b)) agrees qualitatively with those shown in Fig. 11.3 for Josephson junctions.

The dependence of the maximum supercurrent density on normal metal thickness is shown in Fig. 21.3 and on applied magnetic field in Fig. 21.4. The results in both cases are similar to those expected for Josephson tunnelling. The magnetic field dependence is in fact very similar to that shown in Fig. 12.5 calculated for $L/\lambda_J = 10$. There is, however, a major difference between SNS and SIS junctions as far as the temperature dependence of j_J is concerned, as may be seen by contrasting Fig. 21.5 with Fig. 9.3.

Near the critical temperature the theoretical prediction is that j_J should vary as

$$j_J \sim (T - T_c)^2. \tag{21.1}$$

The experimental data are shown in Fig. 21.6. The observed dependence of j_J is indeed linear in the range $6.6°$K to $7°$K. The sudden decrease after that may be attributed to the experimental realisation of the superconductors in the form of films of 0.7 μm thickness. When the temperature is sufficiently high the coherence length becomes comparable to that figure and the order parameter will be depressed everywhere in the superconductor, leading to a more rapid decrease in j_J.

Similar experiments on S_1–N–S_2 junctions (S_1 = Sn, N = Cu, S_2 = Pb) were recently performed by Kobayashi et al. [541]. The theory modified for this case predicts a variation

$$j_J \sim (1 - T/T_c) \tag{21.2}$$

in the vicinity of the lower critical temperature. The experiments (Fig. 21.7) bear this out, though again at the end of the region there seems to be a small discontinuity in j_J.

According to the measurements of Bondarenko et al. [542], however, there is little difference between the behaviour of S_1–N–S_1 and S_1–N–S_2 junctions (S_1 = Sn, N = Ag, S_2 = Ta).

An interesting effect in the I–V characteristic was reported by Mitani et al.

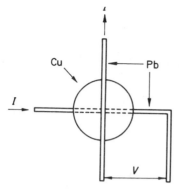

Fig. 21.1 Experimental realisation of a Pb–Cu–Pb junction. After Clarke [258].

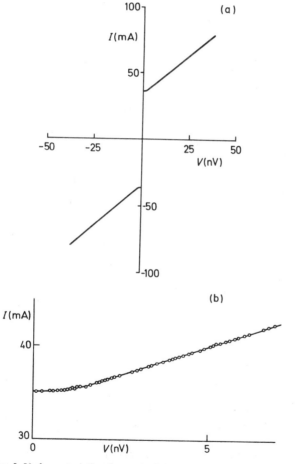

ig. 21.2 (a) The I–V characteristic of a typical SNS junction. $T = 2 \cdot 98°$K, $w = 552$ nm, $= 14$ nm, where w is the width of the junction and l is the mean free path. (b) Enlarged ⧫ortion of the characteristic. After Clarke [362].

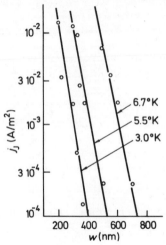

Fig. 21.3 Maximum supercurrent density as a function of barrier thickness in a Pb–Cu–P junction. After Clarke [362].

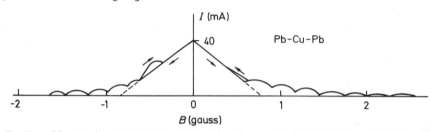

Fig. 21.4 The maximum supercurrent as a function of applied magnetic field. Measure period in flux is $(1{\cdot}95\ 0{\cdot}2)10^{-7}$ gauss cm^2, $w = 225$ nm, $l = 8$ nm, $T = 6{\cdot}3°$K, $L/\lambda_J = 8{\cdot}5$ After Clarke [362].

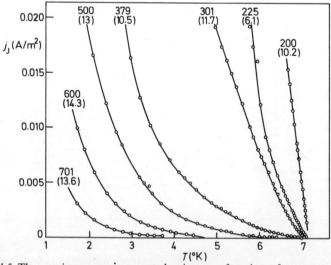

Fig. 21.5 The maximum supercurrent density as a function of temperature for variou barrier thicknesses in Pb–Cu–Pb junctions. The numbers in brackets refer to the mean fre path of the Cu. All dimensions in nm. After Clarke [362].

[543, 544]. They found that the junction switched to the resistive state at a certain value of the current but switched back to the zero voltage state when the current was further increased. They explained the effect with the aid of the self magnetic field (higher current causes higher magnetic field for which the maximum permissible supercurrent is higher).

SNS junctions were also used in double junction interferometers [456] showing the same interference effects as other types of junctions.

Finally, we mention here theoretical calculations by Yeh and Mechetti [545,

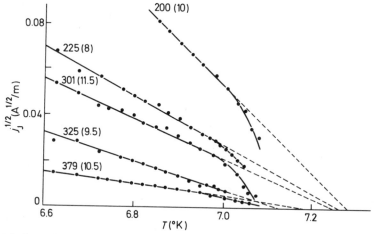

Fig. 21.6 $j_J^{\frac{1}{2}}$ as a function of T in the vicinity of the critical temperature. Numbers refer to barrier thickness and those in brackets to its mean free path. All dimensions in nm. After Clarke [362].

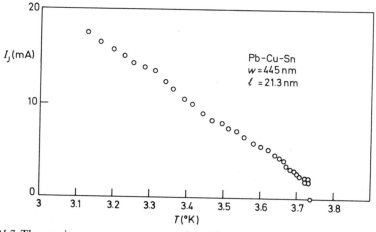

Fig. 21.7 The maximum supercurrent as a function of temperature in the vicinity of the critical temperature of Sn. After Kobayashi *et al.* [541].

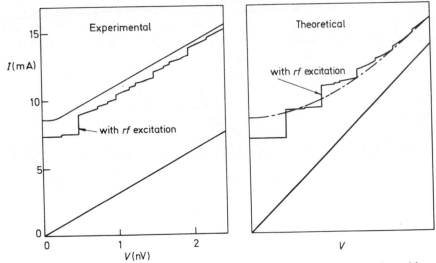

Fig. 21.8 Experimental and theoretical *I–V* characteristics for a wide junction with and without rf excitation. The experimental results refer to a square junction of side $L = 0.2$ mm for which λ_J is 0.031 mm. The exciting frequency is about 250 kHz. The theoretical results refer to a one-dimensional junction of the same width. Its area is chosen so as to make the critical currents equal, which in this case gives the theoretical junction a substantially smaller resistance than the real junction. After Waldram *et al.* [397].

545a] and Aslamazov *et al.* [545b] on the magnitude of the supercurrent, by Galaiko [545c] and Gogadze and Kulik [545d] on resonance effects due to the presence of a magnetic field, by Ishii [545e and f] on the current–phase relationship, and by Svidzinskii *et al.* [545g] on the maximum supercurrent in wide junctions.

The effect of an a.c. signal upon the *I–V* characteristic was studied by Clarke *et al.* [396] and Waldram *et al.* [397] both for small and large junctions. Their results for small junctions were discussed in Section 11.6. For large junctions in the viscous limit (second temporal derivative is neglected in Equation (12.41)) they obtained numerical solutions which are compared with experimental results for square junctions in Fig. 21.8. The agreement seems to be very good both for the positions and magnitudes of the subharmonic and harmonic steps.

21.3 Mechanical analogues

The analogy between a small Josephson junction and a simple pendulum was first noted by Anderson [329] and has been several times evoked since for drawing qualitative conclusions [365, 439, 498, 546].

The simplest form of the analogy may be obtained by considering a small tunnel junction which has a capacitance C. Its energy (disregarding the spatial variation and the constant in Equation (12.36)) may be written in the form

$$E = -\frac{\hbar}{q}j_{J}\cos\phi + \tfrac{1}{2}CV^2$$

$$= -\frac{\hbar}{q}j_{J}\cos\phi + \frac{1}{2}\left(\frac{\hbar}{q}\right)^2 C\left(\frac{d\phi}{dt}\right)^2. \qquad (21.3)$$

The energy of a pendulum of length l and mass m has the analogous form

$$E = -mgl\cos\phi + \tfrac{1}{2}ml^2\left(\frac{d\phi}{dt}\right)^2 \qquad (21.4)$$

where ϕ is now the angular displacement and the first and second terms correspond to potential and kinetic energies respectively. Writing further the formula for the torque

$$T = mgl\sin\phi + ml^2\frac{d\phi}{dt} \qquad (21.5)$$

it may be immediately seen that the analogues are

phase difference $\rightarrow$ angular displacement
voltage $\rightarrow$ angular velocity
current $\rightarrow$ torque
supercurrent $\rightarrow$ vertical displacement

This simplest form can be easily generalised for more complicated junction circuits. If, for instance, the junction is part of a ring of inductance L we need to add a term $\tfrac{1}{2}LI^2$ to the energy in Equation (21.5). Remembering Equations (13.7) and (14.17) we may write this term in the form

$$\tfrac{1}{2}LI^2 = \tfrac{1}{2}K(\phi - \phi_x)^2 \qquad (21.6)$$

where K and ϕ_x are constants. The right-hand side of Equation (21.6) may be regarded in the mechanical analogy as representing a horizontal torsion bar of torque constant K, the other end of which is attached to a rigid support whose angular displacement ϕ_x can be varied.

Similarly, a constant conductance in the junction is equivalent to mechanical damping, a battery of zero internal resistance to a constantly applied angular velocity and the resistance (in a partly resistive ring) to a viscous clutch in the torsion bar which permits an additional nonconservative phase slippage between ϕ and ϕ_x.

A mechanical analogue based on the above principles was built by Sullivan and Zimmerman (reported by Zimmerman [365]). The equivalence of the various components is shown in Fig. 21.9.

For large Josephson junctions the spatial variation of ϕ must also be taken into account. The mechanical analogue consists then of a series of coupled pendula as shown by Scott [547] who demonstrated nonlinear wave propagation on his model. Coupled pendula placed in a viscous medium provide an analogue for flux flow in a large SNS junction. Both the static and dynamic behaviour were studied on such a model by Waldram et al. [397].

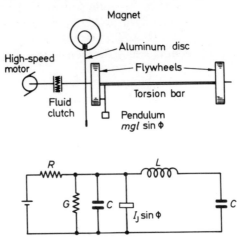

Fig. 21.9 Schematic representation of mechanical analogue of Josephson junction coupled to a resonant circuit and driven by a constant current source. After Zimmerman [365].

21.4 Superfluid analogues

Analogies between the supercurrent in a superconductor and the flow of superfluid helium were discussed by Anderson [548] with particular reference to the validity of Josephson equations. He suggested that the analogue of a junction (or rather weak link) is a small orifice connecting two superfluid helium reservoirs and the analogue of the electric potential is the gravitational potential. Thus superfluid helium may be expected to display both d.c. interference phenomena and the a.c. Josephson effect.

The easiest experiment to carry out seems to be to detect the presence of a.c. induced (by an ultrasonic wave) difference in superfluid levels.* The equation to be satisfied is

$$mg(z_2 - z_1) = \frac{n_1}{n_2}\hbar\omega \qquad (21.7)$$

where m is the mass of a helium atom, g is gravitational acceleration, $z_2 - z_1$ is

---

* This is analogous to the experiments of Langenberg et al. [498] on tunnel junctions mentioned in Section 19.1.

the level difference of the two reservoirs, n_1, n_2 are integers, ω is the angular frequency of the impressed ultrasonic wave. The first experiment was performed by Richards and Anderson [549]. We shall discuss here later experimental results by Khorana and Chandrashekar [550] who realised the two reservoirs in the form of a beaker inside a completely enclosed chamber. The bottom of the beaker was a nickel foil of thickness 40 μm with a hole of 8 μm diameter in it. This hole provided the only link between the beaker and the chamber. The ultrasonic wave was produced by a quartz transducer; its frequency being 99·722 kHz. For $n_1/n_2 = 1$, Equation (21.7) yields $z_2 - z_1 = 1\cdot01$ mm. The level of superfluid helium in the beaker as a function of time is shown in Fig. 21.10.

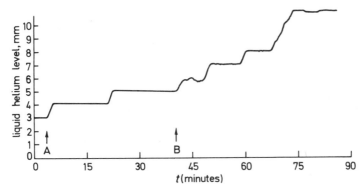

Fig. 21.10 Recorder trace showing the liquid helium level as a function of time, due to the pumping action of the sound generator, at 1·34°K. The sound generator was switched on at point A, and the intensity was momentarily increased at point B in order to induce a level change. After Khorana and Chandrashekar [550].

The ultrasonic generator was switched on at point A. The level remained at the same value for a fairly long time (in this particular case for about 15 minutes; in general it could be anything between 1 minute and 45 minutes) and then switched spontaneously to another value differing from the previous one by about 1 mm as expected. Transitions between the stable levels could be induced by increasing the intensity of the ultrasonic waves. An example is shown at point B where the change in level is caused by a momentary increase in intensity.

In the absence of the ultrasonic wave the level of superfluid helium decreased by about 0·8 mm per minute; thus the stabilisation of the level for periods of the order of ten minutes is a clear indication of the occurrence of a Josephson type effect. For further measurements see Khorana [551].

Note that m, the mass of the helium atom in Equation (21.7) may be expressed as m^*/N where N is Avogadro's number and m^* is an accurately known quantity from mass spectroscopy measurements. Hence measuring $z_2 - z_1$ and ω one can determine the product of fundamental constants hN [552, 553].

The analogue of the d.c. Josephson effect was measured by Hulin *et al.* [554].

The experimental apparatus consisted again of two superfluid helium baths communicating through a small orifice. They disturbed the level of one of the baths by inserting an aluminium cylinder at varying speeds. It was found that up to a certain critical speed (that is up to a certain critical superfluid velocity) there was no difference between the levels. Above the critical speed the superfluid current through the orifice stays at its critical value, vortices are created and a level difference between the baths appears.

Finally, we shall mention some calculations on Josephson junction analogues by Mamaladze and Cheishvili [555–558] which are based on the phenomenological theory of Ginzburg and Pitaevskii [559].

21.5 Microwave enhancement of superconductivity in weak links

We have seen in Section 10.2 that the zero voltage current ($n = 0$ in Fig. 10.7) of a point contact junction decreases as a function of microwave input power, and similar results were found for infrared input in Section 17.5. Interestingly, the zero voltage current of thin film bridges has been found to *increase* when irradiated by microwaves. This was first found by Wyatt *et al.* [452, 560] in experiments conducted at 9·6 GHz. The relative increase in I_J is more pronounced near the critical temperature as shown in Fig. 21.11. The phenomenon was investigated in more detail by Dayem and Wiegand [383]. They conclude that the enhancement appears at a frequency of about 2 GHz and becomes stronger as the frequency increases (data given up to 11 GHz). The same effect was also found by Seraphim [425], Wyatt and Evans [560a], Gregers–Hansen and Levinsen [561] and Dmitriev *et al.* [562] for thin film bridges and reported for some tunnel junctions both in the presence [563] and absence [564] of magnetic fields.

Two explanations have been suggested so far but neither of them is suitable for detailed comparison with the experimental results. Hunt and Mercereau [565] use a thermodynamical argument, based on measured results, that the critical temperature of the bridge section is below that of wider portions of the same film. They argue that when the bridge becomes superconducting one 'degree of freedom' (the freedom to assign arbitrary values to the phases on both sides) is lost and an energy of the order of kT is released. Hence the bridge remains normal until this gain in thermal energy is compensated by the loss of Gibbs free energy. In the presence of microwaves the phase is determined anyway so no freedom is lost when the bridge becomes superconducting. According to this explanation all that happens is that in the presence of the microwaves the critical temperature of the bridge section becomes identical with that of the rest of the film which appears experimentally as an increase in the critical temperature of the whole sample.

The other explanation is due to Eliashberg [566]. He gives a proof starting with the BCS theory that when $\omega\tau \gg 1$ (τ is the relaxation time of the order parameter) the presence of an electromagnetic field does lead to an increase in T_c. This is a much more general explanation suggesting that the increase in critical temperature is by no means limited to configurations containing a weak link.

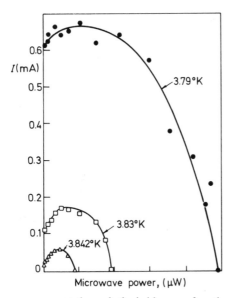

Fig. 21.11 Maximum supercurrent through the bridge as a function of applied microwave power. The critical temperature of the film is 3·845°K. After Wyatt *et al* . [452].

21.6 Applications

The term 'applications' is used in this section in two meanings, firstly what we may call *diagnostic* applications (we use the phenomenon of Josephson tunnelling to learn something about something else) and secondly when we use the specific properties of Josephson junction *devices* to measure the properties of something else.

A list of those in the first category is as follows: Experiments by Baixeras and Pech [567, 568] on point contact junctions for observing the motion of vortices (bundles rather than individual ones). The penetration depth in super-conducting tin was derived by Matisoo [569] from measurements of self-excited resonances in Josephson junctions. Another measurement aimed again at determining the penetration depth was performed by Meservey [570] with the aid of a superconducting Al ring containing two weak links. Jaklevic *et al.* [347] used a double junction interferometer to show that there is modulation of the maximum supercurrent in a magnetic field free region due to the finite vector potential. Weinberg [571] used Josephson tunnelling as an indicator of film quality when investigating image force effects in normal tunnelling. Greenspoon and Smith [572] deduced conclusions about the proximity effect by measuring the supercurrent across Pb–I–Cu–Pb sandwiches.

Those in the second category do not strictly belong to the subject of this book. Once a magnetometer has been built it may be used for thousands of different purposes which bear no relevance to the original construction of the magnetometer. However, Josephson devices are fairly new and they usually have rather specific applications so it seems justified to mention a few of them (we mentioned one in Section 19.2 where induced voltage steps on different SNS junctions were compared and the current, or rather the absence of current, was measured by a SLUG).

An early application was that of Zimmerman and Mercereau [573] who measured quantised flux pinning by drawing slowly a superconducting wire through a double junction interferometer. Resistivity was determined by Hanabusa and Silver [574] by measuring the decaying magnetic field caused by eddy currents. Low field magnetic susceptibilities were measured by Hanabusa et al. [575] and Gollub et al. [576]. Detection of nuclear magnetic resonance (by measuring the change in magnetic field caused by the change in static nuclear susceptibility) was suggested by Kamper et al. [449].

The sensitivity of the developed magnetometers is so high that it proved possible to measure Johnson noise [577, 578], the fluctuations of the Earth's magnetic field [493], and the magnetic field produced by the human heart [579, 580].

A further interesting application is that of Hebard and Fairbank [581] whose experiments are designed to detect quarks. They use a Josephson magnetometer to measure small displacements (less than 1 μm) of a Nb sphere from which the charge on the sphere (hopefully one or two thirds of that of an electron) is to be determined.

As detectors of electromagnetic waves we shall mention in conclusion the 1 mm measurements of Ulrich [582] in radioastronomy.

21.7 Parametric amplification

We have shown in Section 11.7 that a Josephson junction possesses a nonlinear inductance, and some measured results were discussed in Section 14.6. In the present section we shall outline the possible use of a Josephson junction as a parametric amplifier.

The first attraction of Josephson junctions for this purpose comes from the fact that the junction not only serves as a nonlinear inductance but can provide the pump power as well if biased to a certain d.c. voltage. In Electronic Engineering terms we may call a Josephson junction a d.c. pumped parametric amplifier.

The basic circuit arrangement treated by Russer [583] is shown in Fig. 21.12 for the case when the signal circuit (at ω_1) and the idler circuit (at ω_2) may be

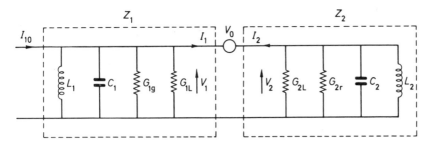

Fig. 21.12 Circuit for parametric amplification.

regarded independent of each other $(Z_1(\omega_2) = Z_2(\omega_1) = 0)$. Note that G_{1g} is the generator conductance, G_{1L} and G_{2L} are the primary and secondary load conductances and G_{2r} represents the losses of the secondary circuit. I_{10} is the signal current source at ω_1 and the d.c. voltage bias V_0 is chosen in such a way that

$$\frac{qV_0}{\hbar} = \omega_0 = \omega_1 + \omega_2 \tag{21.8}$$

The power gain g_{p1} and the conversion power gain g_{p2} are defined as

$$g_{p1} = \frac{P_{L2}}{P_{a1}}, \qquad g_{p2} = \frac{P_{L2}}{P_{a1}} \tag{21.9}$$

where

$$P_{a1} = \frac{I_{10}^2}{8G_{1g}} \tag{21.10}$$

is the available input power and

$$P_{L1} = \tfrac{1}{2}V_1^2 G_{1L}, \qquad P_{L2} = \tfrac{1}{2}V_2^2 G_{2L} \tag{21.11}$$

are the output powers in circuits 1 and 2 respectively.

Formulae for the gains were derived by Russer by investigating the response of a Josephson junction to a voltage input

$$V = V_0 + V_1 \cos(\omega_1 t + \alpha_1) + V_2 \cos(\omega_2 t + \alpha_2) \tag{12.12}$$

same, in fact, as assumed in Equation (17.5).

The gain–bandwidth product is also calculated giving for narrow band amplifiers

$$\Delta\omega(g_{p1\,max})^{1/2} = \left(\frac{Q_1}{\omega_1} + \frac{Q_2}{\omega_2}\right)^{-1} \tag{21.13}$$

$$\Delta\omega(g_{p2\,max})^{1/2} = 2\left(\frac{\omega_2}{\omega_1}\right)^{1/2}\left(\frac{Q_1}{\omega_1} + \frac{Q_2}{\omega_2}\right)^{-1} \tag{21.14}$$

which are in agreement with formulae derived for other parametric amplifiers [584].

General energy relations for a Josephson junction were also derived by Russer [585] in a form similar to the Manley–Rowe equations [584] but with an additional term for d.c. power.

We mention here some further work by Zimmer [586], Vystavkin et al. [586a] and Kanter and Silver [794], the last authors obtaining a gain of 11 db.

21.8 Relaxation oscillations

We shall describe here an interesting effect found by Vernon and Pedersen [587] in which the joint action of the junction and of the external circuit result in continuous oscillation. The circuit is shown in Fig. 21.13(a) where R and L are lumped elements of the external circuit and the real junction is represented by the parallel combination of an ideal junction with a capacitor.

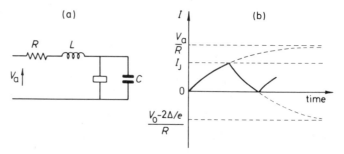

Fig. 21.13 Relaxation oscillations; (a) circuit, (b) waveform.

The oscillatory effect occurs when V_a, the applied d.c. voltage, satisfies the inequality

$$RI_J < V_a < 2\Delta/e \qquad (21.15)$$

An oscillation period can be divided into the following stages:

(i) V_a is applied at $t = 0$; the current increases up to I_J with a time constant L/R as shown in Fig. 21.13(b).

(ii) When the current reaches I_J the junction turns normal and the capacitor (assumed to be small) is charged up to $2\Delta/e$ in a time much faster than the time constant.

(iii) The junction acts as a battery of voltage $2\Delta/e$ (as discussed in Section 16.5). The current decays along the normal tunnelling characteristic from I_J towards $(V_0 - 2\Delta/e)/R$ as shown again in Fig. 21.13(b).

(iv) The voltage is reduced to zero as the capacitor discharges through the resistor and the inductance (time again negligible) and the whole process starts again.

22.   Statistical notes

Josephson tunnelling is one of those few disciplines which has both a clear beginning (Josephson's paper in 1962) and a fairly well defined boundary. In the large majority of cases it is relatively easy to say whether the main theme of a given paper is Josephson tunnelling or not. Hence it seems an ideal field to see how a new branch of Science starts up and expands, how the papers are distributed between different journals, and in what languages the publications are written, etc. The statistics will be done for the years* 1962 to 1970 for publications appearing in the open literature. Conference publications are included (except those for which the Abstract is only given) but not patents, unpublished reports and theses. The total number coming into this category was 422, out of which 36 were rejected as not belonging to Josephson tunnelling 'proper', like applications (in the sense used in Section 21.6) and some publications only vaguely connected with the main theme.

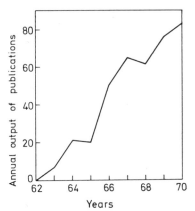

Fig. 22.1 The annual output of publications for the years between 1962 and 1970.

First of all let us see how the annual number of publications grew starting with one in 1962. This is shown in Fig. 22.1 which looks more like a linear curve than an exponential one usually found for growth phenomena. On this

---

*Conference publications are classified by the date of the conference and not by the publication date of the conference proceedings. Papers written in the Soviet Union are dated by the publication of the English translation.

basis we could, for example, predict that the number of publications in 1971 will be about 90.

The distribution between papers, letters,* conference publications** and books*** is shown in Fig. 22.2. Taking together the papers and letters it is interesting to see (Fig. 22.3) that out of 295 publications 163 were published by no more than five journals. The remaining 132 is shared by 48 other journals.

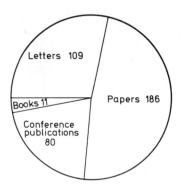

Fig. 22.2 The distribution of publications between letters, papers, conference proceedings and books.

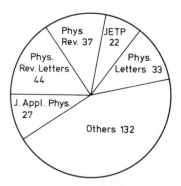

Fig. 22.3 The distribution of letters and papers among the various journals.

*Only those are counted under this heading which appeared in journals specialising in quick publishing; the correspondence sections of 'ordinary' journals are not included.
**Only those are counted which have come out in special conference editions. The conferences reported in, for example, Reviews of Modern Physics or Journal of Applied Physics are taken as papers. The only exceptions are the lectures delivered at the 'Science of Superconductivity' Conference, Stanford, 1969, which are regarded as conference publications (dated 1969) although they have been published in Physica.
***Contributions published in books are regarded here as separate. Thus for example Tunneling Phenomena in Solids counts as 4 because it contains four relevant chapters written by different authors.

Even higher concentration is found if we do the statistics according to the country in which the contributors worked at the time the publication was submitted. As may be seen in Fig. 22.4 five countries are responsible for 348 (90%) of the publications. The rest came from the following countries: France 8, Germany 8, Canada 6, Italy 5, India 2, Hungary 2, Denmark 2, Austria 2, Australia 1, Israel 1, Finland 1. Finally, if the classification is according to the languages in which the publications are written, we may see (Fig. 22.5) that two languages, English and Russian, account for 373, that is 96·8%, of the publications. The rest is divided between German 6, French 3, Ukrainian 3, Japanese 1, Italian 1.

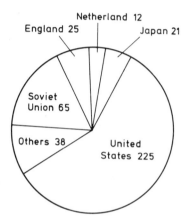

Fig. 22.4 The distribution of publications according to the country in which the contributor worked at the time of submission.

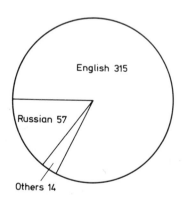

Fig. 22.5 The distribution of publications according to languages.

IV.   Additional material

Note:

The aim of Part IV is twofold, partly to discuss subjects which did not fit into any of the preceding chapters and partly to present new material published up to October 1971. The additional chapter on normal tunnelling (Chapter 23) differs in design from that on Josephson tunnelling (Chapter 24). It was considered that the basic phenomena of tunnelling and the potentialities of diagnostic applications had already been described in sufficient detail and that therefore no further details would be of general interest. For this reason, instead of referring to all publications, a few tables were compiled which summarise the experimental work done and enable the reader to find easily those publications of most interest. The structure of Chapter 24 follows the order of Part III and an attempt is made to emphasise the latest advances in the art section by section. Regrettably there was not room for a detailed description of the wealth of new material on Fluctuations so only the prominent experimental results are described.

23.   To normal tunnelling

23.1 A review of the reviews

There are only three major reviews concerned with normal electron tunnelling.

Duke [293] regard superconductive tunnelling as part of the general theory of tunnelling in solids and makes frequent comparisons between superconductive and other types of tunnelling. There is a fair number of experimental results included in Duke's book but naturally there is more room devoted to topics (e.g. zero bias anomalies) which have more general implications and are not restricted to the superconducting state.

'Tunneling Phenomena in Solids', edited by Burstein and Lundquist, contains a large number of reviews by various authors covering again all aspects of tunnelling in solids. The contributions (updated versions of lectures delivered at the Advanced Study Institute, Risö, Denmark, 19–30 June 1967) concerned with normal electron tunnelling are as follows: 'Tunneling between Super-conductors' by Giaever [588], 'Tunneling Density of States—Experiment' by Rowell [589], 'Single-Particle Tunneling in Superconductors' by Schrieffer [105], 'Geometrical Resonances in the Tunneling Characteristics of Thick Superconducting Films' by Tomasch [150], 'Multiparticle Tunneling' by Wilkins [125], 'Photon-Assisted Single-Particle Tunneling between Super-conductors' by Everett [590], 'Phonon Generation and Detection by Single-Particle Tunneling in Superconductors' by Eisenmenger [591], 'Gapless Superconducting Tunneling—Theory' by Fulde [592], and 'Gapless Super-conductor Tunneling—Experiment' by Claeson [593]. The standard is much higher than in the present book; in the majority of contributions familiarity with microscopic theory is assumed.

'Superconductivity', edited by Parks, is also a collection of contributions by various authors. Those of interest for our purpose are 'Equilibrium Properties: Comparison of Experimental Results with Predictions of the BCS Theory' by Meservey and Schwartz [594] and 'Tunneling and Strong-Coupling Super-conductivity' by McMillan and Rowell [165]. The treatment is again at a high standard.

We shall further mention here the reviews of Fiske and Giaever [595] (aimed at Electronic Engineers), of Douglass and Falicov [33] (on the super-conducting energy gap), of Rowell [287] (an approach emphasising the physical fundamentals) and of Guyon [277] (on gapless superconductors).

23.2 On basic characteristics, special tunnelling effects and devices

Semiconductors were mentioned in the main text as barriers only, although they merit consideration in two further aspects, first as superconducting semiconductors and second as one of the electrodes in superconductor–semiconductor junctions.

The first observation of a superconducting energy gap in a semiconductor was made by Stiles *et al.* [596] with the aid of an Al–I–GeTe junction by measuring the differential resistance. A theory, taking account of the low carrier density, was developed by Koonce [597].

Superconductor–semiconductor junctions were first investigated by Rowell and Chynoweth [598] using mercury frozen onto degenerated silicon, and a number of other publications [120, 190, 599–601] were also concerned with the problem. Molnar *et al.* [450] developed a point contact technique where a degenerate GaAs point is pressed against a clean Pb surface. The energy gap and density of states of Pb may be determined by this method. The *I–V* characteristic is asymmetric which is taken as a proof that the barrier is provided by the existing Schottky barrier and not by an oxide layer. A more detailed report on point-contact tunnelling, including measurements on tin, vanadium and intermetallic lanthanum compounds, was published by Thompson and Molnar [602]. The effect of magnetic field was investigated by Tsuda [603, 604], the phonon induced structure for GaAs–Pb [605] and for Si–Pb [606] junctions by Tsui and the effect of pressure by Guetin and Schreder [606a]. A theoretical paper discussing phonon assisted tunnelling was published by Duke and Kleiman [607].

Photon assisted tunnelling in the S–I–S configuration was discussed in Section 5.2 in fair detail commenting on the controversy between the Tien–Gordon [58] and Cook–Everett [129] theories. An interesting further development is a proof by Heidam [608] showing that the expression of Cook and Everett [129] (discussed also by Everett [590]) is almost identical with that of Hamilton and Shapiro [135]. We shall further mention here another experiment by Hamilton and Shapiro [609] where the junction is irradiated by microwaves of two different frequencies (69 GHz and 80 GHz) simultaneously and current steps are detected corresponding to sum and difference frequencies.

Phonon assisted tunnelling was dealt with rather curtly in Section 5.3 referring only to a few selected publications. We shall complete the list below. There are a number of further experimental studies [610–614] on phonon interaction, a theoretical one considering the effect of the complex gap parameter [615] and publications concerned with the excitation of electromagnetic waves by phonons making a small angle with the plane of the junction [616–618]. Calculations for the tunnelling current based on microscopic theory are also available [619, 619a].

Subharmonic structure was observed by a number of authors [620–625] besides those already mentioned. Zawadowsky [626] made an attempt to explain the experimental results of Rochlin and Douglass [622] by assuming a process where p Cooper pairs break up and n electrons tunnel across the junction with a resulting structure at $2p\Delta/ne$ in the I–V characteristic.

On the *Tomasch effect* we shall mention here further experimental work with superconductive and ferromagnetic overlays [627] and theoretical work by Wolfram and Lehman [628, 629], Wolfram and Einhorn [630], Wolfram [631] and McMillan [632]. More detailed experimental results are reported in a recent paper by Lykken *et al.* [633] who used single crystals of lead films. Some of their results imply the presence of carriers (coming from different Brillouin zones) with two different Fermi velocities in agreement with Bennett's [634] calculations. Fermi velocities in *rf* sputtered superconducting Nb films were measured by Smith and Miller [635].

Another rather special tunnelling effect (where the special part occurs at room temperature) was discussed by Pedersen and Vernon [636]. They have shown that room temperature measurements for low resistance junctions may be misleading because of the finite resistance of the metal films. The apparent resistance decreases as the temperature goes up and may even turn negative.

Among *devices* we shall mention further work on neuristor propagation [636a] and the application of the nonlinear tunnelling characteristics for detection and mixing of electromagnetic waves [637, 638]. Signals in the ranges 2 MHz, 9·36 GHz and 24 GHz were detected, and a 30 MHz *if* output was obtained from two microwave inputs.

Finally we shall mention two recent papers concerned with the generation and detection of phonons. The spectrum of phonons emitted from a Sn–I–Sn junction into a solid (Ge doped by Sb which served as the detector as well) was determined by Dynes *et al.* [639] and found to be approximately monochromatic ($\hbar\omega \cong 2\Delta$). The power output was of the order of milliwatts. Phonon detection in the THz region (the experimental arrangement differs from that of Ref. [310] to the extent that the junction is not in thermal contact with the He II bath) was reported by Schulz and Weis [640].

23.3 Summary of the experiments

The aim of this section is to show the reader in tabular form where the various experimental results can be found. Table 23.1 gives the references for two-electrode junctions containing an insulating barrier (excluding those designed for device applications).

Columns 1 and 2 give the material of the respective electrodes (the order of the electrodes has no significance; it is *not* implied that the insulating layer is the oxide of the first electrode). In Column 3 are listed the papers in which the energy gap of at least one of the electrodes is determined. Columns 4, 5 and 6 are concerned with *I–V* characteristics and with their first and second derivatives. If at least one of the electrodes was prepared by low temperature condensation that is shown in Column 7, and the fact that a magnetic field was used is noted in Column 8. The references in Columns 9–12 give information about multiparticle and subharmonic tunnelling, the effect of crystal orientation, pressure and paramagnetic impurities respectively. The phonon spectrum (by the gap inversion technique) is given in the papers of Column 13 whereas Columns 14 and 15 give references to experiments on phonon and photon assisted tunnelling. Finally, Column 16 lists the papers in which point-contact junctions were used.

A list of two-electrode structures containing a semiconductor barrier is given in Table 23.2 where only the electrodes, the barrier material and the references are given.

Experiments on three-electrode sandwiches (insulating layer is between electrodes 1 and 2), nearly all concerned with the proximity effect, are summarised in Table 23.3.

23.4 Further notes

A new diagnostic application of tunnelling is concerned with the spin of the electron. The splitting of the normal electron states was observed by Meservey *et al.* [679] by measuring the normalised tunnelling conductance, $\sigma(V)$, of Al–I–Ag tunnel junctions in the presence of high (parallel) magnetic fields. The experimental results were explained by simply splitting the BCS density of states into spin-up and spin-down parts displaced in energy by $\pm\mu_m H$ where μ_m is the magnetic moment of the electron. In fact, some mixing of the spin states (due to spin–orbit interaction) does take place and was detected in the form of extra tunnelling conductance at $V = (\Delta_1 + \Delta_2 - 2\mu_m H)/e$ by Tedrow and Meservey [690]. The same authors also observed [691] spin dependent tunnelling on Al–I–Ni junctions.

As further diagnostic applications we mention here the spectroscopic studies (observing the optical phonon band of PbO and SnO_2) of Yanson [722], the measurements of Ziemba and Bergman [710] on the mean free path dependence of the energy gaps of strong coupling (Pb and Hg) superconductors and the tunnelling determination of the critical temperature by Feldman and Rowell [723]. The effect of the measuring current on the zero bias conductivity of Al–I–Sn junctions was studied by Paterno *et al.* [724], the dependence of normal electron recombination on film thickness by Levine and Hsieh [725] and the effect of vortices on the tunnelling characteristic by Gogadze [725a].

Table 23.1

Electrode 1	Electrode 2	Energy gap	I-V characteristics	First derivative measurements	Second derivative measurements	Amorphous	Magnetic field	Multiparticle and subharmonic tun.	Gap anisotropy	Pressure	Paramagnetic impurities	Phonon spectrum	Phonon assisted tunnelling	Photon assisted tunnelling	Point-contact
1	2	3	4	5	6	7	8	9	10	11	12	13	14	15	16
Al	Al	43, 45, 69, 99, 209, 702, 709	43, 45, 61, 66, 69, 99, 209	66, 110, 690, 703, 709	92, 239		690		209, 702			238			
	Pb	43, 45, 46, 69, 70, 77, 164, 171, 191, 192, 209, 237, 249, 623, 654, 660, 672, 677, 683, 716	43, 44, 45, 46, 60, 66, 69, 80, 102, 109, 164, 171, 192, 209, 249, 610, 666, 670, 672, 673, 677, 682, 694, 716	43, 44, 65, 69, 80, 113, 139, 209, 219, 221, 234, 246, 247, 251, 261, 611, 612, 623, 642, 658, 660, 663, 664, 665, 670, 671, 673, 676, 682, 683, 688, 692, 709	65, 156, 209, 219, 221, 234, 246, 247, 612, 656, 660, 673, 676	234, 246, 673, 677, 716	66, 70, 191, 192, 654, 663, 664, 665, 682, 683, 688	623	209, 623	77, 246, 247, 249, 251, 252, 660		234, 246, 252, 656, 673, 677, 716	59, 60, 139, 610, 611, 612, 618		
	Sn	69, 193, 209, 237, 709	69, 209, 666	86, 139, 146, 147, 193, 209, 237, 642, 709	146, 147, 241, 246	241, 246, 673	86, 193		209			246, 673	139		
	In	45, 69, 209, 650, 702	45, 57, 69, 129, 209	57, 110, 129, 145, 151, 209, 642, 643, 650, 709	113, 145, 151, 246, 650	246			209, 650, 702					57, 129	

Table 23.1 (cont.)

1 Electrode 1	2 Electrode 2	3 Energy gap	4 I–V characteristics	5 First derivative measurements	6 Second derivative measurements	7 Amorphous	8 Magnetic field	9 Multiparticle and subharmonic tun.	10 Gap anisotropy	11 Pressure	12 Paramagnetic impurities	13 Phonon spectrum	14 Phonon assisted tunnelling	15 Photon assisted tunnelling	16 Point-contact
Al	Ga	236, 672, 677	672, 677	168, 236, 677		168, 648, 677, 717		672				648, 677, 717			
	Bi	207, 677, 716	72, 207, 677, 716	207, 240, 673	240	207, 240, 648, 673, 677, 716, 717						240, 648, 673, 677, 716, 717			
	Tl	649, 680	680	113, 649, 660	91, 649, 660					660					
	Hg	225	225, 226	225, 226	225, 226							226			
	Cd	675	675	675											
	Nb			635	635										
	Ag			86, 679			86, 679								
	La	172, 173, 175, 657	172	173, 175, 176											

Electrode 1	Electrode 2	Energy gap	I–V characteristics	First derivative measurements	Second derivative measurements	Amorphous	Magnetic field	Multiparticle and subharmonic tun.	Gap anisotropy	Pressure	Paramagnetic impurities	Phonon spectrum	Phonon assisted tunnelling	Photon assisted tunnelling	Point-contact
	Ni			691			691								
	Zn	108					108								
	PbBi	230, 233, 677, 716	233, 661, 677, 716	200, 233, 692	230, 726	677, 716, 717	200, 646, 692					677, 716, 717			
	PbTl	82, 83, 230		656	229, 230, 656	673	82, 83					656			
	PbCu											673			
	PbGd	186, 187				186, 187					186, 187				
Al	PbMn	186				186					186				
	PbIn	230		227, 228	230, 228		84, 646					227			
	SnIn			160, 200, 201, 692, 701			160, 200, 201, 646, 692, 701								
	SnCu					673						673			

Table 23.1 (cont.)

1 Electrode 1	2 Electrode 2	3 Energy gap	4 I–V characteristics	5 First derivative measurements	6 Second derivative measurements	7 Amorphous	8 Magnetic field	9 Multiparticle and subharmonic tun.	10 Gap anisotropy	11 Pressure	12 Paramagnetic impurities	13 Phonon spectrum	14 Phonon assisted tunnelling	15 Photon assisted tunnelling	16 Point-contact
	InTl			85, 667								243			
	InBi	78, 186				78, 186					78, 186				
	InFe						66,								
	BiTl			656	656							656			
	BiCu					673						673			
	BiSb			673		673						673			
	LaCe	189		188, 189											
	LaLu	175, 657		175											
	Nb₃Sn	179				179									
Pb	Pb	153, 163, 622, 623, 624, 633, 644, 714, 718	61, 62, 107, 141, 142, 163, 192, 620, 621, 622, 623, 644, 714, 718	66, 67, 139, 142, 153, 611, 620, 623, 633, 644, 705, 718	67, 90, 113, 153, 221, 633, 644, 718, 722		90, 142, 192, 621, 622, 623, 624	61, 62, 107, 141, 142, 620, 621, 622, 623, 624	153, 623, 633, 644, 718			67	139, 141, 611		179

1 Electrode 1	2 Electrode 2	3 Energy gap	4 I-V characteristics	5 First derivative measurements	6 Second derivative measurements	7 Amorphous	8 Magnetic field	9 Multiparticle and subharmonic tun.	10 Gap anisotropy	11 Pressure	12 Paramagnetic impurities	13 Phonon spectrum	14 Phonon assisted tunnelling	15 Photon assisted tunnelling	16 Point-contact
Pb	Sn	102, 685, 686, 696, 700, 710	62, 102, 129, 131, 134, 294, 686, 698, 700	131, 614	722	710	685, 686	62					139, 141, 611, 614		
	In	700	700												
	Ga	706, 707, 708, 236	707	236					706, 708						
	Tl	231		231											
	Ta	167, 651, 653, 689	167, 651, 653, 689		231				653						
	Nb	166, 693	166, 661, 681, 693	166											
	Mg			661											
	SnCu	687	687												
	Nb₃Sn	178, 180	178, 180	178, 180			180								
	NbZr	652	652	652											
	NbN	674	674												
	NbC	684													

Table 23.1 (cont.)

Electrode 1	Electrode 2	Energy gap	I-V characteristics	First derivative measurements	Second derivative measurements	Amorphous	Magnetic field	Multiparticle and subharmonic tun.	Gap anisotropy	Pressure	Paramagnetic impurities	Phonon spectrum	Phonon assisted tunnelling	Photon assisted tunnelling	Point-contact
1	2	3	4	5	6	7	8	9	10	11	12	13	14	15	16
Sn	Sn	73, 75, 76, 164, 700	64, 66, 73, 74, 75, 132, 135, 164, 636, 700	73, 74, 75, 76, 135	113, 609, 722		64	64	73, 75, 76					132, 134, 135, 609	
	In	102, 710	102	142, 202		710	202							130, 645	
Sn	Tl	710	61			710		61							
	Hg		710												
	Mg			66											
	Nb	167, 232, 689	130, 167, 232, 645, 689												
In	BiPb		232, 661												
	Nb₃Sn	177	177	178											
	Nb₃Sn	177	177, 180												
	Mg			66											
	Nb	212, 213, 215, 216, 668, 678	211–215, 678, 695	211, 213–215, 668, 678			695		212, 213, 215, 216, 668						

Electrode 1	Electrode 2	Energy gap	I–V characteristics	First derivative measurements	Second derivative measurements	Amorphous	Magnetic field	Multiparticle and subharmonic tun.	Gap anisotropy	Pressure	Paramagnetic impurities	Phonon spectrum	Phonon assisted tunnelling	Photon assisted tunnelling	Point-contact
Tl	In			649	722										
Ta	Mg				91, 649										
	Ag				704										
	Nb₃Sn		122												122
	NbC	684	625	625				625							684
Cu	Nb	177	177												625
Nb	Nb₃Sn		625	122, 625				625						137	137, 625
	Nb														
	La		174												174
	V₃Si	122	122												122
	V₃Ge	122	122												122
	NbC	684													684
NbTa	NbTa	625	625	625				625							625
	TiMo			625				625							625
PbBi	PbBi	210	210						210						

Table 23.2 *Table 23.3*

Electrode [1]	Barrier	Electrode [s]	References
Pb	C	P	121
	C	Sn	121
	C	Cu	121
	Ge	Sn	144
	Ge	Nb	715
	Si	Nb	715
	CdS	Pb	118, 120, 144
	CdS	Tl	144
	CdS	PbIn	144
	ZnS	Pb	119
	InSb	Nb	715
	GaAs	Nb	715
	Ge–C	Pb	121
Sn	Ge	Sn	119, 120
	CdS	Sn	118–120, 144
	ZnS	Sn	119, 120
	CdS	Al	119
	ZnS	Al	119
	Ge	Sn	713
Re	C	In	711, 712

Electrode [1]	Electrodes 2 and 3	References
Al	Al–Pb	260, 267
	Al–Sn	264, 265
	Pb–Ag	259, 267, 697
	Pb–Al	80, 259, 267
	Pb–Cu	111, 641, 655
	Pb–Fe	261, 697, 719
	Pb–Mn	187, 267
	Pb–Ni	260, 261, 670
	Pb–Pt	260, 261
	Pb–CuFe	655
	Sn–Al	699
	Sn–Pb	262
	Sn–Mn	187
	In–Al	627
	In–Fe	78
	In–Pb	627
	In–Fe_3O_4	669, 721
	Ag–Pb	155, 259, 262, 267
	Cu–Pb	111, 262, 641, 720
	Cd–Pb	260
	Pt–Pb	260
	Zn–InBi	266, 269, 647
	InBi–Zn	266, 269, 647
Pb	Sn–Pb	662
	Ag–Pb	268
	Ag–Sn	696, 697
	Fe–Sn	696, 697
Mg	Cu–Pb	263, 659
	Pb–Cu	263, 659
Sn	Ag–Pb	79

Fluctuation effects on NIS junctions (N just above its critical temperature) were investigated by Cohen *et al.* [726] and Cheishvili [727].

A numerical calculation by Pike [728] gives the relationship (for NIS junctions using the BCS density of states) between $2\Delta(T)/e$ and the junction voltage V_m at which the differential conductance is maximum. The tunnelling characteristics for Ta–I–Nb and V–I–Nb junctions for several values of the temperature were computed by Blume [729].

Kummel [730] proposed tunnelling experiments for observing the energy bands due to the periodic structure of the intermediate state. Kaiser and Zuckerman [731] generalised McMillan's tunnelling model of the proximity effect to include magnetic impurities. NIS tunnelling where the superconductor contains magnetic impurities was studied by Urushadze [732]. An explanation for Hoffstein and Cohen's failure [179] in getting the right value for the energy gap of Nb_3Sn was given by Weger [733].

Further theoretical studies were conducted by Vashista and Carbotte [734, 735], Maki [736], Maki and Griffin [737], Griffin and Demers [738], Radhakrishnan [739], Juranek et al. [740], Nakajima and Nakao [741] and Kulik [742].

24.   To Josephson tunnelling

24.1 A review of the reviews

The fascination of the Josephson effect phenomena has led to a planthora of reviews.

The reviews vary enormously from high brow to low brow, from general to specific, from abstract to practical. There are reviews of the microscopic theory [329, 349, 368, 369, 533, 743, 744] and of the phenomenological theory [366], and there are popular reviews [441, 745, 746, 768]. There are reviews for chemists [747] and for electronic engineers [748–751], for East Germans [751, 752] and West Germans [753], for the French [748], the Finns [754] and the Russians [755], and there is one for the Czechs [371] written in English by a Hungarian. There are reviews on double junction point-contact devices [414, 765], on magnetometers [364, 488, 490, 491], on infrared detectors [482, 483], on transmission lines [757], on fabrication [758] and on devices in general [100, 759–762], on the determination of e/h [763–767], on the a.c. Josephson effect [473, 769], on weak links [770], on quantum interference [434, 771, 772], on the infrared microwave gap [773] and on submillimeter generation [379]. There are historical reviews [774, 775] (with titles 'Brian Josephson and macroscopic quantum interference' and 'How Josephson discovered his effect') and optimistic reviews [750] (with titles like 'Josephson junctions—all-purpose components of the future').

There is a book by Kulik and Yanson [776] devoted to the high frequency properties of Josephson junctions, there is Duke's book [293] with emphasis on microscopic theory, and de Gennes' book [367] containing his usual mixture of high mathematics and physical intuition.

Josephson junctions are both difficult and simple depending on the type of approach and on the man looking at them. In 1965 when Feynman included them into his undergraduate lectures [21] the subject seemed pretty advanced but with the passage of time this is no longer so. Undergraduates have already made experiments [777] on Josephson junctions and newly published textbooks [778, 779] designed for final year undergraduates (and first year graduates) also discuss the subject.

24.2 The d.c. Josephson effect

Temperature dependence. The maximum supercurrent, I_J, as a function of temperature was measured by Fiske [323] and was shown in Fig. 9.3. The agreement with the theory of Ambegaokar and Baratoff [332] is quite good which is largely borne out by other experiments as well [261, 780–782]. It is interesting to note that even gapless superconductors (as shown by Hauser [261] on Cr–Pb–I–Pb–Cr junctions) display the same temperature dependence.

A theory for Pb–I–Pb junctions taking into account strong-coupling effects was developed (on the lines of the Fulton–McCumber [378] paper) by Lim et al. [782]. They find a somewhat different temperature dependence which shows better agreement with their experimental results.

We shall further mention here a theory by Gorbonosov and Kulik [783] who relate the $I_J(T)$ curve to the gap anisotropy of the electrodes.

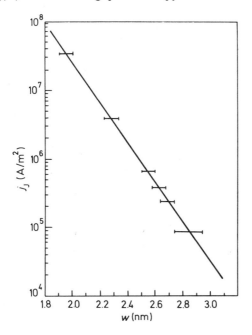

Fig. 24.1 The maximum supercurrent density as a function of barrier thickness. After Eldridge and Matisoo [786].

Barrier thickness dependence. The maximum supercurrent as a function of barrier thickness was investigated by Schwidtal and Finnegan [784, 785] and Eldridge and Matisoo [786] on Pb–I–Pb junctions. The former authors derived the thickness from R_{NN}, the normal tunnelling resistance whereas the latter authors measured the thickness of the oxide layer in situ. An experimental curve by Eldridge and Matisoo [786] is shown in Fig. 24.1. The data are well fitted by the expression

$$j_J = 1 \cdot 5 \; 10^{13} \exp(-w/0 \cdot 15); \qquad A/m^2 \qquad (24.1)$$

where w, the thickness of the barrier, is to be substituted in nanometers. For detailed comparison between theory and experiment one would need an independent determination of the barrier height but at least the exponential dependence on barrier thickness is confirmed.

Magnetic field dependence. I_J as a function of magnetic field should follow a diffraction pattern; a neat experimental result was shown in Fig. 10.1. That result is true only for small junctions, uniform barriers and small currents. If the current across the junction is large then self field effects must also be taken into account. The self magnetic field must be added to the applied magnetic field resulting in a tilt in the I_J versus B curves. This effect was investigated in detail by Yamashita and Onodera [787], Yamashita *et al.* [788, 789] and Schwidtal and Finnegan [784]. A typical curve by Yamashita and Onodera [787] is shown in Fig. 24.2.

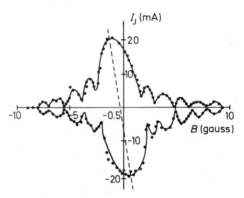

Fig. 24.2 The maximum supercurrent as a function of applied magnetic field. After Yamashita and Onodera [787].

Another reason for deviations from the $I_J - B$ curve of Fig. 10.1 is the nonuniformity of the barrier. Then j_J in Equation (10.4) cannot be taken as a constant. In a practical case we have very little information about the nonuniformity of j_J so a comparison between theory and experiment for that case would be rather difficult. We can, however, invert the problem (take the inverse Fourier transform) and derive j_J from the measured $I_J - B$ curve as done by Zappe [790] and Dynes and Fulton [791]. The latter authors conclude that for plasma grown oxides of Sn and thermally grown oxides of Pb the assumption of uniform current density is fairly well justified; there is, however, significant spatial variation in current density for light-sensitive Sn–CdS–Sn junctions.

Theoretical calculations on the magnetic field dependence of I_J for thin film bridges were done by Likharev [792] by taking account of vortex motion. He found only minor deviations from the $I_J - B$ curve valid for tunnel junctions.

Film thickness dependence. Our basic formulae in Chapter 9 were derived on the basis that the superconductors on both sides of the barrier extend well beyond the penetration depth. In practice it often occurs that the thickness of one of the electrodes is comparable with the penetration depth. Modifications of the formulae for this case were derived by Weihnacht [793]. He finds (among others) that λ_J, the Josephson penetration depth, decreases with decreasing film thickness.

Superconducting barriers. We mention here some experimental work by Mitani *et al.* [543] and further theoretical work by Yamafuji *et al.* [795].

24.3 *I–V* characteristics

In this section we shall briefly summarise a fair amount of further work done on the *I–V* characteristics of tunnel junctions and weak links. Most of the results are extensions of those discussed in Chapter 11.

The treatments of McCumber [335] and Stewart [336] are generalised by Warman and Blackburn [796] by including the effect of finite source resistance, and by Fack and Kose [797] by including a load resistance and calculating the output power at the Josephson frequency. Current–voltage characteristics of double (solder drop) junctions are discussed by Clarke and Fulton [798] and Jaoul [799]. The former authors explain the measured asymmetry by self field effects, the latter one explains the measured hysteresis by examining the energy stored in the junctions and in the inductance of the loop.

The effect of microwave excitation was studied by a number of authors [800–809]. In particular, Fack *et al.* [800] investigated the effect of finite source resistance, Volkov and Nad' [802] examined the *I–V* characteristics in the vicinity of an induced step, Aslamazov and Larkin [803] developed a theory of irradiated point contacts. Arrays of point contacts were investigated by Clark [804]. He obtained steps at voltages $V = nm\hbar\omega/q$ where n is an integer and m is the number of junctions in an array. The first current step due to incident laser radiation at 2·5 THz was observed by McDonald *et al.* [805].

We shall discuss here in a little more detail the measurements of Gregers–Hansen *et al.* [561, 808] on small (0·2 μm × 0·5 μm) Dayem bridges. The heights of the microwave induced steps were deduced from the simple model of Section 11.6. This is shown in Fig. 24.3 where the experimental results are compared with the analogue computer calculations (based on the circuit of Fig. 11.8 where an ohmic resistance in parallel with an ideal Josephson junction is excited by a d.c. and a.c. current source) of Russer [809].

Self induced steps were also widely investigated [811–818]. Kulik [811] came to the conclusion that the height of steps initially increases with increasing Q (of the tunnel junction resonance) but decays again after reaching a maximum. The variation of a self induced step with temperature was investigated by

Dmitrenko and Yanson [811], resonances in a cylindrical structure by Bermon and Mesak [812]. Numerical solutions by Blackburn *et al.* [813] show that when calculating the relative magnitudes of steps the damping constant (B in Equation (9.35)) rather than the Q must be taken constant. Experiments by Angadi and Graham [814] on Nb wires inserted into Hg also showed a step structure which the authors attributed to high frequency resonances in the presence of a self magnetic field.

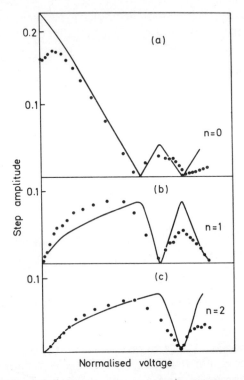

Fig. 24.3 Amplitude of microwave-induced steps for Dayem bridges as a function of normalised voltage for $\hbar\omega G/qI_J = 0.22$. After Gregers-Hansen and Levinson [561].

We shall also mention here an effect, strictly speaking, outside the field of Josephson junctions. Fiory [819] measured the I–V characteristics of super-conducting Al films and obtained steps when the frequency of the applied a.c. current satisfied the relationship

$$f = nv/d \qquad (n = 1, 2, 3, ...) \tag{24.2}$$

where v is the average vortex velocity and d is the magnitude of a two-dimensional lattice vector of the vortex structure. The exact mechanism of the effect is not clear but there is no doubt that it is some manifestation of a weak line effect.

Finally, we shall describe briefly a new effect found by Fulton and Dynes [820]. They have shown both experimentally and theoretically (on the basis of the model of Section 11.2) that a possible mode of operation for the junction is to skip the $V = 0$ part of the characteristic as shown in Fig. 24.4. As the driving current is reduced from positive to negative values the characteristic may follow the dotted lines of the normal tunnelling characteristic instead of switching over to the $V = 0$ Josephson characteristic. Note that the effect is more likely to occur for small junctions.

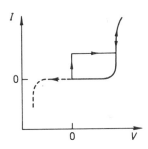

Fig. 24.4 Possible I–V characteristic of a Josephson junction. As the current decreases there is no switching to the $V = 0$ line.

24.4 Long junctions and vortex solutions

There has been further experimental work on the dependence of I_J on the length of the junction. The results depend to a certain extent on the geometry of the junction but in general good agreement was obtained [821–823] with the theory of Owen and Scalapino [333], as plotted in Fig. 12.4. The dependence of I_J on length and width is discussed by Stuehm and Wilmsen [824] on the basis of the experimental results of Pritchard and Schroen [463].

Detailed experimental results on the I_J versus B curve were reported by Schwidtal [825]. Again good agreement is found with the theory of Owen and Scalapino [333] as plotted in Fig. 12.5.

Wave propagation in the junction was studied by Scott [826, 827], Renard [828], Gorbonosov and Kulik [828a] and Kulik [829], the last author being concerned also with the case of finite voltage across the junction. Propagation in an S–I–S–I–S sandwich was studied by Owen and Scalapino [830] with a view to couple in and out of a junction. The coupling in this case arises from the magnetic field penetration through the common superconductor.

Special effects in the I–V characteristics, related to flux motion, were found by Scott and Johnson [831], Barone [832] and Goldman [833].

24.5 Phenomena in rings containing Josephson junctions

Metastable current carrying states. The question is here that what are the permitted values of persistent current in a ring [834–836]. In a simple model taking into account only the energy of the junction (assumed small in comparison with the Josephson penetration depth) and the magnetic energy stored in the ring we get

$$E = -\frac{\hbar I_J}{q}\cos\left(\frac{q}{\hbar}LI_{\text{circ}}\right) + \tfrac{1}{2}LI_{\text{circ}}^2 \tag{24.3}$$

where Equations (12.36), (14.17) and (13.7) (with $\Phi_{\text{ext}} = 0$) have been used. Differentiating E with respect to I_{circ} we get the condition of minimum energy

$$-\frac{I_{\text{circ}}}{I_J} = \sin\left(\frac{q}{\hbar}LI_{\text{circ}}\right) \tag{24.4}$$

which may be solved graphically as shown in Fig. 24.5 (*b*). Wherever the straight line $-I_{\text{circ}}/I_J$ intersects the sinusoidal, there is a local energy minimum (Fig. 24.5 (*a*)) which may serve as a metastable state for a persistent current.

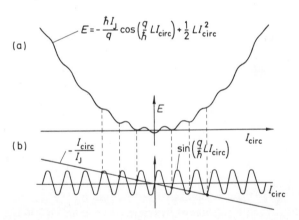

Fig. 24.5 (a) Local energy minima for metastable states, (b) solution of Equation (25.4). After Goldman *et al.* [835].

The lifetime of the states against thermal excitation was calculated by Goldman [837] who came to the conclusion that the lifetime is strongly dependent on I_J. Taking reasonable values for the parameters of the ring he obtained a very wide range of lifetimes from 10^{-9} sec to 10^{12} years. For sufficiently low temperatures thermal excitation becomes negligible and then lifetime is determined by tunnelling out of the metastable states. This was calculated by Scalapino [369, 533] coming also to the conclusion that the lifetime may vary in a very wide range.

Experimental work on rings containing two Josephson junctions was carried out by Goldman et al. [834–836]. The presence of persistent currents was deduced either by measuring the magnetic field created or by measuring the critical current when one of the junctions turned resistive.

High-order quantum transitions. The experiment, performed by Zimmerman and Silver [838] was set up in the following manner. A point contact junction in a resistive ring was voltage biased by a d.c. current and coupled to a resonant circuit at $\omega/2\pi = 27$ MHz. The output at this frequency was measured as a function of the d.c. current. An output was found whenever the relationship

$$\omega = \frac{q}{\hbar} \frac{V_0}{N} \qquad (N = 1, 2, 3, \ldots) \qquad (24.5)$$

was satisfied. N could be a large number up to 200. The effect was explained by the passage across the junction of N flux quanta per cycle.

Computer study. A ring containing a junction and coupled to a resonant circuit was investigated by Simmonds and Parker [839] by generalising the theory (both the linear and the nonlinear one) presented in Chapter 14. The resultant curves (obtained by an analogue computer) bear strong resemblance to those shown in Chapter 14 proving that the interaction with the resonant circuit gives rise to no new phenomena.

24.6 Fabrication

There is not much new work on tunnel junctions; there is a study of thermal oxidisation [840], a report on plasma oxidisation of Nb [841] and further reports on the use of semiconducting [842] and formvar [714, 843] insulators.

On point contacts we shall mention here a control mechanism developed by Contaldo [844], a contact between a solder coated Cu wire and a Nb block [844a] and a stable (can be cycled between room- and liquid helium temperatures) structure by Buhrman et al. [845].

There has been a little more attention paid to the fabrication of thin film bridges. Cerdonio et al. [846] report on work with photomasks and Nad' and Polyanski [847] on direct evaporation through metal masks. An interesting process resulting in small dimensions has been developed by Gregers–Hansen and Levinson [561]. They cut the surface of a glass substrate by a razor blade and immerse it for a short time in dilute hydrofluoric acid resulting in a groove of semicircular cross section of about 0·5 μm radius. Next a 0·1 μm thick Sn layer is evaporated and another cut, this time across the groove, is made. The material removed is often only 0·2 μm wide leaving the tin at the bottom of the groove to serve as a bridge between the two sides.

A technique using single crystal $NbSe_2$ is reported by Consadori et al. [848].

The crystal can be made a few unit cell thick by repeated cleaving. The bridge is obtained by cutting away sections of the crystal and bringing it to the form shown in Fig. 24.6. Typical dimensions were 5 μm × 5 μm × 20 nm.

A sputtering technique for making bridges was developed by Janocko et al. [849, 850]. First a 100 nm layer of NbN was deposited on a sapphire substrate, then an Al layer of the same thickness. Using further a photoresist technique (see Section 15.5) to form the bridge in the Al film, the remaining structure (the Al bridge on the top of the NbN film) is sputter etched. Since the NbN layer is sputtering away at the same rate as the Al on the top, the NbN bridge (with a little Al on top) remains there when the substrate becomes exposed. The limit of this technique is set to about 1 μm by the wavelength of the light used for illuminating the photoresist. However if the sputter etching continues further, then the edge of the bridge starts to sputter away and the result is a very narrow bridge (sometimes down to a width of a mere 60 nm).

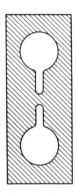

Fig. 24.6 A superconducting bridge obtained by the technique of Consadori et al. [848].

An exciting new development was reported recently by Goodkind and Dundon [851]. They were trying to make bridges on a cylindrical structure (see Fig. 14.10(a)) by using Nb as the superconductor. Since it is difficult to evaporate Nb they used a sputtering technique (the Nb sputtered away was deposited upon a sapphhire rod) and subsequent anodisation. They found that the whole film behaved as a weak link with a definite I_j, adjustable by the number of anodisations. Mechanical strength is one of the obvious advantages of this technique but more importantly this opens the road for mass production by a simple technique. The physical mechanism is not clear; quite likely there are a large number of small weak links in series.

24.7 Computer elements

The possible failure of a junction to return to the $V = 0$ state when the current is suddenly reduced from positive to negative values has already been mentioned in Section 24.3. The significance of the effect (as noted by Fulton [851a]) is that it is is inadvisable to try to speed up switching by 'overdrive', that is to reduce the current to a negative value.

The same effects also influence the operation of the flip-flop (discussed in Section 16.4). When the capacitor discharges current, oscillations occur in the LC circuit which cause current overshoot and may lead to accidental locking to the $V = 0$ state. The remedy is either to introduce more damping (accepting longer switching and steering times) or use larger junctions (a square junction with $I_J = 10$ mA may already be regarded as 'large' for this purpose).

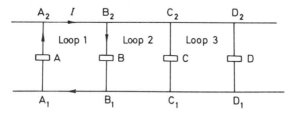

Fig. 24.7 A 'ladder' of junctions which may be used as a counter shift register. After Clarke [759].

A new development in the field of computer elements was mentioned by Anderson [774, 775] reported by Anderson et al. [851b] and discussed by Clarke [759]. It consists of a series of junctions in the configuration of Fig. 24.7. The maximum supercurrent, I_J, is chosen for each junction in such a way that each loop can maintain no more than one flux quantum in the absence of applied fields, that is

$$\Phi_0 < LI_J < 2\Phi_0. \tag{24.6}$$

Assume that initially no currents flow in this ladder circuit and that at $t = 0$ a current pulse $I_J < I < 2I_J$ is applied to A_1A_2. The current flowing across junction A will momentarily exceed I_J so the junction will turn normal and a flux quantum will be admitted into loop 1. When the current pulse is over a circulating current, maintaining one flux quantum, will flow in loop 1. A second current pulse will cause a second flux quantum admitted into loop 1 but since that cannot be maintained junction B will turn normal admitting a flux quantum into loop 2, etc. Hence the flux quanta move up the ladder depending on the number of input pulses. The presence of a flux quantum in the nth loop can be sensed by a fluxmeter and so the circuit may serve as a counter.

It is also possible to use the circuit of Fig. 24.7 as a shift register. Assuming that a flux quantum is maintained in loop 1 and no current flows in any of the other loops, a current pulse of less than I_J is applied to B_1B_2. The current pulse is in the same direction as the loop current and it is sufficiently large so that the sum of the pulse and loop current will drive junction B normal and admit the flux quantum into loop 2. When the pulse is removed the flux quantum will stay in loop 2 so the effect of the pulse is to transfer the flux quantum from loop 1 to loop 2. If loop 1 contained no flux quantum initially (or both loops 1 and 2 contained flux quanta) then the pulse incident upon $B_1 B_2$ is insufficient to drive the junction normal and no transfer of flux occurs. Now if the ladder is filled with flux quanta up to the nth and we apply successively pulses to the nth, n–lth, etc. junctions, the whole content of the ladder can be shifted, that is we have a shift register.

24.8 Generation, mixing and detection of electromagnetic waves

Further work on generation of microwaves in tunnel junctions was done by Dmitrenko et al. [852] and by Galkin and Svistunov [853]. A theoretical analysis of the radiation properties of point contacts in various cavity configurations was given by McCumber [854]; the radiation properties of an array of point contacts were investigated by Tilley [855].

An experimental proof of the existence of the Riedel-peak (see Fig. 17.3) was given by Hamilton and Shapiro [856] and Buckner et al. [964].

Mixers were analysed by DiNardo and Sard [857] who came to the conclusion that for optimum conversion efficiency the junction should be biased half-way between the zeroth and the first step. The point-contact junction as a mixer in the configuration of Fig. 17.7 was discussed by Longacre [858]. Current structure at voltages $(n\omega_1 + m\omega_2)\hbar/q$ was reported by Chen [859] where n and m are integers, $\omega_1/2\pi$ is the frequency of applied microwave radiation and $\omega_2/2\pi$ is the fundamental self-resonant frequency of the tunnel junction used in the experiment.

McDonald et al. [860] mixed the 84th harmonic of 10·60358 GHz with the fundamental laser output at 890·761 GHz in a point contact junction, and detected the if at 60 MHz. Similar measurements were done by Blaney et al. [861] who applied signals at 36 GHz from a klystron, and at 891 and 964 GHz from a laser and detected the if output in the range of 1 to 100 MHz. It should be noted that in the latter application the Josephson junction proved superior both to silicon and to metal–oxide–metal (MOM) point contact mixers.

A theoretical paper on detection was published by Likharev [862] investigating the case when the junction is part of a resonant circuit at the same frequency as that of the input radiation (the same problem was treated by Richards and Sterling [355] as mentioned in Section 17.5). The voltage across the junction is written in the form

$$V(t) = V_0(t) + V_1(t)\cos\left[\omega t + \phi(t)\right]_j, \qquad V_0 \cong \frac{\hbar\omega}{q} \tag{24.7}$$

and an analytical solution is obtained by assuming that $V_0(t)$, $V_1(t)$ and $\phi(t)$ are slowly varying functions of time.

24.9 Measurement of magnetic field, current, voltage, resistance and inductance

The basic physics of the single junction magnetometer was described in Chapter 14 and the device itself in Section 18.5. The heart of the device is a ring containing a single junction coupled to a resonant circuit. A new mode of operation was reported by Goodkind and Stolfa [863] who rely on the change in the kinetic inductance of the ring (realised in the form of a thin film cylinder as shown in Fig. 14.10(a)) caused by the rf input for providing an rf output. They also discuss the dissipative mode where the bridge is driven into its resistive (or resistive-superconductive) state by the rf input. For both modes of operation high Q and large coupling between the ring and the coil of the resonant circuit are required. One, rather unexpected, way of improving the coupling is offered by the fabrication process (see Section 24.6) of Goodkind and Dundon [851] who produced weak links without the need for cutting bridges (the cut in the cylindrical film reduces coupling).

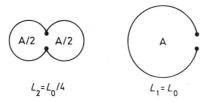

Fig. 24.8 Two loops of the same area; loop 2 may be regarded as having fractional turns, $N = \frac{1}{2}$.

Another new development is the use of fractional turn loops by Zimmerman [864]. It is based on the relationships that (i) the output power of rf biased single junction devices is inversely proportional (see Equation (17.3)) to the inductance of the ring and (ii) the inductance of a coil is $L = \mu_0 N^2 A/l$ where N = number of turns, A = enclosed area and l = length. Hence the device performance can be improved if $N \ll 1$. Interestingly this is quite a possible proposition. The symmetric double-hole device of Fig. 15.3(e) has in fact $N = \frac{1}{2}$ if compared with a single turn loop of the same area. This is shown in Fig. 24.8, the inductance of the single loop is $L_1 = L_0 = \mu_0 A/l$ whereas the double loop is the parallel combination of two loops of area $A/2$ yielding $L_2 = L_0/4$, that is $N = \frac{1}{2}$. Extending further the same principle Zimmerman [864] used a 12 hole device, and did indeed obtain increased sensitivity. Considering further the possibility of increasing the rf bias frequency above 30 MHz (Zimmerman and Frederick [580] used 300 MHz with corresponding improvement in sensitivity) Zimmerman [864] predicts a sensitivity above 10^{-11} gauss.

A new development promising significant improvement in the sensitivity of double junction devices was reported by Clarke and Paterson [865]. Feeding the ring in an asymmetric manner they obtained I_{max} versus Φ curves as shown

in Fig. 24.9. Note that the value of $dI_{max}/d\Phi$ on the steep part of the characteristics is by about two orders of magnitude larger than that for the sinusoidal
curves discussed in Chapter 13.

The magnetometer can also be used for accurate measurement of current as
was proposed by Meservey [866] some time ago. The basic arrangement is as
follows. A thin film cylinder containing two junctions is placed (coaxially)
inside a solenoid (made also of thin film) in which the current to be measured
is passed. The number of flux quanta entering the ring may then be counted as
the current increases. The accuracy of the current measurement depends on the
accuracy with which the various geometrical data can be determined.

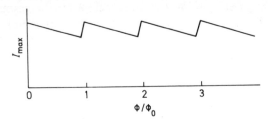

Fig. 24.9 I_{max} as a function of flux for asymmetrically fed double junctions. After Clarke
[759].

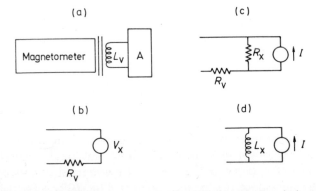

Fig. 24.10 Circuit A coupled to a magnetometer with the aid of an inductance L_v; (b, c
and d), circuits for measurements of voltage (V_x), resistance (R_x), and inductance (L_x)
respectively.

An application of the magnetometer for measuring voltage, resistance and
inductance was outlined by Lukens et al. [867]. The basic circuit connection is
shown in Fig. 24.10(a) where the network A may take the forms shown on
Figs. 24.10b, c and d). In all three cases the magnetic field created by the current
flowing through L_v is measured from which the values of V_x, R_x and L_x may
be dervived.

24.10 Determination of *e/h*

There has been further work on the determination of *e/h* [868–872]. The accuracy of the best result, reported by Finnegan *et al.* [868] is 0·12 ppm where the main contributions to the uncertainty come from (i) lack of stability of the local electrochemical-standard-cell and (ii) the intercomparison between the local volt and the volt as maintained by the NBS. The improved accuracy was achieved by improved methods of comparison (better potentiometers [873]) between the voltage of the Josephson junction (or junctions if there were several connected in series) and the voltage of the standard cell. In fact the stage has been reached where the accuracy of the comparison is considerably higher than the accuracy with which the standard can be maintained.

Summarising the state of art, the present uncertainties in the value of *q/h* showing the four most accurate determinations, are plotted in Fig. 24.11.

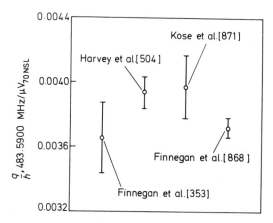

Fig. 24.11 The value of *q/h* according to the four closest determinations to date.

On the theoretical front Nordtvedt [874] presented an argument that the Josephson frequency–voltage relationship is not exact but accurate only to about 1 part in 10^{10} because of quantum electrodynamical corrections. The reason is that in a metal virtual excitation of electron–positron pairs is prevented by Pauli's principle for occupied states below the Fermi energy whereas these states are available in vacuo. As a consequence the charge of an electron in a metal is different. This argument was refuted by Langenberg and Schrieffer [875] and Hartle *et al.* [876] who claimed that (i) there are no quantum electrodynamical corrections to the charge of an electron in a metal and (ii) that the question of charge renormalisation is not directly relevant to the *e/h* measurements.

For a review related to the theoretical problems arising from the Josephson frequency–voltage relation see Scalapino [767].

24.11 Fluctuations

Fluctuations in Josephson junctions represent a field currently explored with great intensity. Some of the simpler concepts and a number of experimental results were described in Chapter 20. We shall mention here some further experimental results and give references to theoretical works.

A detailed study of impressed low frequency (up to 20 kHz) noise was carried out recently by Galkin et al. [877]. They investigate the effect of noise upon (i) $I–V$ characteristics in general and the value of I_J in particular, (ii) magnetic field dependence of I_J, (iii) shape of an induced current step, (iv) temperature dependence of I_J. The experimental curves are compared with their theory and good agreement is found. They suggest that low frequency noise is practically always present and may be responsible for some of the anomalous experimental results (e.g. those in Reference [780]).

The shape of induced steps in the $I–V$ characteristics of irradiated point contact junctions was measured by Henkels and Webb [878] who found good agreement (using a one parameter fit) with the theories of Stephen [514] and Lee [879].

The linewidth of Josephson radiation at centre frequencies of 10 MHz and 34 MHz was measured by Kirschman et al. [880, 881] on Notarys bridges (see Fig. 14.10(b)). Good agreement with theory is obtained by assuming the simple model (discussed in Section 20.2) that fluctuations in the supercurrent are caused indirectly by fluctuations in the normal current.

The linewidth for Nb–Nb point contacts was measured by Vernet and Adde [882] who obtain good agreement with the theory mentioned in Section 20.9. They claim that the discrepancy of a factor of 2 found by Dahm et al. [534] is a consequence of using self-induced steps in tunnel junctions where the rf fields are fairly large in contrast to the small rf field assumption of the theory.

The effect of noise in resistive rings was discussed by Harding and Zimmerman [883] and Zimmerman [884], formulae for vortex fluctuations in thin film bridges were dervived by Thiene and Zimmerman [885], phase fluctuations were treated by Eck et al. [886], experimental results on impressed current noise were given by Kanter and Vernon [887], some results on $I–V$ characteristics were reported by Galkin et al. [888]. For further threoretical work see References [889–904].

There is finally the problem of fluctuations in 'one-dimensional' superconductors which has some bearing on various properties (e.g. the microwave enhancement mechanism discussed in Section 21.5) of superconducting bridges but falls outside the scope of the present section.

24.12 Miscellaneous effects

In this section we shall discuss a few further effects which could not be easily fitted into any of the other sections of this chapter.

A new approach. Leplae *et al.* [905, 906] have applied their theory, the so-called 'boson theory of superconductivity' to the problem of Josephson junctions. Since the relevant equations have a certain simplicity and the results for vortex solutions [905] and a.c. effects [906] agree closely with those derived by conventional means, this new approach seems to have considerable potential.

Granular superconductors. A possible model for a granular superconductor consists of grains of superconducting material separated from each other by insulating barriers exhibiting Josephson tunnelling. A theoretical treatment on the basis of this model was undertaken by Parmenter [907, 908]. Controlled experiments may be conducted on 'artificial' granular superconductors [909, 910] which consist of small but macroscopic (about 0·1 mm diameter) grains pressed together. The experiments of Clark [474, 804, 911, 912] and the theoretical work of Clark and Tilley [913] on arrays of superconducting balls belong to the same category.

Weak link induced by irradiation. The possibility of realising a weak link whose 'weakness' is controllable by an incident electromagnetic wave, has been recently studied by Volkov [914]. The maximum supercurrent I_J is calculated as a function of incident power and of the dimensions of the illuminated region.

Surface plasmons. A metal–insulator–metal structure is capable to support surface plasma modes which is still true when the metals become superconducting. There is hardly any difference in the dispersion relation but there is radical decrease in damping because of the scarcity of normal electrons. The influence of these surface plasmon modes upon the Josephson effects (and vice versa) have been investigated by Economou and Ngai [915] and Ngai [916]. An outcome of the study is a new set of current steps and resonance peaks as well as an alternative explanation for Giaever's experiments [380] on coupled tunnel junctions (see Section 10.2).

Generation of phonons. When the junction is stabilised on a self-induced step the power absorbed is likely to be converted into phonons. The intensity of emitted phonons at $\omega \geqslant 2\Delta/\hbar$ was measured by Kinder [917] as a function of junction voltage (it was possible to move the self-induced step at a constant value of current by varying slowly the magnetic field). By comparing the measured intensity with that obtained from normal tunnelling Kinder concludes that the dominant loss mechanism in the a.c. Josephson effect must be the excitation of normal electrons (which create the phonons by recombination).

A theoretical study of some other aspect of phonon generation was conducted by Ivanchenko and Medvedev [918] who show that due to electron–phonon coupling phonons at the Josephson frequency are generated at the junction.

We shall mention here some further work on *cylindrical junctions* by

Cheishvili [919] and Tilley [920], on *negative self inductance* by Christiansen *et al.* [921], on junctions in *series* by Kao [922], on *nucleation* of superconductivity at a barrier by Boyd [923, 924], on *pressure contacts* by Pankove [925, 926], on *singularities* in the *I–V* characteristics by Yanson [927] and Yanson and Albegova [928], and on *layered structures* by Lawrence and Doniach [928a].

24.13 Microscopic theory

A number of publications concerned with microscopic theory have already been mentioned at the relevant places. There will be no attempt made here to discuss the rest but for completeness we shall list them in References [929–961].

V.   Appendices

Appendix 1

In deriving Equation (2.22) it is assumed that:
   (i) only a very small proportion of the electrons will make the transition so that $q \approx 1$ and $r \approx 0$.
   (ii) the electrons must be either in ψ_1 or in ψ_r so that

$$\frac{d}{dt}[|q|^2 + |r|^2] = 0 \qquad (A1.1)$$

whence it follows that $dq/dt \approx 0$.
   Substituting Equation (2.20) into (2.21), using the above results, multiplying both sides of the equation by ψ_r^* and integrating, Equation (2.22) is obtained.

Appendix 2

First it should be recognised that $H\psi_1 = E\psi_1$ for $x < x_2$ and $H\psi_r = E\psi_r$ for $x > x_1$. Hence the integral

$$-\int_{x_B}^{\infty} \psi_1(H - E_r)\psi_r^* \, dx \qquad x_B \geq x_1 \qquad (A2.1)$$

is zero and could be added to the integral in Equation (2.24) yielding the symmetric form

$$T_{rl} = \int_{x_B}^{\infty} [\psi_r^*(H - E_1)\psi_1 - \psi_1(H - E_r)\psi_r^*] \, dx \qquad x_1 \leq x_B \leq x_2 \qquad (A2.2)$$

Integrating by parts we get

$$T_{rl} = -\frac{\hbar^2}{2m}\left(\psi_r^*\frac{d\psi_1}{dx} - \psi_1\frac{d\psi_r^*}{dx}\right)_{x_B}. \qquad (A2.3)$$

Substituting ψ_r and ψ_1 from Equation (2.19) into the above equation, we obtain

$$T_{rl} = \frac{\hbar^2 \kappa}{m} C_r^* C_1. \qquad (A2.4)$$

Appendix 3

We shall perform here the integration indicated in Equation (4.8)

$$F(T, \Delta) = \int \frac{|E|}{(E^2 - \Delta^2)^{1/2}} [f(E) - f(E - eV)] \, dE. \qquad (A3.1)$$

Integrating for all possible energies

$$F(T, \Delta) = \int_{\Delta}^{\infty} \frac{E}{(E^2 - \Delta^2)^{1/2}} [f(E - eV) - f(E)] \, dE$$

$$-\int_{-\infty}^{-\Delta} \frac{E}{(E^2-\Delta^2)^{1/2}} [f(E-eV)-f(E)]\, dE. \tag{A3.2}$$

Introducing $x+\Delta = E$ in the first integral and $x+\Delta = -E$ in the second integral, Equation (A3.2) is transformed into

$$F(T,\Delta) = \int_0^\infty \frac{x+\Delta}{[x(x+2\Delta)]^{1/2}} [f(x+\Delta-eV)-f(x+\Delta)]\, dx$$

$$+ \int_0^\infty \frac{x+\Delta}{[x(x+2\Delta)]^{1/2}} [f(-x-\Delta-eV)-f(-x-\Delta)]\, dx. \tag{A3.3}$$

Using further the general relationship

$$f(-\xi) = 1-f(\xi) \tag{A3.4}$$

valid for the Fermi function, we get

$$F(T,\Delta) = \int_0^\infty \frac{x+\Delta}{[x(x+2\Delta)]^{1/2}} [f(x+\Delta-eV)-f(x+\Delta+eV)]\, dx. \tag{A3.5}$$

Expanding now the Fermi function

$$f(\xi) = [1+\exp\xi]^{-1} = \exp(-\xi)[1+\exp(-\xi)]^{-1}$$

$$= \exp(-\xi) \sum_{m=0}^\infty (-1)^m \frac{\exp(-m\xi)}{m!} \tag{A3.6}$$

which is valid for $\xi > 0$, substituting $\xi = x+\Delta-eV$ and $x+\Delta+eV$ respectively, and performing the operations within the square bracket in Equation (A3.5) we get

[handwritten margin notes: $x+\Delta-eV>0$; $x+\Delta+eV>0$; $\Rightarrow -(x+\Delta)<eV<(x+\Delta)$; $|eV|<(x+\Delta)$]

$$F(T,\Delta) = 2\sum_{m=0}^\infty (-1)^{m+1} \exp\left(-\frac{m\Delta}{kT}\right) \sinh\frac{meV}{kT}$$

$$\times \int_0^\infty \frac{x+\Delta}{[x(x+2\Delta)]^{1/2}} \exp\left(-\frac{mx}{kT}\right) dx. \tag{A3.7}$$

The above integral is tabulated among the known Laplace transforms; it may be found in the form [962]

$$\int_0^\infty \frac{t+a}{[t(t+2a)]^{1/2}} \exp(-pt)\, dt = a\exp(ap)K_1(ap) \tag{A3.8}$$

where K_1 is the first order modified Bessel function of the second kind. Equation (A3.7) reduces then to the final form

$$F(T,\Delta) = 2\Delta \sum_{m=0}^\infty (-1)^{m+1} K_1(m\Delta/kT) \sinh(meV/kT). \tag{A3.9}$$

Appendix 4

We shall perform here the integration indicated in Equation (4.14) for the case when $T = 0$ and $\Delta_1 = \Delta_2 = \Delta$.

$$G(\Delta) = \int \frac{|E - eV|}{[(E - eV)^2 - \Delta_1^2]^{1/2}} \frac{|E|}{(E^2 - \Delta^2)^{1/2}} [f(E - eV) - f(E)] \, dE. \quad (A4.1)$$

The difference of Fermi functions is given by Equation (4.4) and the limits of integration extend from Δ, the top of the gap on the right, to $eV - \Delta$, the bottom of the gap on the left. Thus

$$G(\Delta) = 0 \quad \text{for} \quad V < 2\Delta/e \quad (A4.2)$$

and for $V \geq 2\Delta$ the integral to solve is as follows

$$G(\Delta) = \int_\Delta^{eV - \Delta} \frac{eV - E}{[(E - eV)^2 - \Delta^2]^{1/2}} \frac{E}{(E^2 - \Delta^2)^{1/2}} \, dE. \quad (A4.3)$$

Introducing the new variable

$$t = \frac{E - eV/2}{\Delta - eV/2} \quad (A4.4)$$

the above integral may be transformed into

$$G(\Delta) = \int_{-1}^{1} \frac{(eV/2)^2 - t^2(\Delta - eV/2)^2}{(\Delta + eV/2)(t^2 - 1)^{1/2}(\alpha^2 t^2 - 1)} \, dt \quad (A4.5)$$

which can be expressed as a linear combination of the complete elliptic integrals [101]

$$K(\alpha) = \int_0^1 \frac{dt}{(1 - t^2)^{1/2}(1 - \alpha^2 t^2)^{1/2}}, \qquad E(\alpha) = \int_0^1 \frac{(1 - \alpha^2 t^2)^{1/2}}{(1 - t^2)^{1/2}} \, dt \quad (A4.6)$$

in the form [66]

$$G(\Delta) = (2\Delta + eV)E(\alpha) - 4\frac{\Delta(\Delta + eV)}{2\Delta + eV} K(\alpha) \quad (A4.7)$$

where

$$\alpha = \frac{eV - 2\Delta}{eV + 2\Delta}. \quad (A4.8)$$

Appendix 5

The proof that Equation (5.16) reduces to (5.17) when $\hbar\omega \to 0$ was given both by Hamilton and Shapiro [135] and by Sweet [963]. We shall follow here the derivation of the former authors (though the latter one is more rigorous) because it leads more quickly to the desired relationship.

When $\hbar\omega \to 0$ and $\alpha \to \infty$ Equation (5.16) may be written in integral form

$$I = \int_{-\infty}^{\infty} J_{\eta}^{2}(\alpha) I_{0}(V + \eta V_{\rm rf}/\alpha)\, d\eta \tag{A5.1}$$

Since $J_{\eta}^{2}(\alpha)$ decreases rapidly for $\eta > \alpha$ the limits of integration may be changed to $\pm\alpha$. Noting further that for large values of α it is possible to replace $J_{\eta}^{2}(\alpha)$ in the integrand by its average value over a small range of η which, using standard expansions, may be obtained in the form

$$\langle J_{\eta}^{2}(\alpha)\rangle = \frac{1}{\pi\alpha}\left[1 - \left(\frac{\eta}{\alpha}\right)^{2}\right]^{-1/2} \tag{A5.2}$$

and substituting

$$\frac{\eta}{\alpha} = \sin\theta \tag{A5.3}$$

Equation (A5.1) reduces finally to

$$I = \frac{1}{\pi}\int_{-\pi/2}^{\pi/2} I_{0}(V + V_{\rm rf}\sin\theta)\, d\theta. \tag{A5.4}$$

Appendix 6

We shall derive here the $\eta(\theta)$ function starting with the differential equation

$$\frac{d\phi}{\kappa - \sin\phi} = d\theta \tag{A6.1}$$

Using the $x = \tan\phi/2$ substitution Equation (A6.1) may be integrated to give

$$\phi = 2\tan^{-1}\left\{\frac{1}{\kappa} + \left(1 - \frac{1}{\kappa^{2}}\right)^{1/2}\tan\left[\frac{\theta}{2}(\kappa^{2}-1)^{1/2}\right]\right\} \tag{A6.2}$$

whence

$$\eta(\theta) = \frac{d\phi}{d\theta} = \frac{\kappa^{2}-1}{\kappa + \sin[(\kappa^{2}-1)^{1/2}\theta + \tan^{-1}(\kappa^{2}-1)^{-1/2}]}. \tag{A6.3}$$

The average value of $\eta(\theta)$ may be obtained either with the aid of Equation (11.10) or by integrating Equation (A6.3) over a period, yielding

$$\langle\eta(\theta)\rangle = (\kappa^{2}-1)^{1/2}. \tag{A6.4}$$

Note that $\eta(\theta)$ is always positive for $\kappa > 1$; its maximum value is $\kappa+1$ and its minimum value $\kappa-1$.

Appendix 7

We shall derive here the fundamental component of the emf in a ring containing a Josephson junction when it is excited by a flux

$$\Phi_{ext} = \Phi_x^0 + \Phi_x^1 \sin \omega t \qquad (A7.1)$$

and the condition $LI_c \ll \Phi_0$ is satisfied. The expressions needed are those of Equations (14.19) and (14.25) as follows*

$$\Phi_{int} + LI_c \sin\left(\frac{q}{\hbar}\Phi_{int}\right) = \Phi_{ext} \qquad (A7.2)$$

and

$$V_i(t) = -\omega\Phi_x^1 \frac{d\Phi_{int}}{d\Phi_{ext}} \cos \omega t. \qquad (A7.3)$$

Since the flux caused by the ring itself is small in comparison with the flux quantum we may replace Φ_{int} by Φ_{ext} in the argument of the sine function in Equation (A5.2). Φ_{int} obtained this way may then be differentiated with respect to Φ_{ext} and substituted into Equation (A5.3), yielding

$$V_i(t) = -\omega\Phi_x^1 \cos \omega t \left\{ 1 - \frac{q}{\hbar}LI_c \cos\left[\frac{q}{\hbar}(\Phi_x^0 + \Phi_x^1 \sin \omega t)\right] \right\}$$

$$= -\omega\Phi_x^1 \cos \omega t \left\{ 1 - \frac{q}{\hbar}LI_c \cos\left(\frac{q}{\hbar}\Phi_x^0\right) \left[J_0\left(\frac{q}{\hbar}\Phi_x^1\right) + 2\sum_{n=1}^{\infty} J_{2n}\left(\frac{q}{\hbar}\Phi_x^1\right)\cos 2n\omega t \right] \right.$$

$$\left. + 2\frac{q}{\hbar}LI_c \sin\left(\frac{q}{\hbar}\Phi_x^0\right) \sum_{n=1}^{\infty} J_{2n-1}\left(\frac{q}{\hbar}\Phi_x^1\right) \sin(2n-1)\omega t \right\} \qquad (A7.4)$$

where the Fourier–Bessel expansion [381] of the trigonometric functions is used. The fundamental component may be obtained by selecting the terms in ωt leading to

$$V_{i1} = -\omega\Phi_x^1 + 2\omega LI_c \cos\left(\frac{q}{\hbar}\Phi_x^0\right) J_1\left(\frac{q}{\hbar}\Phi_x^1\right) \qquad (A7.5)$$

where the relationship between Bessel functions [381] $z[J_0(z)+J_2(z)] = 2J_1(z)$ is used.

---

*As we are no longer interested in quantum transitions the term $(q/\hbar)k\Phi_0 = 2k\pi$ in the argument of the sine function is omitted.

Appendix 8

We shall solve here the differential equation

$$\omega'(\eta \sin \phi - 1) + (\xi \cos \phi + 1)\frac{d\phi}{dt} = 0. \tag{A8.1}$$

Separating the variables we get

$$\omega' \int dt = \int \frac{\xi \cos \phi + 1}{1 - \eta \sin \phi} d\phi. \tag{A8.2}$$

The integral on the right-hand side may be divided into two parts; the first one may be directly integrated

$$\xi \int \frac{\cos \phi \, d\phi}{1 - \eta \sin \phi} = -\frac{\xi}{\eta} \ln (1 - \eta \sin \phi) \tag{A8.3}$$

and the second one may be integrated after the substitution of $x = \tan(\phi/2)$ yielding at the end

$$\int \frac{d\phi}{1 - \eta \sin \phi} = \frac{2}{(1 - \eta^2)^{1/2}} \tan^{-1} \frac{\tan(\phi/2) - \eta}{(1 - \eta^2)^{1/2}}. \tag{A8.4}$$

Hence the solution of Equation (A6.1) is

$$\omega'(1 - \eta^2)^{1/2}t = 2 \tan^{-1} \frac{\tan(\phi/2) - \eta}{(1 - \eta^2)^{1/2}} - \frac{\xi}{\eta}(1 - \eta^2)^{1/2} \ln (1 - \eta \sin \phi). \tag{A8.5}$$

The integration constant has not been included as it would affect only the zero of the time axis.

Next we shall calculate the average value of the voltage across the Josephson junction

$$V_0 = \frac{\hbar}{q}\left\langle \frac{d\phi}{dt} \right\rangle = \frac{\hbar}{q} \lim_{t \to \infty} \frac{1}{t}\left[\phi(t) - \phi(0)\right] \tag{A8.6}$$

which with the aid of Equation (A8.5) comes to

$$V_0 = \frac{\hbar \omega_0}{q} = \frac{\hbar}{q}\omega'(1 - \eta^2)^{1/2} = [V'^2 - (RI_c)^2]^{1/2}. \tag{A8.7}$$

With the above defined quantities Equation (A8.5) may be written as

$$\tan \frac{\phi}{2} = \eta + (1 - \eta^2)^{1/2} \tan \frac{F}{2} \tag{A8.8}$$

where

$$F = \omega_0\left[t + \frac{L}{R} \ln\left(1 - \eta \frac{I}{I_c}\right)\right]. \tag{A8.9}$$

The aim is now to eliminate ϕ and get an expression in terms of I and t. It may be proven by substitution that apart from a phase factor in the sine function, Equations (A8.8) and (A8.9) are identical with

$$\frac{I - \eta I_c}{I_c - \eta I} = \sin\left\{\omega_0\left[t + \frac{L}{R}\ln\left(1 - \eta\frac{I}{I_c}\right)\right]\right\}$$

(A8.10)

which is the required form.

Appendix 9

The inertia of charge carriers (usually neglected in electric circuits) leads to the same type of relation between current and voltage as that due to the self-magnetic field. We shall now work out this so-called kinetic inductance for a conductor of length l and cross-section A. We start with the equation of motion in the form

$$m\left(\frac{dv}{dt} + \frac{v}{\tau}\right) = e\mathscr{E}$$

(A9.1)

where τ is a phenomenological relaxation time. Further, using the definitions of current

$$I = Ae\rho v$$

(A9.2)

(where ρ is the charge density) and assuming the relation

$$V = \mathscr{E}l$$

Equation (A9.1) modifies to

$$V = \frac{lm}{Ae^2\rho}\left(\frac{dI}{dt} + \frac{I}{\tau}\right).$$

(A9.3)

Comparing the above expression with the familiar inductive and resistive relations between current and voltage we get the resistance in the usual form

$$R = \frac{lm}{Ae^2\rho\tau}$$

(A9.4)

and may now define the kinetic inductance as

$$L = \frac{lm}{Ae^2\rho}.$$

(A9.5)

For the hollow cylinder dicussed in Section 20.2

$$l = 2\pi r \quad \text{and} \quad A = ad$$

(A9.6)

Distinguishing further between the normal and the superfluid by the subscripts N and S we obtain Equation (20.2).

Appendix 10

We shall derive here the condition for minimum energy. First Equation (20.3) must be expressed as a function of the single variable Φ, for a given value of the fluxoid Φ_c. From Equations (20.6) and (20.7)

$$I_S = \frac{\Phi - \Phi_c}{L_S} \quad \text{and} \quad I_N = \frac{1}{L_S}\left[\Phi\left(1 + \frac{L_S}{L}\right) - \Phi_c\right]. \tag{A10.1}$$

Substituting I_N and I_S into Equation (20.3) we get an expression for the energy in terms of Φ. The energy is a minimum with respect to Φ when

$$\Phi = \frac{\Phi_c}{1 - L_S/L} \tag{A10.2}$$

which leads to

$$I_N = 0 \quad \text{and} \quad I_S = \frac{\Phi_c}{L + L_S}. \tag{A10.3}$$

References

SPECIAL ABBREVIATIONS

LT7   Proceedings of the 7th International Conference on Low Temperature Physics, (G. M. Graham and A. C. Hollis, Eds.), University of Toronto Press, 1961.

LT8   Proceedings of the 8th International Conference on Low Temperature Physics, (R. O. Davies, Ed.), Butterworths, London, 1963.

LT9   Proceedings of the 9th International Conference on Low Temperature Physics, (J. G. Daunt, D. O. Edwards, F. J. Milford and M. Yaqub), Plenum Press. New York, 1965.

LT10   Proceedings of the 10th International Conference on Low Temperature Physics, (M. P. Malkov, Ed.), Viniti, Moskva, 1967.

LT11   Proceedings of the 11th International Conference on Low Temperature Physics, (J. F. Allen, D. M. Finlayson and D. M. McCall, Eds.), University of St Andrews, 1968.

LT12   Proceedings of the 12th International Conference on Low Temperature Physics (E. Kanda, Ed.), Keigaku Publishing Co. Ltd., Tokyo, 1971.

TPS   Tunneling Phenomena in Solids, (E. Burstein and S. Lundquist, Eds.), Plenum Press, New York, 1969.

PSPSD   Proceedings of the Symposium on the Physics of Superconducting Devices, April 28–29, 1967, University of Virginia, Charlottesville.

PCFS   Proceedings of the Conference on Fluctuations in Superconductors, (W. S. Goree and F. Chilton, Eds.), March 13–15, 1968, Asilomar Conference Grounds, Pacific Grove, California.

PCPMFC   Proceedings of the International Conference on Precision Measurement and Fundamental Constants (D. N. Langenberg and B. N. Taylor, Eds.), National Bureau of Standards Special Publication 343, U.S. GPO, Washington, D.C., 1971.

PCSS   Proceedings of the International Conference on the Science of Superconductivity, Stanford, 1969 (published as volume 55 of Physica).

[1] H. Kamerlingh Onnes, *Comm. Leiden Suppl.* **58**, 16 (1924).
[2] W. Tuyn, *Comm. Leiden* **198**, 16 (1929).
[3] W. Meissner and R. Ochsenfeld, *Naturwissenschaften* **21**, 787 (1933).
[4] C. J. Gorter and H. B. G. Casimir, *Phys. Z.* **35**, 963 (1934).
[5] F. London and H. London, *Proc. Roy. Soc.* **A149**, 71 (1935).
[6] F. London and H. London, *Physica* **2**, 341 (1935).
[7] V. L. Ginzburg and L. D. Landau, *Zh. Eksp. Teor. Fiz.* **20**, 1064 (1950).
[8] A. B. Pippard, *Proc. Roy. Soc.* **A216**, 547 (1953).
[9] H. Frohlich, *Phys. Rev.* **79**, 845 (1950).
[10] C. A. Reynolds, B. Serin, W. H. Wright and L. B. Nesbitt, *Phys. Rev.* **78**, 487 (1950).
[11] E. Maxwell, *Phys. Rev.* **78**, 477 (1950).

[11] B. Maxwell, *Phys. Rev.* **78**, 477 (1950).
[12] L. N. Cooper, *Phys. Rev.* **104**, 1189 (1956).
[13] J. Bardeen, L. N. Cooper and J. R. Schrieffer, *Phys. Rev.* **108**, 1175 (1957).
[14] N. N. Bogoliubov, D. N. Zubarev and Yu. A. Tserkovnik, *Dokl. Akad. Nauk SSSR* **117**, 788 (1957).
[15] L. P. Gorkov, *JETP* **7**, 505 (1958).
[16] M. W. Zemansky, 'Heat and Thermodynamics', McGraw Hill, New York, 1951, (3rd edition) p. 365.
[17] F. London, 'Superfluids', Vol. I, Wiley, New York, 1950.
[18] L. O. Landau and E. M. Lifshitz, 'Statistical Physics', Pergamon Press, London, 1958.
[19] L. P. Gorkov, *JETP* **9**, 1364 (1959).
[20] L. P. Gorkov, *JETP* **10**, 998 (1960).
[21] R. P. Feynman, R. B. Leighton and M. Sands, 'The Feynman lectures on physics', Vol. III, Chapter 21, Addison-Wesley, Reading, Massachusetts, 1965.
[22] H. Frohlich, *Proc. Phys. Soc.* **87**, 330 (1966).
[23] R. Doll and M. Näbauer, *Phys. Rev. Letts.* **7**, 51 (1961).
[24] B. S. Deaver, Jr. and W. M. Fairbank, *Phys. Rev. Letts.* **7**, 43 (1961).
[25] A. A. Abrikosov, *JETP* **5**, 1174 (1957).
[26] A. A. Abrikosov, *J. Phys. Chem. Solids* **2**, 199 (1957).
[27] W. H. Kleiner, L. M. Roth and S. H. Autler, *Phys. Rev.* **133**, A1226 (1964).
[28] J. Matricon, *Phys. Letts.* **9**, 289 (1964).
[29] D. Cribier, B. Jacrot, R. L. Madhav and B. Farnoux, *Phys. Letts.* **9**, 106 (1964).
[30] U. Essman and H. Träuble, *Phys. Letts.* **24A**, 526 (1967).
[31] M. H. Tinkham, *Phys. Rev.* **129**, 2413 (1963).
[32] M. H. Tinkham, *Rev. Mod. Phys.* **36**, 268 (1964).
[33] D. H. Douglass, Jr. and L. M. Falicov, 'Progress in Low Temperature Physics', Vol. IV (C. J. Gorter, Ed.), North Holland, Amsterdam, 1964, p. 97.
[34] P. W. Anderson, *J. Chem. Phys. Solids* **11**, 26 (1959).
[35] J. R. Oppenheimer, *Phys. Rev.* **31**, 66 (1928).
[36] R. H. Fowler and L. Nordheim, *Proc. Roy. Soc.* **A119**, 173 (1928).
[37] G. Gamow, *Z. Physik* **51**, 204 (1928).
[38] J. Frenkel, *Phys. Rev.* **36**, 1604 (1930).
[39] A. Sommerfeld and H. Bethe, 'Handbuch der Physik', Vol. XXIV, Part 2 (S. Flugge, Ed.), Julius Springer, Berlin, 1933, p. 333.
[40] A. H. Wilson, *Proc. Roy. Soc.* **A136**, 487 (1932).
[41] C. Zener, *Proc. Roy. Soc.* **A145**, 523 (1943).
[42] L. Esaki, *Phys. Rev.* **109**, 603 (1957).
[43] I. Giaever, LT7, p. 327.
[44] I. Giaever, *Phys. Rev. Letts.* **5**, 147 (1960).
[45] I. Giaever, *Phys. Rev. Letts.* **5**, 464 (1960).
[46] J. Nicol, S. Shapiro and P. H. Smith, *Phys. Rev. Letts.* **5**, 461 (1960).
[47] B. D. Josephson, *Phys. Letts.* **1**, 251 (1962).
[48] J. Bardeen, *Phys. Rev. Letts.* **6**, 57 (1961).
[49] E. O. Kane, TPS p. 1.
[50] M. H. Cohen, L. M. Falicov and J. C. Phillips, *Phys. Rev. Letts.* **8**, 316 (1962).
[51] W. A. Harrison, *Phys. Rev.* **123**, 85 (1961).
[52] J. E. Dowman, M. L. A. MacVicar and J. R. Waldram, *Phys. Rev.* **186**, 452 (1969).
[53] W. Franz, TPS p. 13.
[54] R. Holm and W. Meissner, *Z. Phys.* **74**, 715 (1932).
[55] R. Glover and M. Tinkham, *Phys. Rev.* **108**, 243 (1957).

[56] R. W. Schmitt, *Physics Today* **14**, No. 12, 38 (1961).
[57] A. Dayem and R. J. Martin, *Phys. Rev. Letts.* **8**, 246 (1962).
[58] P. K. Tien and J. Gordon, *Phys. Rev.* **129**, 647 (1963).
[59] Y. Goldstein and B. Abeles, *Phys. Letts.* **14**, 78 (1965).
[60] E. Lax and F. L. Vernon, Jr., *Phys. Rev. Letts.* **14**, 256 (1965).
[61] B. N. Taylor and E. Burstein, *Phys. Rev. Letts.* **10**, 14 (1963).
[62] C. J. Adkins, *Phil. Mag.* **8**, 1051 (1963).
[63] J. M. Rowell, *Rev. Mod. Phys.* **36**, 215 (1964) (discussion).
[64] I. K. Yanson, V. M. Svistunov and I. M. Dmitrenko, *JETP* **20**, 1404 (1965).
[65] W. J. Tomasch, *Phys. Rev. Letts.* **15**, 672 (1965).
[66] I. Giaever, H. R. Hart and K. Megerle, *Phys. Rev.* **126**, 941 (1962).
[67] W. L. McMillan and J. M. Rowell, *Phys. Rev. Letts.* **14**, 108 (1965).
[68] G. M. Eliashberg, *JETP* **11**, 696 (1960).
[69] I. Giaever and K. Megerle, *Phys. Rev.* **122**, 1101 (1961).
[70] D. H. Douglass, Jr., *Phys. Rev. Letts.* **7**, 14 (1961).
[71] J. L. Levine, *Phys. Rev. Letts.* **15**, 154 (1965).
[72] J. P. Wilson, *Phys. Letts.* **28A**, 43 (1968).
[73] N. V. Zavaritskii, LT8, p. 175.
[74] N. V. Zavaritskii, *JETP* **16**, 793 (1963).
[75] N. V. Zavaritskii, *JETP* **18**, 1260 (1964).
[76] N. V. Zavaritskii, *JETP* **21**, 557 (1965).
[77] J. P. Franck and W. J. Keeler, *Phys. Rev. Letts.* **20**, 379 (1968).
[78] F. Reif and M. A. Woolf, *Phys. Rev. Letts.* **9**, 315 (1962).
[79] P. H. Smith, S. Shapiro, J. L. Miles and J. Nicol, *Phys. Rev. Letts.* **6**, 686 (1961).
[80] R. Frerichs and J. P. Wilson, *Phys. Rev.* **142**, 264 (1966).
[81] W. J. Tomasch, *Phys. Letts.* **8**, 104 (1964).
[82] W. J. Tomasch, LT9, p. 424.
[83] W. J. Tomasch, *Phys. Rev.* **139**, A746 (1965).
[84] J. Sutton, *Proc. Phys. Soc.* **87**, 791 (1966).
[85] P. Nedellec, E. Guyon and F. Brochard, *J. Low Temp. Phys.* **1**, 519 (1969).
[86] G. B. Donaldson and D. J. Brassington, LT12, p. 429.
[87] D. M. Ginsberg, *Phys. Rev. Letts.* **8**, 204 (1962).
[88] B. I. Miller and A. H. Dayem, *Phys. Rev. Letts.* **18**, 1000 (1967).
[89] J. L. Levine and S. Y. Hsieh, LT11, p. 713.
[90] S. M. Marcus, *Phys. Letts.* **23**, 28 (1966).
[91] T. D. Clark, *Phys. Letts.* **24A**, 459 (1967).
[92] J. Klein and A. Leger, *Phys. Letts.* **30A**, 96 (1969).
[93] J. L. Miles, P. H. Smith and W. Schonbein, *Proc. IEEE* **51**, 937 (1963).
[94] H. T. Yuan and A. C. Scott, *Solid State Electron.* **9**, 1149 (1966).
[95] E. Burstein, D. N. Langenberg and B. N. Taylor, 'Advances in Quantum Electronics', (J. R. Singer, Ed.), Columbia University Press, New York, 1961, p. 480.
[96] W. Eisenmenger and A. H. Dayem, *Phys. Rev. Letts.* **18**, 125 (1967).
[97] W. D. Gregory, L. Leopold, D. Repici and J. Bostock, *Phys. Letts.* **29A**, 13 (1969).
[98] G. H. Wood and B. L. White, *Appl. Phys. Letts.* **15**, 237 (1969).
[99] B. L. Blackford and R. H. March, *Can. J. Phys.* **46**, 141 (1968).
[100] B. N. Taylor, *J. Appl. Phys.* **39**, 2490 (1968).
[101] Jahnke–Emde–Lösch, 'Tables of Higher Functions', McGraw Hill, New York, 1960.
[102] S. Shapiro, P. H. Smith, J. Nicol, J. L. Miles and P. F. Strong, *IBM J. Res. Develop.* **6**, 34 (1962).
[103] B. N. Taylor, E. Burstein and D. N. Langenberg, *Bull. Am. Phys. Soc.* **7**, 190 (1962).
[104] J. Bardeen, *Phys. Rev. Letts.* **9**, 147 (1962).

[105] J. R. Schrieffer, TPS, p. 287.
[106] J. R. Schrieffer, *Rev. Mod. Phys.* **36**, 200 (1964).
[107] C. J. Adkins, *Rev. Mod. Phys.* **36**, 211 (1964).
[108] G. B. Donaldson, LT10, S163, p. 291.
[109] C. J. Gillespie and W. A. Rachinger, *J. Appl. Phys.* **38**, 2615 (1967).
[110] A. C. Thorsen, T. Wolfram and L. E. Valby, *Phys. Letts.* **25A**, 548 (1967).
[111] C. J. Adkins and B. W. Kington, *Phil. Mag.* **13**, 971 (1966).
[112] J. C. Fisher and I. Giaever, *J. Appl. Phys.* **32**, 172 (1961).
[113] J. M. Rowell and L. Kopf, *Phys. Rev.* **137**, A907 (1965).
[114] J. L. Miles and P. H. Smith, *J. Electrochem. Soc.* **110**, 1240 (1963).
[115] W. Schroen, *J. Appl. Phys.* **39**, 2671 (1968).
[116] J. L. Miles and H. O. McMahon, *J. Appl. Phys.* **32**, 1176 (1961).
[117] K. Blodgett and I. Langmuir, *Phys. Rev.* **51**, 964 (1937).
[118] I. Giaever, *Phys. Rev. Letts.* **20**, 1286 (1968).
[119] I. Giaever and H. R. Zeller, *Phys. Rev. Letts.* **21**, 1385 (1968).
[120] I. Giaever and H. R. Zeller, *J. Vac. Sc. Tech.* **6**, 502 (1969).
[121] M. L. A. MacVicar, S. M. Freake and C. J. Adkins, *J. Vac. Sc. Tech.* **6**, 717 (1969).
[122] H. J. Levinstein and J. E. Kunzler, *Phys. Letts.* **20**, 581 (1966).
[123] H. J. Levinstein and J. E. Kunzler, LT10, p. 241.
[124] J. R. Schrieffer and J. W. Wilkins, *Phys. Rev. Letts.* **10**, Z7 (1963).
[125] J. W. Wilkins, TPS, p. 333.
[126] M. H. Cohen, L. M. Falicov and J. C. Phillips, LT8, p. 163.
[127] E. Riedel, *Z. Naturforsch.* **19a**, 1634 (1964).
[128] N. R. Werthamer, *Phys. Rev.* **147**, 255 (1966).
[129] C. F. Cook and G. E. Everett, *Phys. Rev.* **159**, 374 (1967).
[130] D. Bonnet and H. Rabenhorst, *Phys. Letts.* **26A**, 174 (1968).
[131] S. Teller and B. Kofoed, *Solid State Commun.* **8**, 235 (1970).
[132] J. N. Sweet and G. I. Rochlin, *Phys. Rev.* **B2**, 656 (1970).
[133] H. Buttner and E. Gerlach, *Phys. Letts.* **27A**, 226 (1968).
[134] J. N. Sweet and G. I. Rochlin, *Solid State Commun.* **8**, 1341 (1970).
[135] C. A. Hamilton and S. Shapiro, *Phys. Rev.* **B2**, 4494 (1970).
[136] J. C. Swihart, *J. Appl. Phys.* **32**, 461 (1961).
[137] A. Longacre, Jr. and S. Shapiro, Microwave Research Inst. Symp. Series, v. 20, p. 295, Polytechnic Press, Brooklyn, 1971.
[138] J. M. Ziman, 'Electrons and Phonons', Oxford University Press, 1960, p. 191.
[139] Y. Goldstein, B. Abeles and R. W. Cohen, *Phys. Rev.* **151**, 349 (1966).
[140] L. Kleinman, *Phys. Rev.* **132**, 2484 (1963).
[141] L. Kleinman, B. N. Taylor and E. Burstein, *Rev. Mod. Phys.* **36**, 208 (1964).
[142] J. M. Rowell and W. L. Feldmann, *Phys. Rev.* **172**, 393 (1968).
[143] Yu. M. Ivanchenko, *JETP Letts.* **4**, 242 (1966).
[144] I. Giaever and H. R. Zeller, *Phys. Rev.* **B1**, 4278 (1970). PCSS, p. 455.
[145] W. J. Tomasch, *Phys. Rev. Letts.* **16**, 16 (1966).
[146] W. J. Tomasch, *Phys. Letts.* **23**, 204 (1966).
[147] W. J. Tomasch, LT10, S41, p. 266.
[148] W. L. McMillan and P. W. Anderson, *Phys. Rev. Letts.* **16**, 85 (1966).
[149] W. Schattke, *Z. Naturforsch.* **25a**, 189 (1970).
[150] W. J. Tomasch, TPS, p. 315.
[151] W. J. Tomasch and T. Wolfram, *Phys. Rev. Letts.* **16**, 352 (1966).
[152] R. T. Mina and M. S. Khaikin, *JETP* **24**, 42 (1967).
[153] G. I. Lykken, A. L. Geiger and E. N. Mitchell, *Phys. Rev. Letts.* **25**, 1578 (1970).
[154] K. Maki and A. Griffin, *Phys. Rev.* **150**, 356 (1966).

[155] J. M. Rowell and W. L. McMillan, *Phys. Rev. Letts.* **16**, 453 (1966).
[156] D. E. Thomas and J. M. Rowell, *Rev. Sci. Instr.* **36**, 1301 (1965).
[157] J. G. Adler and J. E. Jackson, *Rev. Sci. Instr.* **37**, 1049 (1966).
[158] J. W. T. Dabbs, LT9, p. 428.
[159] J. S. Rogers, J. G. Adler and S. B. Woods, *Rev. Sci. Instr.* **35**, 208 (1964).
[160] A. Gaudefroy-Demonbynes, E. Guyon, A. Martinet and J. Sanchez, *Rev. Phys. Appl.* **1**, 18 (1966).
[161] D. E. Thomas and J. M. Klein, *Rev. Sci. Instr.* **34**, 920 (1963).
[162] A. Longacre Jr., *Rev. Sci. Instr.* **41**, 448 (1970).
[163] R. F. Gasparovic, B. N. Taylor and R. E. Eck, *Solid State Commun.* **4**, 59 (1966).
[164] D. H. Douglass, Jr. and R. Meservey, *Phys. Rev.* **135**, A19 (1964).
[165] W. L. McMillan and J. M. Rowell, in 'Superconductivity', (R. D. Parks, Ed.), Marcel Dekker Inc., New York, 1969, p. 561.
[166] M. D. Sherril and H. H. Edwards, *Phys. Rev. Letts.* **6**, 460 (1961).
[167] P. Townsend and J. Sutton, *Phys. Rev.* **128**, 591 (1962).
[168] H. Wuhl, J. E. Jackson and C. V. Briscoe, *Phys. Rev. Letts.* **20**, 1496 (1968).
[169] S. Bermon, Tech. Rept. 1, University of Illinois, Urbana, National Science Foundation Grant NSF GP1100, 1964.
[170] J. L. Harden and R. S. Collier, *Cryogenics* **2**, 369 (1962).
[171] J. R. Merril and R. W. Christy, *Am. J. Phys.* **37**, 61 (1969).
[172] A. S. Edelstein and A. M. Toxen, *Phys. Rev. Letts.* **17**, 196 (1966).
[173] J. J. Hauser, *Phys. Rev. Letts.* **17**, 921 (1966).
[174] H. J. Levinstein, V. G. Chirba and J. E. Kunzler, *Phys. Letts.* **24A**, 362 (1967).
[175] A. S. Edelstein, *Phys. Rev.* **164**, 510 (1967).
[176] J. S. Rogers and S. M. Khanna, *Phys. Rev. Letts.* **20**, 1284 (1968).
[177] T. Seidel and A. W. Wicklund, LT8, p. 176.
[178] Y. Goldstein, *Rev. Mod. Phys.* **36**, 213 (1964).
[179] V. Hoffstein and R. W. Cohen, *Phys. Letts.* **29A**, 603 (1969).
[180] R. W. Cohen, G. D. Cody and Y. Goldstein, *RCA Rev.* **25**, 433 (1964).
[181] B. T. Matthias, H. Suhl and E. Corenzwit, *Phys. Rev. Letts.* **1**, 92 (1958).
[182] B. T. Matthias, H. Suhl and E. Corenzwit, *J. Phys. Chem. Solids* **13**, 156 (1960).
[183] K. Schwidtal, *Z. Phys.* **158**, 563 (1960).
[184] A. A. Abrikosov and L. P. Gorkov, *JETP* **12**, 1243 (1961).
[185] S. Skalski, O. Betbeder-Matibet and P. R. Weiss, *Phys. Rev.* **136**, A1500 (1964).
[186] M. A. Woolf and F. Reif, *Phys. Rev.* **137**, A557 (1965).
[187] F. Reif and M. A. Woolf, *Rev. Mod. Phys.* **36**, 238 (1964).
[188] A. S. Edelstein, *Phys. Rev. Letts.* **19**, 1184 (1967).
[189] A. S. Edelstein, *Phys. Rev.* **180**, 505 (1969).
[190] N. Tsuda, *J. Phys. Soc. Japan* **27**, 1025 (1969).
[191] D. H. Douglass, Jr., *IBM J. Res. Develop.* **6**, 44 (1962).
[192] R. Meservey and D. H. Douglass, Jr., *Phys. Rev.* **135**, A24 (1964).
[193] R. S. Collier and R. A. Kamper, *Phys. Rev.* **143**, 323 (1966).
[194] P. G. de Gennes, *Phys. Kond. Mat.* **3**, 79 (1964).
[195] K. Maki, *Progr. Theoret. Phys.* (Kyoto) **29**, 10 (1963).
[196] K. Maki, *Progr. Theoret. Phys.* (Kyoto) **29**, 333 (1963).
[197] K. Maki, *Progr. Theoret. Phys.* (Kyoto) **29**, 603 (1963).
[198] K. Maki, *Progr. Theoret. Phys.* (Kyoto) **31**, 731 (1964).
[199] K. Maki and P. Fulde, *Phys. Rev.* **140**, A1586 (1965).
[200] E. Guyon, A. Martinet, J. Matricon and P. Pincus, *Phys. Rev.* **138**, A746 (1965).
[201] J. L. Levine, *Phys. Rev.* **155**, 373 (1967).
[202] J. Millstein and M. Tinkham, *Phys. Rev.* **158**, 325 (1967).

[203] S. Strassler and P. Wyder, *Phys. Rev.* **158**, 319 (1967).

[204] J. Bardeen, *Rev. Mod. Phys.* **34**, 667 (1962).

[205] P. Fulde, LT9, p. 438.

[206] Catalin D. Mitescu, LT10, S181, p. 366.

[207] J. T. Chen, T. T. Chen, J. D. Leslie and H. J. T. Smith, *Phys. Letts.* **25A**, 679 (1967).

[208] A. J. Bennett, *Phys. Rev.* **153**, 482 (1967).

[209] C. K. Campbell and D. G. Walmsley, *Can. J. Phys.* **45**, 159 (1967).

[210] C. K. Campbell, R. C. Dynes and D. G. Walmsley, *Can. J. Phys.* **44**, 2601 (1966).

[211] M. L. A. MacVicar and R. M. Rose, PSPSD, Q1.

[212] M. L. A. MacVicar and R. M. Rose, *Phys. Letts.* **25A**, 681 (1967).

[213] M. L. A. MacVicar and R. M. Rose, *J. Appl. Phys.* **39**, 1721 (1968).

[214] M. L. A. MacVicar and R. M. Rose, *Phys. Letts.* **26A**, 510 (1968).

[215] M. L. A. MacVicar and R. M. Rose, LT11, p. 717.

[216] J. W. Hafstrom, R. M. Rose and M. L. A. MacVicar, *Phys. Letts.* **30A**, 379 (1969).

[217] M. L. A. MacVicar, *Phys. Rev.* **B2**, 97 (1970).

[218] P. Morel and P. W. Anderson, *Phys. Rev.* **125**, 1263 (1962).

[219] J. M. Rowell, A. G. Chynoweth and J. C. Phillips, *Phys. Rev. Letts.* **9**, 59 (1962).

[220] J. R. Schrieffer, D. J. Scalapino and J. W. Wilkins, *Phys. Rev. Letts.* **10**, 336 (1963).

[221] J. M. Rowell, P. W. Anderson and D. E. Thomas, *Phys. Rev. Letts.* **10**, 334 (1963).

[222] B. N. Brockhouse, T. Arase, G. Caglioti, K. R. Rao and A. D. B. Woods, *Phys. Rev.* **128**, 1099 (1962).

[223] D. J. Scalapino and P. W. Anderson, *Phys. Rev.* **133**, A291 (1964).

[224] L. Van Hove, *Phys. Rev.* **89**, 1189 (1953).

[225] S. Bermon and D. M. Ginsberg, *Phys. Rev.* **135**, A306 (1964).

[226] W. N. Hubin and D. M. Ginsberg, *Phys. Rev.* **188**, 716 (1969).

[227] J. M. Rowell, W. L. McMillan and P. W. Anderson, *Phys. Rev. Letts.* **14**, 633 (1965).

[228] J. G. Adler, J. E. Jackson and B. S. Chandrashekar, *Phys. Rev. Letts.* **16**, 53 (1966).

[229] T. Claeson, *Solid State Commun.* **5**, 119 (1967).

[230] J. G. Adler, J. E. Jackson and T. A. Will, *Phys. Letts.* **24A**, 407 (1967).

[231] T. Claeson and G. Grimwall, *J. Phys. Chem. Solids* **29**, 387 (1968).

[232] I. Giaever, LT8, p. 171.

[233] J. G. Adler and S. C. Ng, *Can. J. Phys.* **43**, 594 (1965).

[234] N. V. Zavaritskii, *JETP Letts.* **6**, 155 (1967).

[235] N. V. Zavaritskii, LT11, p. 721.

[236] R. W. Cohen, B. Abeles and G. S. Weisbarth, *Phys. Rev. Letts.* **18**, 336 (1967).

[237] R. W. Cohen and B. Abeles, *Phys. Rev.* **168**, 444 (1968).

[238] A. Leger and J. Klein, *Phys. Letts.* **28A**, 751 (1969).

[239] J. Klein and A. Leger, *Phys. Letts.* **28A**, 134 (1968).

[240] N. V. Zavaritskii, *JETP Letts.* **5**, 352 (1967).

[241] K. Knorr and N. Barth, *Solid State Commun.* **6**, 791 (1968).

[242] W. L. McMillan, *Phys. Rev.* **167**, 331 (1968).

[243] R. C. Dynes, *Phys. Rev.* **B2**, 644 (1970).

[244] J. W. Garland and P. B. Allen, PCSS, p. 669.

[245] N. B. Brandt and N. I. Ginzburg, *Sov. Phys. Uspekhi* **85**, 3 (1965).

[246] N. V. Zavaritskii, E. S. Itskevich and A. I. Voronovskii, LT11, p. 725.

[247] N. V. Zavaritskii, E. S. Itskevich and A. N. Voronovskii, *JETP Letts.* **7**, 211 (1968).

[248] T. F. Smith and C. W. Chu, *Phys. Rev.* **159**, 353 (1967).

[249] A. A. Galkin and V. M. Svistunov, *Phys. Stat. Sol.* **26**, K55 (1968).

[250] B. T. Gelikman and V. Kresin, *Sov. Phys. Solid State* **7**, 2659 (1966).

[251] J. P. Franck and W. J. Keeler, *Phys. Letts.* **25A**, 624 (1967).

[252] J. P. Franck, W. J. Keeler and T. M. Wu, *Solid State Commun.* **7**, 483 (1969).

[253] H. Meissner, *Phys. Rev.* **109**, 686 (1958).
[254] H. Meissner, *Phys. Rev.* **117**, 672 (1960).
[255] P. Fulde and K. Maki, *Phys. Rev. Letts.* **15**, 675 (1965).
[256] P. G. de Gennes and S. Mauro, *Solid State Commun.* **3**, 381 (1965).
[257] W. L. McMillan, *Phys. Rev.* **175**, 537 (1968).
[258] J. Clarke, *J. Phys.* supplement au no. 2–3, Tome 29, C2 (1968).
[259] T. Claeson and S. Gygax, *Solid State Commun.* **4**, 385 (1966).
[260] J. J. Hauser, *Physics* **2**, 247 (1966).
[261] J. J. Hauser, *Phys. Rev.* **164**, 558 (1967).
[262] C. J. Adkins and B. W. Kington, *Phys. Rev.* **177**, 777 (1969).
[263] S. M. Freake and C. J. Adkins, *Phys. Letts.* **29A**, 382 (1969).
[264] J. Vrba and S. B. Woods, LT12, p. 419.
[265] J. Vrba and S. B. Woods, *Phys. Rev.* **B3**, 2243 (1971).
[266] E. Guyon, A. Martinet, S. Mauro and F. Meunier, *Phys. Kondens. Mat.* **5**, 123 (1966).
[267] T. Claeson, S. Gygax and K. Maki, *Phys. Kondens. Mat.* **6**, 23 (1967).
[268] S. M. Marcus, *Phys. Letts.* **20**, 467 (1966).
[269] J. P. Burger, G. Deutscher, E. Guyon and A. Martinet, *Phys. Letts.* **17**, 180 (1965).
[270] P. G. de Gennes and J. P. Hurault, *Phys. Letts.* **17**, 181 (1965).
[271] J. L. Levine and S. Y. Hsieh, *Phys. Rev. Letts.* **20**, 994 (1968).
[272] J. R. Schrieffer and D. M. Ginsberg, *Phys. Rev. Letts.* **8**, 207 (1962).
[273] K. E. Gray, A. R. Long and C. J. Adkins, *Phil. Mag.* **20**, 273 (1969).
[274] A. Rothwarf and B. N. Taylor, *Phys. Rev. Letts.* **19**, 27 (1967).
[275] K. E. Gray, *Phil. Mag.* **20**, 267 (1969).
[276] S. Y. Hsieh and J. L. Levine, *Phys. Rev. Letts.* **20**, 1502 (1968).
[277] E. Guyon, *Advan. Phys.* **15**, 41 (1966).
[278] K. Maki, *Ann. Physik* **34**, 363 (1965).
[279] Yu. V. Sharvin, *JETP Letts.* **2**, 183 (1965).
[280] Yu. V. Sharvin, LT10, S172, p. 323.
[281] B. S. Chandrashekar, D. E. Farrel and S. Huang, *Phys. Rev. Letts.* **18**, 43 (1967).
[282] O. Iwanyshyn, J. D. Leslie and H. J. T. Smith, *Can. J. Phys.* **48**, 470 (1970).
[283] R. C. Jaklevic and J. Lambe, *Phys. Rev. Letts.* **17**, 1139 (1966).
[284] J. Lambe and R. C. Jaklevic, TPS, p. 233 and p. 243.
[285] J. Lambe and R. C. Jaklevic, *Phys. Rev.* **165**, 821 (1968).
[286] J. M. Rowell and W. L. McMillan, *Bull. Am. Phys. Soc.* **12**, 77 (1967).
[287] J. M. Rowell, in 'Superconductivity', The McGill University Advanced Summer Study Institute, Montreal, Canada, 17–29 June 1968, Gordon and Breach, 1969, p. 371.
[288] I. Giaever and H. R. Zeller, *Phys. Rev. Letts.* **20**, 1504 (1968).
[289] H. R. Zeller and I. Giaever, *Phys. Rev.* **181**, 789 (1969).
[290] P. W. Anderson, *J. Phys. Chem. Solids* **11**, 26 (1959).
[291] D. Markowitz, *Physics* **3**, 199 (1967).
[292] C. B. Duke, TPS, p. 405.
[293] C. B. Duke, 'Tunneling in Solids', Academic Press, New York, 1969.
[294] I. Giaever and K. Megerle, *IRE Trans.* **ED-9**, 459 (1962).
[295] J. D. Penney, High Frequency Tunnel Diode Amplifiers, Ph.D. Thesis, University of London, July 1968.
[296] A. C. Scott, *Solid State Electron.* **7**, 137 (1964).
[297] I. Simon, *Phys. Rev.* **77**, 384 (1950).
[298] H. D. Crane, *Proc. IRE* **50**, 2048 (1962).
[299] I. Giaever, 1961 International Solid-State Circuits Conference – Digest of Technical Papers, Lewis Winner, New York, p. 14.

[300] T. Ogushi, T. Goto, T. Mitsuoka, Y. Onodera, J. Oizumi, Y. Muto, T. Fukuroi and Y. Shibuya, *Rep. Res. Inst. Elect. Commun.* **14**, 113 (1963).

[301] E. Burstein, D. N. Langenberg and B. N. Taylor, *Phys. Rev. Letts.* **6**, 92 (1961).

[302] G. S. Picus, E. Burstein and S. Jacobs, Zinc doped germanium as a far infrared detector, Conference of the International Commission for Optics, Stockholm, 1959.

[303] D. C. Mattis and J. Bardeen, *Phys. Rev.* **111**, 412 (1958).

[304] W. D. Gregory, L. Leopold and D. Repici, *Can. J. Phys.* **47**, 1171 (1969).

[305] L. Leopold, W. D. Gregory and J. Bostock, *Can. J. Phys.* **47**, 1167 (1969).

[306] G. J. Lasher, *IBM J. Res. Develop.* **7**, 58 (1963).

[307] L. Leopold, W. D. Gregory, D. Repici and R. F. Averill, *Phys. Letts.* **30A**, 507 (1969).

[308] W. D. Gregory, L. Leopold, J. Bostock, R. F. Averill and D. Repici, LT12.

[309] Y. Goldstein, B. Abeles and K. R. Keller, *Elect. Letters* **1**, 97 (1965).

[310] J. P. Schulz and O. Weis, *Phys. Letts.* **32A**, 381 (1970).

[311] H. Kinder, K. Laszmann and W. Eisenmenger, *Phys. Letts.* **31A**, 475 (1970).

[312] L. Tewordt, *Phys. Rev.* **127**, 371 (1962).

[313] L. Tewordt, *Phys. Rev.* **128**, 12 (1962).

[314] D. B. Sullivan and C. E. Roos, *Phys. Rev. Letts.* **1f**, 212 (1967).

[315] J. W. Bakker, H. Van Kempen and P. Wyder, *Phys. Letts.* **31A**, 290 (1970).

[316] G. B. Donaldson and W. T. Band, Meeting of the International Institute of Refrigeration, Tokyo, September 1970.

[317] I. Dietrich, *Z. Phys.* **133**, 499 (1952).

[318] G. H. Hardy, 'A Mathematician's Apology', Cambridge University Press (2nd edition), 1967, p. 89.

[319] J. Bardeen, *Phys. Rev. Letts.* **9**, 147 (1962).

[320] P. W. Anderson and J. M. Rowell, *Phys. Rev. Letts.* **10**, 230 (1963).

[321] J. M. Rowell, *Phys. Rev. Letts.* **11**, 200 (1963).

[322] S. Shapiro, *Phys. Rev. Letts.* **11**, 80 (1963).

[323] M. D. Fiske, *Rev. Mod. Phys.* **36**, 221 (1964).

[324] R. C. Jaklevic, J. Lambe, A. H. Silver and J. E. Mercereau, *Phys. Rev. Letts.* **12**, 274 (1964).

[325] I. K. Yanson, V. M. Svistunov and I. M. Dmitrenko, *JETP* **21**, 650 (1965).

[326] S. Shapiro, *J. Appl. Phys.* **38**, 1879 (1967).

[327] C. C. Grimes and S. Shapiro, *Phys. Rev.* **169**, 397 (1968).

[328] D. N. Langenberg, W. H. Parker and B. N. Taylor, *Phys. Rev.* **150**, 186 (1966).

[329] P. W. Anderson, in 'Lectures on the Many-Body Problem', Academic Press, New York (E. R. Caianello, Ed.), 1964, p. 113.

[330] R. A. Ferrel and R. E. Prange, *Phys. Rev. Letts.* **10**, 479 (1963).

[331] P. G. de Gennes, *Phys. Letts.* **5**, 22 (1963).

[332] V. Ambegaokar and A. Baratoff, *Phys. Rev. Letts.* **10**, 486 (1963); Erratum, *Phys. Rev. Letts.* **11**, 104 (1963).

[333] C. S. Owen and D. J. Scalapino, *Phys. Rev.* **164**, 538 (1967).

[334] P. Lebwohl and M. J. Stephen, *Phys. Rev.* **163**, 376 (1967).

[335] D. E. McCumber, *J. Appl. Phys.* **39**, 3113 (1968).

[336] W. C. Stewart, *Appl. Phys. Letts.* **12**, 277 (1968).

[337] W. C. Scott, *Appl. Phys. Letts.* **17**, 166 (1970).

[338] J. Clarke, *Phil. Mag.* **13**, 115 (1966).

[339] J. E. Zimmerman and A. H. Silver, *J. Appl. Phys.* **39**, 2679 (1968).

[340] J. Clarke, LT10, p. 211.

[341] J. E. Zimmerman and A. H. Silver, *Phys. Rev.* **141**, 367 (1966).

[342] J. E. Mercereau, PSPSD, U1.

[343] M. Nisenoff, *Rev. Phys. Appl.* **5**, 21 (1970).

[344] J. Matisoo, *Appl. Phys. Letts.* **9**, 167 (1966).
[345] C. C. Grimes, P. L. Richards and S. Shapiro, *Phys. Rev. Letts.* **17**, 431 (1966).
[346] J. M. Rowell, U.S. Patent No. 3,281,602, 25 October 1966.
[347] R. C. Jaklevic, J. Lambe, A. H. Silver and J. E. Mercereau, *Phys. Rev. Letts.* **12**, 274 (1964).
[348] R. C. Jaklevic, J. Lambe, J. E. Mercereau and A. H. Silver, *Phys. Rev.* **140**, A1628 (1965).
[349] P. W. Anderson, in 'Progress in Low Temperature Physics', VI, North Holland, Amsterdam, 1967 (C. J. Gorter, Ed.), p. 1.
[350] J. E. Mercereau, *J. Appl. Phys.* **40**, 1994 (1966) (abstract only).
[351] M. R. Beasley and W. W. Webb, PSPSD, V1.
[352] D. G. McDonald, V. E. Kose, K. M. Evenson, J. S. Wells and J. D. Cupp, *Appl. Phys. Letts.* **15**, 121 (1969).
[353] T. F. Finnegan, A. Denenstein and D. N. Langenberg, *Phys. Rev. Letts.* **24**, 738 (1970).
[354] C. C. Grimes, P. L. Richards and S. Shapiro, *J. Appl. Phys.* **39**, 3905 (1968).
[355] P. L. Richards and S. A. Sterling, *Appl. Phys. Letts.* **14**, 394 (1969).
[356] W. H. Parker, B. N. Taylor and D. N. Langenberg, *Phys. Rev. Letts.* **18**, 287 (1967).
[357] B. D. Josephson, *Rev. Mod. Phys.* **36**, 216 (1964).
[358] W. A. Little and R. D. Parks, *Phys. Rev. Letts.* **9**, 9 (1962).
[359] R. D. Parks, J. M. Mochel and L. V. Surgent, *Phys. Rev. Letts.* **13**, 331 (1964).
[360] R. D. Parks and J. M. Mochel, *Rev. Mod. Phys.* **36**, 284 (1964).
[361] P. W. Anderson and A. H. Dayem, *Phys. Rev. Letts.* **13**, 195 (1964).
[362] J. Clarke, *Proc. Roy. Soc.* **A308**, 447 (1969).
[363] H. A. Notarys and J. E. Mercereau, PCSS, p. 424.
[364] J. E. Mercereau, *Rev. Phys. Appl.* **5**, 13 (1970).
[365] J. E. Zimmerman, *J. Appl. Phys.*
[366] B. D. Josephson, *Advan. Phys.* **14**, 419 (1965).
[367] P. G. de Gennes, 'Superconductivity of Metals and Alloys', W. A. Benjamin, New York, 1966.
[368] B. D. Josephson, in 'Superconductivity', (R. D. Parks, Ed.), Marcel Dekker Inc., New York, 1969, p. 423.
[369] D. J. Scalapino, TPS, p. 477.
[370] A. Th. A. M. de Waele and R. de Bruyn Ouboter, *Physica* **41**, 225 (1969).
[371] A. Zawadowski, *Elektrotech. Casopis* **18**, 528 (1967).
[372] D. A. Jacobson, *Phys. Rev.* **138**, A1066 (1965).
[373] A. Baratoff, CFS, p. 287.
[374] A. Baratoff, J. A. Blackburn and B. B. Schwartz, *Phys. Rev. Letts.* **25**, 1096 (1970).
[375] K. Yamafuji, T. Ezaki, T. Matsushita and F. Irie, LT12, p. 437.
[376] T. A. Fulton, *Solid State Commun.* **8**, 1353 (1970).
[377] T. A. Fulton and R. C. Dynes, *Phys. Rev. Letts.* **25**, 794 (1970).
[378] T. A. Fulton and D. E. McCumber, *Phys. Rev.* **175**, 585 (1968).
[379] D. N. Langenberg, D. J. Scalapino and B. N. Taylor, *Proc. IEEE* **54**, 560 (1966).
[380] I. Giaever, *Phys. Rev. Letts.* **14**, 904 (1965).
[381] I. S. Gradshteyn and I. M. Ryzhik, 'Table of Integrals, Series and Products', Academic Press, New York, 1965.
[382] S. Shapiro, A. R. Janus and S. Holly, *Rev. Mod. Phys.* **36**, 223 (1964).
[383] A. H. Dayem and J. J. Wiegand, *Phys. Rev.* **155**, 419 (1967).
[384] A. Longacre and S. Shapiro,
[385] S. Shapiro, *Phys. Letts.* **25A**, 537 (1967).
[386] A. J. Dahm, A. Denenstein, T. F. Finnegan, D. N. Langenberg and D. J. Scalapino, *Phys. Rev. Letts.* **20**, 859 (1968).

[387] A. J. Dahm, A. Denenstein, T. F. Finnegan, D. N. Langenberg and D. J. Scalapino, LT11, p. 709

[388] A. J. Dahm, PCFS, p. 137.

[389] L. G. Aslamazov, A. I. Larkin and Yu. N. Ovchinnikov, *JETP* **28**, 171 (1969).

[390] D. E. McCumber, Private communications.

[391] J. J. Pankove, *Phys. Letts.* **21**, 406 (1966).

[392] S. A. Buckner, J. T. Chen and D. N. Langenberg, *Phys. Rev. Letts.* **25**, 738 (1970).

[393] S. A. Buckner, J. T. Chen and D. N. Langenberg, LT12, p. 453.

[394] Yu. M. Ivanchenko and L. A. Zilberman, *JETP Letts.* **8**, 113 (1968).

[395] Yu. M. Ivanchenko and L. A. Zilberman, *JETP* **28**, 1272 (1969).

[396] J. Clarke, A. B. Pippard and J. R. Waldram, PCSS, p. 405.

[397] J. R. Waldram, A. B. Pippard and J. Clarke, *Phil. Trans. A*, 268, 265 (1970).

[398] A. H. Silver, R. C. Jaklevic and J. Lambe, *Phys. Rev.* **141**, 362 (1966).

[399] R. E. Eck, D. J. Scalapino and B. N. Taylor, *Phys. Rev. Letts.* **13**, 15 (1964).

[400] R. E. Eck, D. J. Scalapino and B. N. Taylor, LT9, p. 415.

[401] I. O. Kulik, *JETP Letts.* **2**, 84 (1965).

[402] I. M. Dmitrenko, I. K. Yanson and V. M. Svistunov, *JETP Letts.* **2**, 10 (1965).

[403] A. H. Dayem and C. C. Grimes, *Appl. Phys. Letts.* **9**, 47 (1966).

[404] J. Matisoo, *Phys. Letts.* **29A**, 473 (1969).

[405] See for example L. M. Milne-Thomson, 'Jacobian Elliptic Function Tables', Dover Publications Inc., 1950.

[406] A. M. Goldman and P. J. Kreisman, *Phys. Rev.* **164**, 544 (1967).

[407] J. Matisoo, *J. Appl. Phys.* **40**, 1813 (1969).

[408] See for example P. M. Morse and H. Feshbach, 'Methods of Theoretical Physics', McGraw Hill, New York, 1953, p. 277.

[409] C. P. Bean and J. D. Livingston, *Phys. Rev. Letts.* **12**, 14 (1964).

[410] H. Suhl, *Phys. Rev. Letts.* **14**, 226 (1965).

[411] Y. B. Kim, C. F. Hempstead and A. R. Strnad, *Phys. Rev.* **139**, A1163 (1965).

[412] A. Seeger, H. Dorth and A. Kochendorfer, *Z. Phys.* **134**, 173 (1953).

[413] E. T. Whittaker and G. N. Watson, 'Modern Analysis', Cambridge University Press, 1940.

[414] R. de Bruyn Ouboter and A. Th. A. M. de Waele, in 'Progress in Low Temperature Physics', VI, North Holland, Amsterdam, 1970, p. 243.

[415] A. Th. A. M. de Waele, W. H. Kraan, R. de Bruyn Ouboter and K. W. Taconis, *Physica* 37, 114 (1967).

[416] I. M. Dmitrenko and S. I. Bondarenko, *JETP Letts.* **7**, 241 (1968).

[417] R. C. Jaklevic, J. Lambe, A. H. Silver and J. E. Mercereau, LT9, p. 446.

[418] R. A. Ferrel, *Phys. Rev. Letts.* **15**, 527 (1965).

[419] R. de Bruyn Ouboter, M. H. Omar, A. J. P. T. Arnold, T. Guinau and K. W. Taconis, *Physica* 32, 1448 (1966).

[420] M. H. Omar and R. de Bruyn Ouboter, *Physica* 32, 2044 (1966).

[421] R. de Bruyn Ouboter, M. H. Omar, A. J. P. T. Arnold, T. Guinau and K. W. Taconis, LT10, 536, p. 246.

[422] R. de Bruyn Ouboter, W. H. Kraan, A. Th. A. M. de Waele and M. H. Omar, *Physica* 35, 335 (1967).

[423] A. Th. A. M. de Waele and R. de Bruyn Ouboter, *Physica* 42, 626 (1969).

[424] A. Th. A. M. de Waele, W. H. Kraan and R. de Bruyn Ouboter, *Physica* 40, 302 (1968).

[425] G. R. S. Seraphim, Weak Links between Superconductors. D.Phil. thesis, Dept. of Engineering Science, Oxford University, 1970.

[426] L. L. Vant-Hull and J. E. Mercereau, *Phys. Rev. Letts.* **17**, 629 (1966).

[427] L. Vant-Hull, PCFS, p. 47.
[428] T. I. Smith, *Phys. Rev. Letts.* **15**, 460 (1965).
[429] J. E. Zimmerman, Proc. 6th Biennial Gas Dynamics Symposium, Northwestern University Press, Evanston, Ill., 1967, p. 257.
[430] J. E. Zimmerman and A. H. Silver, LT10, S33, p. 233.
[431] J. E. Zimmerman, J. A. Cowen and A. H. Silver, *Appl. Phys. Letts.* **9**, 353 (1966).
[432] A. H. Silver, PSPSD, p. F1.
[433] J. E. Zimmerman and A. H. Silver, *Solid State Commun.* **4**, 133 (1966).
[434] A. H. Silver, *Journ. IEEE* **QE-4**, 738 (1968).
[435] J. E. Zimmerman, P. Thiene and J. T. Harding, *J. Appl. Phys.* **41**, 1572 (1970).
[436] J. E. Zimmerman, *J. Appl. Phys.* **41**, 1589 (1970).
[437] A. H. Silver and H. E. Zimmerman, *Phys. Rev. Letts.* **15**, 888 (1965).
[438] A. H. Silver and J. E. Zimmerman, *Phys. Rev.* **157**, 317 (1967).
[439] D. B. Sullivan, R. L. Peterson, V. E. Kose and J. E. Zimmerman, *J. Appl. Phys.* **41**, 4865 (1970).
[440] J. E. Zimmerman and A. H. Silver, *Phys. Rev.* **167**, 418 (1968).
[441] D. N. Langenberg, D. J. Scalapino and B. N. Taylor, *Scientific American* **251**, 30 (1966).
[442] J. Matisoo, *Proc. IEEE* **55**, 172 (1967).
[443] J. P. Pritchard, Jr., J. T. Pierce and B. G. Slay, *Proc. IEEE* **52**, 1207 (1964).
[444] J. T. Pierce and J. P. Pritchard, Jr., *IEEE Trans.* **CP-12**, 8 (1965).
[445] J. P. Pritchard, Jr., J. T. Pierce and B. G. Slay, *Trans. Met. Soc. AIME* **236**, 359 (1966).
[446] J. P. Pritchard, Jr., *IEEE Spectrum* **3**, 46 (1966).
[447] J. E. Nordman, *J. Appl. Phys.* **40**, 2111 (1969).
[448] L. O. Mullen and D. B. Sullivan, *J. Appl. Phys.* **40**, 2115 (1969).
[449] R. A. Kamper, L. O. Mullen and D. B. Sullivan, Report NBS-TN-381, Nat. Bur. Stand., Washington D.C., USA (Oct. 1969).
[450] S. von Molnar, W. A. Thompson and A. S. Edelstein, *Appl. Phys. Letts.* **11**, 163 (1967).
[451] J. Lambe, A. H. Silver, J. E. Mercereau and R. C. Jaklevic, *Phys. Letts.* **11**, 6 (1964).
[452] A. F. G. Wyatt, V. M. Dmitriev, W. S. Moore and F. W. Sheard, *Phys. Rev. Letts.* **16**, 1166 (1966).
[453] A. N. Broers, Private communications.
[454] S. I. Bondarenko, E. I. Balanov, L. E. Kolinko and T. P. Narbut, PTE no. 1, 1970.
[455] S. I. Bondarenko, E. I. Balanov, L. E. Kolinko and T. P. Narbut, *Cryogenics* **9**, 508 (1970).
[456] I. M. Dmitrenko, S. I. Bondarenko and T. P. Narbut, *JETP* **30**, 817 (1970).
[457] D. A. Buck, *Proc. IRE* **44**, 482 (1956).
[458] P. A. Walker, in 'A Guide to Superconductivity', MacDonald, London, 1969, (D. Fishlock, Ed.), p. 111.
[459] A. E. Brennemann, *Proc. IEEE* **51**, 442 (1963).
[460] J. Matisoo, *Proc. IEEE* **55**, 2052 (1967).
[461] J. Matisoo, *Trans. IEEE* **MAG-4**, 324 (1968).
[462] J. Matisoo, PSPSD, p. N1.
[463] J. P. Pritchard and W. H. Schroen, *Trans. IEEE* **MAG-4**, 320 (1968).
[464] G. R. S. Seraphim and L. Solymar, *Electr. Letters* **6**, 289 (1970).
[465] W. Anacker, *Trans. IEEE* **MAG-5**, 968 (1969).
[466] T. D. Clark and J. P. Baldwin, *Electr. Letts.* **3**, 178 (1967).
[467] N. D. Richards, *Trans. IEEE* **MAG-2**, 394 (1966).

[468] D. J. Dumin and J. F. Gibbons, *J. Appl. Phys.* **34**, 1566 (1963).

[469] W. C. Stewart, *Appl. Phys. Letts.* **14**, 392 (1969).

[470] I. M. Dmitrenko and I. K. Yanson, *JETP Letts.* **2**, 154 (1965).

[471] I. M. Dmitrenko and I. K. Yanson, LT10, S32, p. 228.

[472] D. N. Langenberg, D. J. Scalapino, B. N. Taylor and R. E. Eck, *Phys. Rev. Letts.* **15**, 294 (1965).

[473] B. N. Taylor, LT10, p. 59.

[474] T. D. Clark, *Phys. Letts.* **27A**, 585 (1968).

[475] I. Ya. Krasnopolin and M. S. Khaikin, *JETP Letts.* **6**, 129 (1967).

[476] A. H. Silver and J. E. Zimmerman, *Phys. Rev.* **158**, 423 (1967).

[477] W. H. Higa, *Proc. Conf. El. Dev.* Montreal, 1967.

[478] G. K. Gaule, R. L. Ross and K. Schwidtal, PSPSD, p. P1.

[479] F. Rothwarf, H. M. Krisch and D. Ford, *J. Appl. Phys.* **39**, 2683 (1968).

[480] G. R. S. Seraphim and R. C. McDermott, *Phys. Letts.* **32A**, 35 (1970).

[481] A. H. Silver and J. E. Zimmerman, *Appl. Phys. Letts.* **10**, 142 (1967).

[482] S. Shapiro, PSPSD, p. I1.

[483] S. Shapiro, Spring Superconducting Symposia, NRL Report 7023, 1969, p. 15.

[484] P. L. Richards, *J. Opt. Soc. Am.* **54**, 1474 (1964).

[485] N. R. Werthamer and S. Shapiro, *Phys. Rev.* **164**, 523 (1967).

[486] R. Y. Chiao, *Phys. Letts.* **33A**, 177 (1970); LT12, p. 865.

[487] J. Clarke, PSPSD, p. D1.

[488] J. Clarke, *Rev. Phys. Appl.* **5**, 32 (1970).

[489] A. H. Silver, J. E. Zimmerman and R. A. Kamper, *Appl. Phys. Letts.* **11**, 20 9 (1967).

[490] W. S. Goree, Spring Superconducting Symposia, NRL Report 7023, 1970.

[491] W. S. Goree, *Rev. Phys. Appl.* **5**, 3 (1970).

[492] R. L. Forgacs and A. Warnick, *Rev. Sci. Instrum.* **38**, 214 (1967).

[493] R. L. Forgacs and A. Warnick, *Trans. IEEE* **IM-15**, 113 (1966).

[494] J. W. McWane, J. E. Neighbor and R. S. Newbower, *Rev. Sci. Instrum.* **37**, 160 2 (1966).

[495] D. A. Zych, *Rev. Sci. Instrum.* **39**, 1058 (1968).

[496] J. Clarke, W. E. Tennant and D. Woody, *J. Appl. Phys.* **42**, 5194 (1971).

[497] J. E. Zimmerman, Tech. Report No. U4604, Aeronautronic Division, Philco-Ford Corp., Newport Beach, California, 1969.

[498] D. N. Langenberg, D. J. Scalapino, B. N. Taylor and R. E. Eck, *Phys. Letts.* **20**, 563 (1966).

[499] D. N. Langenberg, W. H. Parker and B. N. Taylor, *Phys. Letts.* **22**, 259 (1966).

[500] W. H. Parker, D. N. Langenberg, A. Denenstein and B. N. Taylor, *Phys. Rev.* **177**, 639 (1969).

[501] B. W. Petley and K. Morris, *Phys. Letts.* **29A**, 289 (1969).

[502] B. W. Petley and K. Morris, *Metrologia* **6**, 46 (1970).

[503] A. Denenstein, T. F. Finnegan, D. N. Langenberg, W. H. Parker and B. N. Taylor, *Phys. Rev.* **B1**, 4500 (1970).

[504] I. K. Harvey, J. C. MacFarlane and R. B. Frenkel, *Phys. Rev. Letts.* **25**, 853 (1970).

[505] J. Clarke, LT11, p. 95.

[506] J. Clarke, *Phys. Rev. Letts.* **21**, 1566 (1968).

[507] T. F. Finnegan, A. Denestein, D. N. Langenberg, J. C. McMenamin, D. E. Novoseller and L. Cheng, *Phys. Rev. Letts.* **23**, 229 (1969).

[508] M. O. Scully and P. A. Lee, *Phys. Rev. Letts.* **22**, 23 (1969).

[509] M. J. Stephen, *Phys. Rev. Letts.* **21**, 1629 (1968).

[510] M. J. Stephen, *Phys. Rev.* **182**, 531 (1969).

[511] D. E. McCumber, *Phys. Rev. Letts.* **23**, 1228 (1969); PCSS, p. 421.

[512] F. Bloch, *Phys. Rev. Letts.* **21**, 1241 (1968).

[513] F. Bloch, *Phys. Rev.* **B2**, 109 (1970).
[514] M. J. Stephen, *Phys. Rev.* **186**, 393 (1969).
[515] V. E. Kose and D. B. Sullivan, *J. Appl. Phys.* **41**, 169 (1970).
[516] J. L. Stewart, *Proc. IRE* **42**, 1539 (1954).
[517] W. H. Parker, PCFS, p. 121.
[518] B. N. Taylor, W. H. Parker and D. N. Langenberg, *Rev. Mod. Phys.* **41**, 375 (1969).
[519] B. W. Petley and K. Morris, *J. Sci. Instrum. Series* 2, **2**, 649 (1969).
[520] E. R. Cohen and J. W. M. Dumond, *Rev. Mod. Phys.* **37**, 537 (1965).
[521] B. N. Taylor, W. H. Parker, D. N. Langenberg and A. D. Denenstein, *Metrologia* 3, 89 (1967).
[522] R. E. Burgess, PSPSD, p. H1.
[523] R. E. Burgess, PCFS, p. 47.
[524] V. Ambegaokar and B. I. Halperin, *Phys. Rev. Letts.* **22**, 1364 (1969).
[525] J. T. Anderson and A. M. Goldman, *Phys. Rev. Letts.* **23**, 128 (1969); PCSS, p. 256.
[526] M. Simmonds and W. H. Parker, *Phys. Rev. Letts.* **24**, 876 (1970).
[527] H. Kanter and F. L. Vernon, Jr., *Phys. Letts.* **32A**, 155 (1970).
[528] J. Kurkijärvi and V. Ambegaokar, *Phys. Letts.* **31A**, 314 (1970).
[529] B. T. Ulrich, *Phys. Rev. Letts.* **20**, 381 (1968).
[530] R. A. Ferrel, *J. Low Temp. Phys.* **1**, 423 (1969); PCSS, p. 265.
[531] Y. Yoshihiro and K. Kajimura, *Phys. Letts.* **32A**, 71 (1970).
[532] Y. Yoshihiro and K. Kajimura, LT12, p. 269.
[533] D. J. Scalapino, PSPSD, p. G1.
[534] A. J. Dahm, A. Denenstein, D. N. Langenberg, W. H. Parker, D. Rogovin and D. J. Scalapino, *Phys. Rev. Let.* **22**, 1416 (1969).
[535] H. Kanter and F. L. Vernon, Jr., *Appl. Phys. Letts.* **16**, 115 (1970).
[536] H. Kanter and F. L. Vernon, Jr., *Phys. Rev. Letts.* **25**, 588 (1970); LT12, p. 455.
[537] R. A. Kamper, PSPSD, p. M1.
[538] R. A. Kamper and J. E. Zimmerman, *J. Appl. Phys.* **42**, 132 (1971).
[539] P. G. de Gennes and E. Guyon, *Phys. Letts.* **3**, 168 (1963).
[540] P. G. de Gennes, *Rev. Mod. Phys.* **36**, 225 (1964).
[541] S. Kobayashi, M. Sato and W. Sasaki, LT12.
[542] S. I. Bondarenko, I. M. Dmitrenko and E. I. Balanov, *Sov. Phys. Solid State* **12**, 1113 (1970).
[543] M. Mitani, K. Aihara and N. Hara, LT12, p. 461.
[544] M. Mitani, K. Aihara and N. Hara, *Appl. Phys. Letts.* **18**, 489 (1971).
[545] R. H. T. Yeh and H. Mechetti, *Phys. Stat. Sol.* **25**, K65 (1968).
[545a] R. H. T. Yeh and H. Mechetti, *Phys. Stat. Sol.* **28**, 683 (1968).
[545b] L. G. Aslamazov, A. I. Larkin and Yu. N. Ovchinnikov, *JETP* **28**, 171 (1969).
[545c] V. P. Galaiko, *JETP* **30**, 514 (1970).
[545d] G. A. Gogadze and I. O. Kulik, *JETP* **33**, 984 (1971).
[545e] C. Ishii, *Progr. Theor. Phys.* (Japan) **44**, 1525 (1970).
[545f] C. Ishii, LT12, p. 435.
[545g] A. V. Svidzinskii, T. N. Antsigina and E. N. Bratus, *JETP*
[546] E. E. Shin and B. B. Schwartz, *Phys. Rev.* **152**, 207 (1966).
[547] A. C. Scott, *Am. J. Phys.* **37**, 52 (1969).
[548] P. W. Anderson, *Rev. Mod. Phys.* **38**, 298 (1966).
[549] P. L. Richards and P. W. Anderson, *Phys. Rev. Letts.* **14**, 540 (1965).
[550] B. M. Khorana and B. S. Chandrashekar, *Phys. Rev. Letts.* **18**, 230 (1967).
[551] B. M. Khorana, *Phys. Rev.* **185**, 299 (1969).
[552] B. M. Khorana and D. H. Douglass, Jr., LT11, p. 169.
[553] B. M. Khorana, LT12, p. 85.

[554] J. P. Hulin, B. Perrin, C. Laroche and A. Libchaber, LT12, p. 83.

[555] Yu. G. Mamaladze and O. D. Cheishvili, *JETP Letts.* **2**, 76 (1965).

[556] O. D. Cheishvili and Yu. G. Mamaladze, *Phys. Letts.* **18**, 278 (1965).

[557] Yu. G. Mamaladze and O. D. Cheishvili, *JETP* **23**, 112 (1966).

[558] Yu. G. Mamaladze and O. D. Cheishvili, *JETP* **25**, 117 (1967).

[559] V. L. Ginzburg and L. P. Pitaevskii, *JETP* **7**, 858 (1958).

[560] A. F. G. Wyatt, V. M. Dmitriev and W. S. Moore, LT10, S35, p. 242.

[560a] A. F. G. Wyatt and D. H. Evans, PCSS, p. 288.

[561] P. E. Gregers-Hansen and M. T. Levinsen, *Phys. Rev. Letts.* **27**, 847 (1971).

[562] V. M. Dmitriev, E. V. Khristenko, A. V. Trubitsyn and F. F. Mende, *Ukr. Fiz. Zh.* **15**, 1611 (1970).

[563] P. E. Gregers-Hansen and M. T. Levinsen, *Solid State Commun.* **7**, 1215 (1969).

[564] I. K. Yanson, *JETP* **31**, 800 (1970).

[565] T. K. Hunt and J. E. Mercereau, *Phys. Rev. Letts.* **18**, 551 (1967).

[566] G. M. Eliashberg, *JETP Letts.* **11**, 114 (1970).

[567] J. Baizeras and T. Pech, LT12, p. 409.

[568] J. Baizeras and T. Pech, *Solid State Commun.* **8**, 1055 (1970).

[569] J. Matisoo, *J. Appl. Phys.* **40**, 2091 (1969).

[570] R. Meservey, LT9, p. 455.

[571] I. Weinberg, *J. Appl. Phys.* **38**, 3036 (1967).

[572] S. Greenspoon and H. J. T. Smith, *Can. J. Phys.* **49**, 1350 (1971).

[573] J. E. Zimmerman and J. E. Mercereau, *Phys. Rev. Letts.* **13**, 125 (1964).

[574] M. Hanabusa and A. H. Silver, *Rev. Sci. Instrum.* **41**, 1235 (1970).

[575] M. Hanabusa, A. H. Silver and T. Kushida, *Phys. Rev.* **B2**, 1293 (1970).

[576] J. P. Gollub, M. R. Beasley and M. Tinkham, *Phys. Rev. Letts.* **25**, 1646 (1970).

[577] L. Vant-Hull, R. A. Simpkins and J. T. Harding, *Phys. Letts.* **24A**, 736 (1967).

[578] J. T. Harding and J. E. Zimmerman, *Phys. Letts.* **27A**, 670 (1968).

[579] D. Cohen, E. A. Edelsack and J. E. Zimmerman, *Appl. Phys. Letts.* **16**, 278 (1970).

[580] J. E. Zimmerman and N. V. Frederick, *Appl. Phys. Letts.* **19**, 16 (1971).

[581] A. F. Hebard and W. M. Fairbank, LT12, p. 855.

[582] B. T. Ulrich, LT12, p. 464.

[583] P. Russer, *Arch. El. Ubertr.* **23**, 417 (1969).

[584] W. H. Louisell, 'Coupled Mode and Parametric Electronics', John Wiley and Sons, Inc., New York, 1960.

[585] P. Russer, *Proc. IEEE* **59**, 282 (1971).

[586] H. Zimmer, *Appl. Phys. Letts.* **10**, 193 (1967).

[586a] A. N. Vystavkin, V. N. Gubankov, G. F. Leschenko, K. K. Likharev, V. V. Migulin, *Radiotekhnika i Elektronika*, **15**, 2404 (1970).

[587] F. L. Vernon, Jr. and R. J. Pedersen, *J. Appl. Phys.* **39**, 2661 (1968).

[588] I. Giaever, TPS, p. 255.

[589] J. M. Rowell, TPS, p. 273.

[590] G. E. Everett, TPS, p. 353.

[591] W. Eisenmenger, TPS, p. 371.

[592] P. Fulde, TPS, p. 427.

[593] T. Claeson, TPS, p. 443.

[594] R. Meservey and B. B. Schwartz, in 'Superconductivity', (R. D. Parks, Ed.), Marcel Dekker Inc., New York, 1969, p. 117.

[595] M. D. Fiske and I. Giaever, *Proc. IEEE* **52**, 1155 (1964).

[596] P. J. Stiles, L. Esaki and J. F. Schooley, *Phys. Letts.* **23**, 206 (1966).

[597] C. S. Koonce, *Phys. Rev.* **182**, 568 (1969).

[598] J. M. Rowell and A. G. Chynoweth, *Bull. Am. Phys. Soc.* **7**, 473 (1962).

[599] F. Gschwend, H. P. Kleinknecht, W. Neft and K. Seiler, *Z. Naturforsch.* **18a**, 1366 (1963).

[600] B. M. Vul, E. I. Zavaritskaya and N. V. Zavaritskii, *Sov. Phys. Solid State* **8**, 710 (1966).

[601] J. W. Conley and J. J. Tiemann, *J. Appl. Phys.* **38**, 2880 (1967).

[602] W. A. Thompson and S. von Molnar, *J. Appl. Phys.*, **41**, 5218 (1970).

[603] N. Tsuda, *Jap. J. Appl. Phys.* **8**, 582 (1969).

[604] N. Tsuda, *J. Phys. Soc. Japan* **27**, 874 (1969).

[605] D. C. Tsui, *J. Appl. Phys.* **41**, 2651 (1970).

[606] D. C. Tsui, *Phys. Rev. Letts.* **27**, 574 (1971).

[606a] P. Guetin and G. Schreder, *Solid State Commun.* **9**, 591 (1971).

[607] C. B. Duke and G. G. Kleiman, *Phys. Rev.* **B2**, 1270 (1970).

[608] N. Z. Heidam, *Phys. Letts.* **35A**, 378 (1971).

[609] C. A. Hamilton and S. Shapiro, *Phys. Letts.* **32A**, 223 (1970).

[610] B. Abeles and Y. Goldstein, *Phys. Letts.* **18**, 11 (1965).

[611] R. W. Cohen and B. Abeles, LT10, S165, p. 297.

[612] B. Abeles and Y. Goldstein, *Phys. Rev. Letts.* **14**, 595 (1965).

[613] F. L. Vernon and E. Lax, LT10, S166, 302.

[614] N. Z. Heidam, *Phys. St. Sol.* **A5**, 645 (1971).

[615] R. W. Cohen, *Phys. Letts.* **33A**, 271 (1970).

[616] B. Abeles, LT11, B1.2, 705.

[617] Y. Goldstein, M. Cohen and B. Abeles, *Phys. Rev. Letts.* **25**, 1571 (1970).

[618] M. Cohen, Y. Goldstein and B. Abeles, *Phys. Rev.* **B3**, 2223 (1971).

[619] G. I. Urushadze, *Sov. Phys. Solid State* **9**, 1208 (1967).

[619a] G. I. Urushadze, *Phys. Letts.* **27A**, 381 (1968).

[620] S. M. Marcus, *Phys. Letts.* **19**, 623 (1966).

[621] S. M. Marcus, *Phys. Letts.* **20**, 236 (1966).

[622] G. I. Rochlin and D. H. Douglass, *Phys. Rev. Letts.* **16**, 359 (1966).

[623] G. I. Rochlin, *Phys. Rev.* **153**, 513 (1967).

[624] A. A. Bright and T. R. Merril, *Phys. Rev.* **184**, 446 (1969).

[625] L. J. Barnes, *Phys. Rev.* **184**, 434 (1969).

[626] A. Zawadowsky, *Phys. Letts.* **23**, 225 (1966).

[627] W. J. Tomasch, *Phys. Letts.* **26A**, 379 (1968).

[628] T. Wolfram and G. W. Lehman, *Phys. Letts.* **21**, 631 (1966).

[629] T. Wolfram and G. W. Lehman, *Phys. Letts.* **24A**, 101 (1967).

[630] T. Wolfram and M. B. Einhorn, *Phys. Rev. Letts.* **17**, 966 (1966).

[631] T. Wolfram, *Phys. Rev.* **170**, 481 (1968).

[632] W. L. McMillan, *Phys. Rev. Letts.* **175**, 559 (1968).

[633] G. I. Lykken, A. L. Geiger, K. S. Dy and E. N. Mitchell, *Phys. Rev.* **B4**, 1523 (1971).

[634] A. J. Bennett, *Phys. Rev.* **140**, A1902 (1965).

[635] C. W. Smith and N. C. Miller, *Phys. Letts.* **34A**, 147 (1971).

[636] R. J. Pedersen and F. L. Vernon, *Appl. Phys. Letts.* **10**, 29 (1967).

[636a] R. D. Parmentier, *Solid State Electr.* **12**, 287 (1969).

[637] S. Shapiro and A. R. Janus, LT8, p. 321.

[638] S. Shapiro and E. J. McNiff, 'Superconductive effects in thin films, Technical Documentary Report No. AL-TDR-64-46', 1964 (AF Avionics Lab., Wright-Patterson Base, Ohio).

[639] R. C. Dynes, V. Narayanamurti and M. Chin, *Phys. Rev. Letts.* **26**, 181 (1971).

[640] J. P. Schulz and O. Weis, LT12, p. 433.

[641] C. J. Adkins and B. W. Kington, LT10, S24, p. 202.

[642] J. G. Adler, J. S. Rogers and S. B. Woods, *Can. J. Phys.* **43**, 557 (1965).

[643] J. G. Adler and J. S. Rogers, *Phys. Rev. Letts.* **10**, 217 (1963).

[644] B. L. Blackford and R. H. March, *Phys. Rev.* **186**, 397 (1969).

[645] D. Bonnet, S. Erlenkamper, H. Germer and H. Rabenhorst, *Phys. Letts.* **25A**, 452 (1967).

[646] J. P. Burger, G. Deutscher, E. Guyon and A. Martinet, *Phys. Rev.* **137**, A853 (1965).

[647] J. P. Burger, G. Deutscher, J. P. Hurault and A. Martinet, LT10, S20, p. 190.

[648] T. T. Chen, J. T. Chen, J. D. Leslie and H. J. T. Smith, *Phys. Rev. Letts.* **22**, 526 (1969).

[649] T. D. Clark, *J. Phys.* **C1**, 732 (1968).

[650] T. D. Clark, *Phys. Letts.* **27A**. 608 (1968).

[651] I. Dietrich, *Z. Naturforsch.* **17a**, 94 (1962).

[652] I. Dietrich, *Phys. Letts.* **9**, 221 (1964).

[653] I. Dietrich, LT8, p. 173.

[654] D. H. Douglass, Jr. and R. H. Meservey, LT8, p. 180.

[655] L. Dumoulin, E. Guyon and P. Nedellec, *Solid State Commun.* **8**, 885 (1970).

[656] R. C. Dynes, J. P. Carbotte, D. W. Taylor and C. K. Campbell, *Phys. Rev.* **178**, 713 (1969).

[657] A. S. Edelstein and A. M. Toxen, LT10, S158, p. 270.

[658] J. P. Franck and W. J. Keeler, *Phys. Rev.* **163**, 373 (1967).

[659] S. M. Freake, *Phil. Mag.* **24**, 319 (1971).

[660] A. A. Galkin, V. M. Svistunov, A. P. Dikii and V. N. Taranenko, *JETP* **32**, 44 (1971).

[661] I. Giaever, 'Proc. of the American Institute of Mining, Metallurgical and Petroleum Engineers', Wiley, New York.

[662] A. Gilabert, J. P. Romagnan and E. Guyon, *Solid State Commun.* **9**, 1295 (1971).

[663] Y. Goldstein, LT9, p. 400.

[664] Y. Goldstein, *Phys. Letts.* **12**, 169 (1964).

[665] Groupe de Supraconductivite d'Orsay, *Phys. Kondens. Mat.* **5**, 141 (1966).

[666] C. J. van Gurp and A. P. van Gelder, *Philips Res. Repts.* **19**, 400 (1964).

[667] E. Guyon, F. Meunier and R. S. Thompson, *Phys. Rev.* **156**, 452 (1967).

[668] J. W. Hafstrom and M. L. A. MacVicar, *Phys. Rev.* **B2**, 4511 (1970).

[669] J. J. Hauser, *Phys. Rev.* **B1**, 3624 (1970).

[670] J. J. Hauser, LT10, S23, p. 195.

[671] S. M. Khanna and S. B. Woods, *Phys. Letts.* **20**, 335 (1966).

[672] J. C. Keister, L. S. Straus and W. D. Gregory, *J. Appl. Phys.* **42**, 642 (1971).

[673] K. Knorr and N. Barth, *J. Low Temp. Phys.* **4**, 469 (1971).

[674] K. Komenou, T. Yamashita and Y. Onodera, *Phys. Letts.* **28A**, 335 (1968).

[675] P. Kumbhare, P. M. Tedrow and D. M. Lee, *Phys. Rev.* **180**, 519 (1969).

[676] A. Leger and J. Klein, *Phys. Rev.* **B3**, 3968 (1971).

[677] J. D. Leslie, J. T. Chen and T. T. Chen, *Canad. J. Phys.* **48**, 2783 (1970).

[678] R. B. Laibowitz and J. J. Cuomo, *J. Appl. Phys.* **41**, 2748 (1970).

[679] R. Meservey, P. M. Tedrow and P. Fulde, *Phys. Rev. Letts.* **25**, 1270 (1970).

[680] J. C. Neel, A. Ballonoff and G. J. Santos, Jr., *Appl. Phys. Letts.* **6**, 2 (1965).

[681] J. E. Nordman and W. H. Keller, *Phys. Letts.* **36A**, 52 (1971).

[682] H. Rothman, *Physica* **44**, 353 (1969).

[683] H. Rothman and A. A. Hirsch, *Physica* **36**, 310 (1967).

[684] L. W. Shacklette, L. G. Radosevich and W. S. Williams, *Phys. Rev.* **B4**, 84 (1971).

[685] T. Shigi, T. Okuda, K. Uchico and S. Nakaya, LT10, S162, p. 286.

[686] T. Shigi, K. Uchico and S. Nakaya, *J. Phys. Soc. Japan* **24**, 82 (1968).

[687] E. Strieder, *Ann. Physik* **22**, 151 (1968).

[688] M. Strongin, A. Paskin, O. F. Kammerer and M. Garber, *Phys. Rev. Letts.* **14**, 362 (1965).

[689] J. Sutton and P. Townsend, LT8, p. 182.
[690] P. M. Tedrow and R. Meservey, *Phys. Rev. Letts.* **27**, 919 (1971).
[691] P. M. Tedrow and R. Meservey, *Phys. Rev. Letts.* **26**, 192 (1971).
[692] D. R. Tilley and R. Ward, *J. Phys.* **C3**, 2119 (1970).
[693] P. Townsend and J. Sutton, *Proc. Phys. Soc.* **78**, 309 (1961).
[694] P. Townsend and J. Sutton, *Phys. Rev. Letts.* **11**, 154 (1963).
[695] N. Tsuda, *J. Phys. Soc. Japan* **27**, 856 (1969).
[696] H. Tsuya, *J. Phys. Soc. Japan* **21**, 1011 (1966).
[697] H. Tsuya, *J. Phys. Soc. Japan* **23**, 975 (1967).
[698] K. Uchico, S. Nakaya and T. Shigi, *J. Phys. Soc. Japan* **24**, 476 (1968).
[699] J. Vrba and S. B. Woods, *Phys. Rev.* **B4**, 87 (1971).
[700] D. G. Walmsley and C. K. Campbell, *Canad. J. Phys.* **45**, 1541 (1967).
[701] R. Ward and D. R. Tilley, *J. Phys.* **C3**, 1718 (1970).
[702] G. L. Wells, J. E. Jackson and E. N. Mitchell, *Phys. Rev.* **B1**, 3636 (1970).
[703] S. B. Woods and J. S. Rogers, LT10, S167, p. 303.
[704] A. F. G. Wyatt, *Phys. Rev. Letts.* **13**, 160 (1964).
[705] P. W. Wyatt and A. Yelon, *Phys. Rev.* **B2**, 4461 (1970).
[706] K. Yoshihiro and W. Sasaki, *J. Phys. Soc. Japan* **28**, 262 (1970).
[707] K. Yoshihiro and W. Sasaki, *J. Phys. Soc. Japan* **26**, 860 (1969).
[708] K. Yoshihiro and W. Sasaki, *J. Phys. Soc. Japan* **24**, 26 (1968).
[709] N. V. Zavaritskii, *JETP* **14**, 470 (1962).
[710] G. Ziemba and G. Bergmann, *Z. Phys.* **237**, 410 (1970).
[711] S. I. Ochiai, M. L. A. MacVicar and R. M. Rose, LT12, p. 421.
[712] S. I. Ochiai, M. L. A. MacVicar and R. M. Rose, *Solid State Commun.* **8**, 1031 (1970).
[713] S. G. Lipson and M. M. Stupel, *Phys. Letts.* **33A**, 493 (1970).
[714] G. Faraci, G. Giaquinta and N. A. Mancini, *Phys. Letts.* **30A**, 400 (1969).
[715] W. H. Keller and J. E. Nordman, *J. Appl. Phys.* **42**, 137 (1971), Abstract only.
[716] T. T. Chen, J. D. Leslie and H. J. T. Smith, PCSS, p. 439.
[717] J. E. Jackson, C. V. Briscoe and H. Wuhl, PCSS, p. 447.
[718] B. L. Blackford, PCSS, p. 475.
[719] J. M. Rowell and W. L. McMillan, PCSS, p. 718.
[720] D. H. Prothero, S. M. Freake and C. J. Adkins, PCSS, p. 744.
[721] J. J. Hauser, PCSS, p. 733.
[722] I. K. Yanson, *JETP* **33**, 951 (1971).
[723] W. L. Feldmann and J. M. Rowell, *J. Appl. Phys.* **40**, 312 (1969).
[724] G. Paterno, M. V. Ricci and N. Sacchetti, *Phys. Rev.* **B3**, 3792 (1971).
[725] J. L. Levine and S. Y. Hsieh, PCSS, p. 471.
[725a] G. A. Gogadze, *Ukrain. Fiz. Zh.* **11**, 1307 (1966).
[726] R. W. Cohen, B. Abeles and C. R. Fuselier, *Phys. Rev. Letts.* **23**, 377 (1969).
[727] O. D. Cheishvili, *J. Low Temp. Phys.* **4**, 577 (1971).
[728] G. E. Pike, *J. Appl. Phys.* **42**, 883 (1971).
[729] H.-J. C. Blume, *Proc. IEEE* **56**, 1395 (1968).
[730] R. Kummel, *Phys. Rev.* **B3**, 3787 (1971).
[731] A. B. Kaiser and M. J. Zuckermann, *Phys. Rev.* **B1**, 229 (1970).
[732] G. I. Urushadze, *Sov. Phys. Solid State* **9**, 930 (1967).
[733] M. Weger, *Solid State Commun.* **9**, 107 (1971).
[734] P. Vashista and J. P. Carbotte, *Solid State Commun.* **8**, 161 (1970).
[735] P. Vashista and J. P. Carbotte, *Solid State Commun.* **8**, 1661 (1970).
[736] K. Maki, *Phys. Rev. Letts.* **18**, 835 (1967).
[737] K. Maki and A. Griffin, *Phys. Rev. Letts.* **15**, 921 (1965).
[738] A. Griffin and J. Demers, *Phys. Rev.* **B4**, 2202 (1971).

[739] V. Radjakrishnan, *Canad. J. Phys.* **48**, 630 (1970).
[740] H. J. Juranek, L. Neumann and L. Tewordt, *Z. Phys.* **193**, 459 (1966).
[741] S. Nakajima and Y. Nakao, *J. Phys. Soc. Japan* **20**, 8 (1965).
[742] I. O. Kulik, LT11, B1.1, p. 701.
[743] B. D. Josephson, *Proc. Intern. Symp. on Quantum Fluids*, Brighton, 1965.
[744] M. J. Stephen, in 'Superconductivity', The McGill University Advanced Summer Study Institute, Montreal, Canada, 17–29 June 1968, Gordon and Breach, 1969, p. 371.
[745] B. D. Josephson, *Wireless World* **72**, 484 (1966).
[746] J. Clarke, *New Scientist* **29**, No. 486 (1966).
[747] J. Matisoo, *Analyt. Chem.* **41**, 83a and 139a (1969).
[748] J. Maurer, *Onde Elect.* **48**, 633 (1968).
[749] R. A. Kamper, *Trans. IEEE ED*-16, 840 (1969).
[750] L. Solymar, *Electronics and Power* **14**, 317 (1968).
[751] F. Lange, *Nachrichtentechnik* **19**, 129 and 194 (1969).
[752] H. Karras, *Feingerate Tech.* **19**, 436 (1970).
[753] F. Baumann, *Naturwiss.* **55**, 578 (1968).
[754] M. Otala and T. Wilk, *Elektronikka* **24**, 43 (1971).
[755] G. F. Zharkov, *Soviet Physics—Uspekhi* **9**, 198 (1966).
[756] R. de Bruyn Ouboter and A. Th. A. M. de Waele, *Rev. Phys. Appl.* **5**, 25 (1970).
[757] A. Barone and R. D. Parmentier, *Alta Frequenza* **40**, 166 (1971).
[758] D. N. Langenberg, T. F. Finnegan and A. Denenstein, *Electronics* **41**, No. 5, 42 (1971).
[759] J. Clarke, *Phys. Today* **24**, No. 8, 30 (1971).
[760] D. E. McCumber, *J. Appl. Phys.* **39**, 2503 (1968).
[761] J. Matisoo, *Trans. IEEE* **MAG-5**, 848 (1970).
[762] R. A. Kamper, *Cryogenics* **9**, 20 (1969).
[763] B. W. Petley, *Contemporary Physics* **10**, 139 (1969).
[764] B. W. Petley, *Contemporary Physics* **12**, 453 (1971).
[765] D. N. Langenberg, W. H. Parker and B. N. Taylor, 'Proc. 3rd Intern. Conf. on atomic masses', Winnipeg, Canada, 28 Aug.–1 Sept. 1967, p. 439.
[766] J. Clarke, *Am. J. Phys.* **38**, 1071 (1970).
[767] D. J. Scalapino, PCPMFC,
[768] O. Doyle, *Electronics* **41**, 38 (1971).
[769] D. N. Langenberg, TPS, p. 519.
[770] A. F. G. Wyatt, TPS, p. 541.
[771] J. E. Mercereau, in 'Superconductivity', (R. D. Parks, Ed.), Marcel Dekker Inc., New York, 1969, p. 393.
[772] J. E. Mercerau, TPS, p. 461.
[773] K. Rose, *Cryogenics* **9**, 227 (1969).
[774] P. W. Anderson, LT12, p. 1.
[775] P. W. Anderson, *Phys. Today* **23**, No. 11, 23 (1970).
[776] I. O. Kulik and I. K. Yanson, 'Josephson effect in high-frequency tunnel structures' (in Russian), Nauka, Moskva, 1970.
[777] P. L. Richards, S. Shapiro and C. C. Grimes, *Am. J. Phys.* **36**, 690 (1968).
[778] A. C. Rose-Innes and E. H. Rhoderick, 'Introduction to superconductivity', Pergamon Press, 1969.
[779] C. Kittel, 'Introduction to solid state Physics', 4th Edition, Wiley, New York, 1971.
[780] T. Shigi, S. Nakaya and T. Aso, *J. Phys. Soc. Japan* **21**, 2418 (1966).
[781] I. Taguchi and H. Yoshioka, *J. Phys. Soc. Japan* **29**, 371 (1970).
[782] C. S. Lim, J. D. Leslie, H. J. T. Smith, P. Vashista and J. P. Carbotte, *Phys. Rev.* **B2**, 1651 (1970).

[783] A. E. Gorbonosov and I. O. Kulik, *Fiz. Met. i Metalloved.* **23**, 803 (1967).
[784] K. Schwidtal and R. D. Finnegan, *J. Appl. Phys.* **40**, 2123 (1969).
[785] K. Schwidtal and R. D. Finnegan, *Phys. Rev.* **B2**, 148 (1970).
[786] J. M. Eldridge and J. Matisoo, LT12, p. 427.
[787] T. Yamashita and Y. Onodera, *J. Appl. Phys.* **38**, 3523 (1967).
[788] T. Yamashita, M. Kunita and Y. Onodera, *Jap. J. Appl. Phys.* **7**, 288 (1968).
[789] T. Yamashita, M. Kunita and Y. Onodera, *J. Appl. Phys.* **39**, 5396 (1968).
[790] H. H. Zappe, IBM Report RC 2974, July 23, 1970.
[791] R. C. Dynes and T. A. Fulton, *Phys. Rev.* **B3**, 3015 (1971).
[792] K. K. Likharev, *JETP*
[793] M. Weihnacht, *Phys. Stat. Sol.* **32**, K169 (1969).
[794] H. Kanter and A. H. Silver, *Appl. Phys. Letts.* **19**, 515 (1971).
[795] K. Yamafuji, T. Ezaki and T. Matsushita, *J. Phys. Soc. Japan* **30**, 965 (1971).
[796] J. Warman and J. A. Blackburn, *Appl. Phys. Letts.* **19**, 60 (1971).
[797] H. Fack and V. Kose, *J. Appl. Phys.* **42**, 322 (1971).
[798] J. Clarke and T. A. Fulton, *J. Appl. Phys.* **40**, 4470 (1969).
[799] O. Jaoul, *Rev. Phys. Appl.* **5**, 885 (1970).
[800] H. Fack, V. Kose and H. J. Schrader, *Messtechnik* **79**, 31 (1971).
[801] H. Fack and V. Kose, *J. Appl. Phys.* **42**, 320 (1971).
[802] A. F. Volkov and F. J. Nad', *JETP Letts.* **11**, 56 (1970).
[803] L. G. Aslamazov and A. I. Larkin, *JETP Letts.* **9**, 87 (1969).
[804] T. D. Clark, LT12, p. 449.
[805] D. G. McDonald, K. M. Evenson, J. S. Wells and J. D. Cupp, *J. Appl. Phys.* **42**, 179 (1971).
[806] I. M. Dmitrenko, S. I. Bondarenko, Yu. G. Bevza and L. E. Kolinko, LT11, p. 729.
[807] T. Yamashita and Y. Onodera, *Jap. J. Appl. Phys.* **6**, 746 (1967).
[808] P. E. Gregers-Hansen, M. T. Levinsen, L. Pedersen and C. J. Sjostrom, *Solid State Commun.* **9**, 661 (1971).
[809] P. Russer, *Acta Phys. Austr.* **32**, 373 (1970).
[810] I. O. Kulik, *Sov. Phys.—Tech. Phys.* **12**, 111 (1967).
[811] I. M. Dmitrenko and I. K. Yanson, *JETP* **22**, 1190 (1966).
[812] S. Bermon and R. M. Mesak, *J. Appl. Phys.* **42**, 4488 (1971).
[813] J. A. Blackburn, J. D. Leslie and H. J. Smith, *J. Appl. Phys.* **42**, 1047 (1971).
[814] M. A. Angadi and G. M. Graham, *Canad. J. Phys.* **49**, 629 (1971).
[815] I. M. Dmitrenko, *Ukr. Fiz. Zh.* **14**, 439 (1969).
[816] I. K. Albegova, B. I. Borodai, I. K. Yanson and I. M. Dmitrenko, *Sov. Phys.—Tech. Phys.* **14**, 681 (1969).
[817] D. D. Coon and M. D. Fiske, *Phys. Rev.* **138**, A744 (1965).
[818] Yu. M. Ivanchenko, A. V. Svidzinskii and V. A. Slyusarev, *JETP* **24**, 131 (1967).
[819] A. T. Fiory, *Phys. Rev. Letts.* **27**, 501 (1971).
[820] T. A. Fulton and R. C. Dynes, *Solid State Commun.* **9**, 1069 (1971).
[821] W. Schroen and J. P. Pritchard, Jr., *J. Appl. Phys.* **40**, 2118 (1969).
[822] W. J. Johnson and A. Barone, *J. Appl. Phys.* **41**, 2958 (1970).
[823] C. K. Mahutte, J. D. Leslie and H. J. T. Smith, *Canad. J. Phys.* **47**, 627 (1969).
[824] D. L. Stuehm and C. W. Wilmsen, *J. Appl. Phys.* **42**, 869 (1971).
[825] K. Schwidtal, *Phys. Rev.* **B2**, 2526 (1970).
[826] A. C. Scott, *Proc. IEEE* **57**, 1338 (1969).
[827] A. C. Scott, *Nuovo Cim.* **69B**, 241 (1970).
[828] M. Renard, LT12, p. 457.
[828a] A. E. Gorbonosov and I. O. Kulik, *JETP* **33**, 374 (1971).
[829] I. O. Kulik, *JETP* **24**, 1307 (1967).

[830] C. S. Owen and D. J. Scalapino, *J. Appl. Phys.* **41**, 2047 (1970).
[831] A. C. Scott and W. J. Johnson, *Appl. Phys. Letts.* **14**, 316 (1969).
[832] A. Barone, *J. Appl. Phys.* **42**, 2747 (1971).
[833] A. M. Goldman, PCFS, p. 277.
[834] A. M. Goldman, LT9, p. 421.
[835] A. M. Goldman, P. J. Kreisman and D. J. Scalapino, *Phys. Rev. Letts.* **15**, 495 (1965).
[836] A. M. Goldman and P. J. Kreisman, LT10, S180, p. 362.
[837] A. M. Goldman, *J. Low Temp. Phys.* **3**, 55 (1970).
[838] J. E. Zimmerman and A. H. Silver, *Phys. Rev. Letts.* **19**, 14 (1967).
[839] M. B. Simmonds and W. H. Parker, *J. Appl. Phys.* **42**, 38 (1971).
[840] J. M. Eldridge and D. Dong, *J. Electrochem. Soc.* (to be published).
[841] R. Graeffe and T. Wiik, *J. Appl. Phys.* **42**, 2146 (1971).
[842] Ph. Cardinne, B. Manhes and M. Renard, LT12, p. 463.
[843] G. Faraci, G. Giaquinta, N. A. Mancini and I. F. Quercia, LT12, p. 423.
[844] A. Contaldo, *Rev. Sci. Instrum.* **38**, 1543 (1967).
[844a] S. C. Smith and A. C. Anderson, *Cryogenics* **11**, 53 (1971).
[845] R. A. Buhrman, S. F. Strait and W. W. Webb, *J. Appl. Phys.* **42**, 4527 (1971).
[846] M. Cerdonio, T. Di Leo and G. Lucano, *Alta Frequenza* **39**, 541 (1970).
[847] F. Ya Nad' and O. Yu. Polyanskii, *Inst. and Exp. Tech.* **6**, 1794 (1970).
[848] F. Consadori, A. A. Fife, R. F. Frindt and S. Gygax, *Appl. Phys. Letts.* **18**, 233 (1971).
[849] M. A. Janocko, J. R. Gavaler, C. K. Jones and R. D. Blaugher, *J. Appl. Phys.* **42**, 182 (1971).
[850] M. A. Janocko, J. R. Gavaler, C. K. Jones and R. D. Blaugher, LT12,
[851] J. M. Goodkind and J. M. Dundon, *Rev. Sci. Instrum.* **42**, 1264 (1971).
[851a] T. A. Fulton, *Appl. Phys. Letts.* **19**, 311 (1971).
[851b] P. W. Anderson, R. C. Dynes and T. A. Fulton, *Bull. Am. Phys. Soc. Series II*, **16**, 399 (1971).
[852] I. M. Dmitrenko, I. K. Yanson and I. I. Yurchenko, *Sov. Phys. Solid State* **9**, 2889 (1968).
[853] A. A. Galkin and V. M. Svistunov, *JET P Letts.* **5**, 323 (1967).
[854] D. E. McCumber, *J. Appl. Phys.* **39**, 297 (1968).
[855] D. R. Tilley, *Phys. Letts.* **33A**, 205 (1970).
[856] C. A. Hamilton and S. Shapiro, *Phys. Rev. Letts.* **26**. 426 (1971).
[857] A. J. DiNardo and E. Sard, *J. Appl. Phys.* **42**, 105 (1971) (Abstract only).
[858] A. Longacre, *Electronics* **41**, No. 5, 44 (1971).
[859] J. T. Chen, *Phys. Letts.* **32A**, 437 (1970).
[860] D. G. McDonald, A. S. Risley, J. D. Cupp and K. M. Evenson, *Appl. Phys. Letts.* **18**, 162 (1971).
[861] T. G. Blaney, C. C. Bradley, G. J. Edwards and D. J. E. Knight, *Phys. Letts.* **36A**, 285 (1971).
[862] K. K. Likharev, Vestnik Moskovskogo Universiteta 83 (1969).
[863] J. M. Goodkind and D. L. Stolfa, *Rev. Sci. Instrum.* **41**, 799 (1970).
[864] J. E. Zimmerman, *J. Appl. Phys.* **42**, 4483 (1971).
[865] J. Clarke and J. L. Paterson, *Bull. Am. Phys. Soc. Series II*, **16**, 399 (1971).
[866] R. Meservey, *J. Appl. Phys.* **39**, 2598 (1968).
[867] J. E. Lukens, R. J. Warburton and W. W. Webb, *J. Appl. Phys.* **42**, 27 (1971).
[868] T. F. Finnegan, A. Denenstein and D. N. Langenberg, *Phys. Rev.* **B4**, 1487 (1971).
[869] F. K. Harris, H. A. Fowler and P. T. Olsen, *Metrologia* **6**, 134 (1970).
[870] B. W. Petley and J. C. Gallop, PCPMFC,
[871] V. Kose, F. Melchert, H. Fack and H.-J. Schrader, *PTB-Mitteilung* **81**, 8 (1971).
[872] H. A. Fowler, T. J. Witt, J. Toots, P. T. Olsen and W. Eicke, PCPMFC,

[873] T. F. Finnegan, A. Denenstein and D. N. Langenberg, PCPMFC,

[874] K. Nordtvedt, Jr., *Phys. Rev.* **B1**, 81 (1970).

[875] D. N. Langenberg and J. R. Schrieffer, *Phys. Rev.* **B3**, 1776 (1971).

[876] J. B. Hartle, D. J. Scalapino and R. L. Sugar, *Phys. Rev.* **B3**, 1778 (1971).

[877] A. A. Galkin, B. I. Borodai, L. A. Zilberman, Yu. M. Ivanchenko and V. M. Svistunov, *JETP* **33**, 354 (1971).

[878] W. H. Henkels and W. W. Webb, *Phys. Rev. Letts.* **26**, 1164 (1971).

[879] P. A. Lee, *J. Appl. Phys.* **42**, 325 (1971).

[880] R. K. Kirschman, H. A. Notarys and J. E. Mercereau, *Phys. Letts.* **34A**, 209 (1971).

[881] R. K. Kirschman and J. E. Mercereau, *Phys. Letts.* **35A**, 177 (1971).

[882] G. Vernet and R. Adde, *Appl. Phys. Letts.* **19**, 195 (1971).

[883] J. T. Harding and J. E. Zimmerman, *J. Appl. Phys.* **41**, 1581 (1970).

[884] J. E. Zimmerman, PCFS, p. 303.

[885] P. Thiene and J. E. Zimmerman, *Phys. Rev.* **177**, 758 (1969).

[886] R. E. Eck, M. Nisenoff, B. T. Ulrich and J. E. Mercereau, PCFS, p. 167.

[887] H. Kanter and F. L. Vernon, Jr., *Phys. Rev.* **B2**, 4694 (1970).

[888] A. A. Galkin, B. I. Borodai, V. M. Svistunov and V. N. Taranenko, *JETP Letts.* **8**, 318 (1968).

[889] Yu. M. Ivanchenko, *JETP Letts.* **6**, 313 (1967).

[890] Yu. M. Ivanchenko, *JETP* **25**, 878 (1967).

[891] A. L. Fetter and M. J. Stephen, *Phys. Rev.* **168**, 475 (1968).

[892] A. I. Larkin and Yu. N. Ovchinnikov, *JETP* **26**, 1219 (1968).

[893] L. Ya. Kobolev and Yu. P. Yarovskikh, *Phys. Mets. & Metallogr.* **27**, 367 (1969).

[894] D. Rogovin and D. J. Scalapino, PCSS, p. 399.

[895] M. J. Stephen, PCSS, p. 24.

[896] D. J. Scalapino, *Phys. Rev. Letts.* **24**, 1052 (1970).

[897] A. C. Biswas and S. S. Jha, *Phys. Rev.* **B2**, 2543 (1970).

[898] L. A. Zilberman and Yu. M. Ivanchenko, *Sov. Phys. Solid State* **12**, 1530 (1971).

[899] L. A. Zilberman and Yu. M. Ivanchenko, *JETP* **33**, 1229 (1971).

[900] I. O. Kulik, *JETP* **32**, 510 (1971).

[901] P. A. Lee and M. O. Scully, *Phys. Rev.* **B3**, 769 (1971).

[902] I. O. Kulik, *JETP Letts.* **10**, 313 (1969).

[903] Y. Ichikawa, *Phys. Letts.* **35A**, 5 (1971).

[904] H. Takayama, LT12, p. 267.

[905] L. Leplae, F. Mancini and H. Umezawa, *Lett. Nuovo Cim.* **4**, 963 (1970).

[906] L. Leplae, F. Mancini and H. Umezawa, *Phys. Letts.* **36A**, 475 (1971).

[907] R. H. Parmenter, *Phys. Rev.* **154**, 353 (1967).

[908] R. H. Parmenter, *Phys. Rev.* **167**, 387 (1968).

[909] H. Cortes, P. Pellan and J. Rosenblatt, LT12, p. 487.

[910] J. Rosenblatt, H. Cortes and P. Pellan, *Phys. Letts.* **33A**, 143 (1970).

[911] T. D. Clark, LT11, p. 686.

[912] T. D. Clark, PCSS, p. 432.

[913] T. D. Clark and D. R. Tilley, *Phys. Letts.* **28A**, 62 (1968).

[914] A. F. Volkov, *JETP* **33**, 811 (1971).

[915] E. N. Economou and K. L. Ngai, *Phys. Rev. Letts.* **20**, 547 (1968).

[916] K. L. Ngai, *Phys. Rev.* **182**, 555 (1969).

[917] H. Kinder, *Phys. Letts.* **36A**, 379 (1971).

[918] Yu. M. Ivanchenko and Yu. V. Medvedev, *JETP*

[919] O. D. Cheishvili, *Sov. Phys. Solid State* **11**, 138 (1969).

[920] D. R. Tilley, *Phys. Letts.* **20**, 117 (1966).

[921] P. V. Christiansen, E. B. Hansen and C. J. Sjostrom, *J. Low Temp. Phys.* **4**, 349 (1971).

[922] Y.-H. Kao, *Phys. Letts.* **26A**, 471 (1968).
[923] R. G. Boyd, *Phys. Rev.* **153**, 444 (1967).
[924] R. G. Boyd, *Phys. Letts.* **24A**, 204 (1967).
[925] J. I. Pankove, LT10, S39, p. 257.
[926] J. I. Pankove, *Phys. Letts.* **22**, 557 (1966).
[927] I. K. Yanson, *JETP* **26**, 742 (1968).
[928] I. K. Yanson and I. Kh. Albegova, *JETP* **28**, 826 (1969).
[928a] W. E. Lawrence and S. Doniach, LT12, p. 361.
[929] G. A. Baramidze, A. G. Kvirikadze and I. O. Kulik, LT10, S45, p. 283.
[930] R. G. Boyd, *Phys. Rev.* **167**, 407 (1968).
[931] S. Donlach and W. E. Lawrence, LT12, p. 166.
[932] V. P. Galaiko, A. V. Svidzinskii and V. A. Slyusarev, *JETP* **29**, 454 (1969).
[933] A. A. Golub, *Sov. Phys. Solid State* **10**, 2501 (1969).
[934] A. E. Gorbonosov and I. O. Kulik, *JETP* **28**, 455 (1969).
[935] B. T. Geilikman, R. O. Zaytsev and V. Z. Kresin, *Phys. Met. & Metallogr.* **23**, No. 5, 26 (1967).
[936] Yu. M. Ivanchenko, LT10, S43, p. 276.
[937] Yu. M. Ivanchenko, *JETP* **24**, 225 (1967).
[938] Yu. M. Ivanchenko, *Phys. Letts.* **23**, 289 (1966).
[939] Yu. M. Ivanchenko and A. V. Svidzinskii, *Phys. Met & Metallogr.* **22**, 619 (1966).
[940] I. O. Kulik, *JETP* **22**, 841 (1966).
[941] I. O. Kulik, LT10, S45, p. 283.
[942] I. O. Kulik, *JETP* **23**, 529 (1966).
[943] V. Radhakrishnan, *Phys. Stat. Sol.* **18**, 113 (1966).
[944] A. I. Larkin and Yu. N. Ovchinnikov, *JETP* **24**, 1035 (1967).
[945] A. I. Larkin, Yu. N. Ovchinnikov and M. A. Fedorov, *JETP* **24**, 120 (1967).
[946] K. Maki, *Progr. Theor. Phys. (Japan)* **30**, 573 (1963).
[947] K. Maki, *Phys. Letts.* **10**, 11 (1964).
[948] K. L. Ngai, J. A. Appelbaum, M. H. Cohen and J. C. Phillips, *Phys. Rev.* **163**, 352 (1967).
[949] M. M. Nieto, *Phys. Rev.* **167**, 416 (1968).
[950] M. M. Nieto, *Nuovo Cim.* **B62**, 95 (1969).
[951] D. J. Scalapino and T. M. Wu, *Phys. Rev. Letts.* **17**, 315 (1966).
[952] A. Schmid, *Z. Phys.* **178**, 26 (1964).
[953] A. Schmid, *Phys. Stat. Sol.* **7**, 3 (1964).
[954] H. Shiba and T. Soda, *Progr. Theor. Phys. (Japan)* **41**, 25 (1969).
[955] M. J. Stephen, PCFS, p. 265.
[956] A. V. Svidzinskii and V. A. Slyusarev, *JETP* **24**, 120 (1967).
[957] A. V. Svidzinskii and V. A. Slyusarev, *Phys. Letts.* **27A**, 22 (1968).
[958] A. V. Svidzinskii and V. A. Slyusarev, LT10, S44, p. 281.
[959] A. V. Svidzinskii and V. A. Slyusarev, *Sov. Phys. Dokl.* **14**, 230 (1969).
[960] P. R. Wallace and M. J. Stavn, *Canad. J. Phys.* **43**, 411 (1965).
[961] A. Zawadowsky, *Phys. Rev.* **163**, 341 (1967).
[962] A. Erdelyi, 'Table of Integral Transforms', I, McGraw Hill, New York, 1954, p. 136.
[963] J. N. Sweet, Ph.D. thesis, Lawrence Radiation Laboratory, University of California, 1970, UCRL-19653.
[964] S. A. Buckner, T. F. Finnegan and D. N. Langenberg, *Phys. Rev. Letts.* **28**, 150 (1972).

Author index

The bracketed numbers refer to specific references (see pages 371 to 392) while the preceding unbracketed numbers refer to the pages on which those references are mentioned. Where *no* bracketed (reference) number follows an unbracketed (page) number, the author's name is mentioned on that page *without* any specific reference to a publication.

Abeles, B., 30[59], 66[139], 67[139], 98[236, 237], 99[237], 124[309], 125[309], 332[610–612, 616–618], 335[59, 139, 237, 610–612, 618], 336[236], 338[139, 611], 339[139, 236, 611], 342[726].
Abrikosov, A. A., 12[25, 26], 82[184], 86[184], 196[26]
Adde, R., 358[882]
Adkins, C. J., 30[62], 45[111], 49[62], 50[121], 53[62, 107], 104[262, 263], 338[62, 107], 339[62], 342[111, 121, 262, 263, 641, 720]
Adler, J. G., 77[157, 159], 95[228], 97[230, 233], 335[642, 643], 337[228, 230, 233]
Aihara, K., 312[543, 544], 347[543]
Albegova, I. K., 347[816], 360[928]
Allen, P. B., 99[244]
Ambegaokar, V., 135[332], 151[332], 152[332], 302[524], 304[528], 345[332]
Anacker, W., 250[465], 251[465], 255[465]
Anderson, A. C., 351[844a]
Anderson, J. T., 302[525]
Anderson, P. W., 18[34], 71[148], 93[218, 221], 94[221, 223], 95[227], 96[227], 114[290], 134[320], 135[329], 137[349], 141[329, 361], 146[349], 193[329], 197[349], 200[349], 240 [361], 241[361], 316[329], 318[548], 319[549], 335[221], 337[227], 338[221], 344[329, 349, 774, 775], 353[774, 775, 851b]
Angadi, M. A., 347[814], 348[814]
Antsigina, T. N., 316[545g]
Appelbaum, J. A., 360[948]
Arase, T., 94[222]
Arnold, A. J. P. T., 209[419, 421]
Aslamazov, L. G., 167[389], 316[545b], 347 [803]
Aso, T., 345[780], 358[780]
Autler, S. H., 12[27]
Averill, A. F., 124[307, 308]

Baixeras, J., 321[567, 568]
Bakker, J. W., 129[315]
Balanov, E. I., 243[454, 455], 312[542]
Baldwin, J. P., 251[466]
Ballonoff, A., 336[680]
Band, W. T., 129[316], 130[316]
Baratoff, A., 135[332], 151[332, 373, 374], 152[332], 345[332].

Bardeen, J., 4[13], 13[13], 19[48], 24[48], 30[13], 41[104], 86[204], 122[303], 134[104]
Barnes, L. J., 333[625], 341[625]
Barone, A., 344[757], 349[822, 832]
Barth, N., 98[241], 335[241, 673], 336[673], 337[673], 338[673]
Baumann, F., 344[753]
Bean, C. P., 197[409]
Beasley, M. R., 138[351], 282[351], 322[576]
Bennett, A. J., 88[208], 333[634]
Bergmann, G., 334[710], 339[710], 340[710]
Bermon, S., 80[169], 95[225], 336[225], 347 [812], 348[812]
Betbeder-Matibet, O., 82[185], 83[185]
Bethe, H., 19[39]
Bevza, Yu. G., 347[806]
Biswas, A. C., 358[897]
Blackburn, J. A. 151[374], 347[796, 813], 348[813]
Blackford, B. L., 17, 34[99], 35[99], 78[99], 335[99], 338[644, 718]
Blaney, T. G., 354[861]
Blaugher, R. D., 352[849, 850]
Bloch, F., 289[512, 513]
Blodgett, K., 49[117]
Blume, H.-J. C., 342[729]
Bogoliubov, N. N., 4[14], 17, 134
Bondarenko, S. I., 205[416], 243[454–456], 312[542], 315[456], 347[806]
Bonnet, D., 62[130], 340[130, 645]
Borodai, B. I., 347[816], 358[877, 888]
Bostock, J., 31[97], 123[97, 305], 124[308]
Boyd, R. G., 360[923, 924]
Bradley, C. C., 354[861]
Brandt, N. B., 99[245]
Brassington, D. J., 31[86], 111[86], 336[86]
Bratus, E. N., 316[545g]
Brenneman, A. E., 245[459]
Bright, A. A., 333[624], 338[624]
Briscoe, C. V., 80[168], 336[168, 717], 337[717]
Brochard, F., 31[85], 111[85], 338[85]
Brockhouse, B. N., 94[222]
Broers, A. N., 243[453]
Buck, D. A., 244[457]
Buckner, S. A., 174[392, 393], 302[392, 393], 303[392], 304[392]· 354[964]
Buhrman, R. A., 351[845]

393

Subject index